NONCOMMUTATIVE DIFFERENTIAL GEOMETRY AND ITS APPLICATIONS TO PHYSICS

MATHEMATICAL PHYSICS STUDIES

VOLUME 23

Noncommutative Differential Geometry and Its Applications to Physics

Proceedings of the Workshop at Shonan, Japan, June 1999

Edited by

Yoshiaki Maeda
Keio University, Yokohama, Japan

Hitoshi Moriyoshi
Keio University, Yokohama, Japan

Hideki Omori
Science University of Tokyo, Noda, Japan

Daniel Sternheimer
CNRS and Université de Bourgogne, Dijon, France

Tatsuya Tate
Keio University, Yokohama, Japan

and

Satoshi Watamura
Tohoku University, Sendai, Japan

KLUWER ACADEMIC PUBLISHERS
DORDRECHT / BOSTON / LONDON

A C.I.P. Catalogue record for this book is available from the Library of Congress.

ISBN 0-7923-6930-0

Published by Kluwer Academic Publishers,
P.O. Box 17, 3300 AA Dordrecht, The Netherlands.

Sold and distributed in North, Central and South America
by Kluwer Academic Publishers,
101 Philip Drive, Norwell, MA 02061, U.S.A.

In all other countries, sold and distributed
by Kluwer Academic Publishers,
P.O. Box 322, 3300 AH Dordrecht, The Netherlands.

Printed on acid-free paper

Printed in the Netherlands.

TABLE OF CONTENTS

PREFACE

A workshop on "noncommutative differential geometry and its applications to physics" was held at Shonan International Village at Hayama, Kanagawa, Japan, from May 31 through June 4, 1999. The aim of the workshop was to enhance international cooperation in various aspects of this field between mathematicians and physicists.

Noncommutative differential geometry is a novel approach to geometry, which promises to be applicable to unsolved geometric phenomena and physical observations. It was founded in the early eighties by 1982 Fields Medalist Alain Connes on the basis of his fundamental works in operator algebras, a subject well developed in Japan, and relies extensively on his notion of cyclic cohomology. It gave, in particular, striking extensions of the Atiyah–Singer index theory, has several common points with deformation quantization, and was applied to a variety of mathematical and physical problems. A 1994 book by Alain Connes takes stock of many developments up to that time. More recently new applications have appeared, including, in physics, to string theory and M-theory.

Nevertheless, noncommutative differential geometry is not sufficiently understood among both mathematicians and physicists. To address this issue, the abovementioned topical workshop was organized. The topics covered included: deformation problems; Poisson groupoids; theory of operads; quantization problems; moduli spaces; D-branes and other applications to quantum field theory. It was attended by a mixture of mathematicians and physicists, and produced interesting discussions on these problems. As hoped, the meeting was extremely fruitful for all participants.

This volume contains papers presented at the workshop. All papers were submitted by conference speakers and participants, and were refereed. The papers are devoted to presentations of new results which have not appeared previously in professional journals or to comprehensive reviews (including an original part) of recent developments in those topics. The volume is accessible to researchers and graduate students interested in them and in the interface with modern theoretical physics.

The workshop was held in cooperation with Keio University and the Mitsubishi foundation. The editors and conference organizers wish to thank these institutions for their generous financial support, and for help and encouragement in the planning phase. The Kluwer Publishing company has been very helpful in the production of this volume. We are pleased to record our gratitude. Special thanks go to Dr. Paul Roos and Mrs. Betty van Herk for their kind help.

Last but not least, thanks are due to the participants and speakers for making the meeting fruitful.

Y. MAEDA, H. MORIYOSHI, H. OMORI

D. STERNHEIMER, T. TATE AND S. WATAMURA

METHODS OF EQUIVARIANT QUANTIZATION

Christian Duval
Université de la Méditerranée and CPT-CNRS
Luminy Case 907, F–13288 Marseille, Cedex 9, FRANCE
duval@cpt.univ-mrs.fr

Pierre B. A. Lecomte
Institut de Mathématiques, Université de Liège
Sart Tilman, Gde, Traverse, 12 (B 37), B–4000 Liège, BELGIUM
plecomte@ulg.ac.be

Valentin Ovsienko
CNRS, Centre de Physique Théorique, CPT-CNRS
Luminy Case 907, F–13288 Marseille, Cedex 9, FRANCE
ovsienko@cpt.univ-mrs.fr

Abstract This article is a survey of recent work [15, 6, 7, 13] developing a new approach to quantization based on the equivariance with respect to some Lie group of symmetries. Examples are provided by conformal and projective differential geometry: given a smooth manifold M endowed with a flat conformal/projective structure, we establish a canonical isomorphism between the space of symmetric contravariant tensor fields on M and the space of differential operators on M. This leads to a notion of conformally/projectively invariant star product on T^*M.

Mathematics Subject Classification (2000): 58B99, 17B56, 17B65, 47E05

Keywords: Quantization, Lie algebras of vector fields, hypergeometric functions star products.

Y. Maeda, et al. (eds.), Noncommutative Differential Geometry and Its Applications to Physics, 1–12.

Proceedings of the workshop "Noncommutative Differential Geometry and its Applications to Physics", Shonan-Kokusaimura, Japan, May 31–June 4, 1999.

1. Introduction

This paper is a survey of recent articles [15, 6, 7, 13] (see also [3]) developing a new viewpoint on quantization.

1.1 Let $\mathcal{D}(M)$ be the space of differential operators on a smooth manifold M and $\mathcal{S}(M)(= \mathrm{Pol}(T^*M))$ the space of symbols, that is, of smooth functions on T^*M polynomial on fibers. Note that $\mathcal{S}(M)$ is naturally isomorphic to the space $\Gamma(STM)$ of symmetric contravariant tensor fields on M.

A *quantization map* (inverse to a *symbol map*) is a linear map

$$\mathcal{Q} : \mathcal{S}(M) \to \mathcal{D}(M) \tag{1.1}$$

satisfying the following natural condition: for every $P \in \mathcal{S}(M)$ the principal symbol of $\mathcal{Q}(P)$ coincides with the homogeneous higher-order part of P.

We will be considering a Lie group G acting on M and impose the G-equivariance condition; we will then seek a map (1.1) commuting with the G-action. We shall be, furthermore, interested in the situation where the equivariance condition determines the quantization map *in a unique fashion.*

1.2 Our main examples of symmetry groups will be:

$$G = SL(n+1, \mathbb{R}) \text{ and } G = SO(p+1, q+1),\ p+q = n\ (= \dim M) \tag{1.2}$$

acting in the standard way on S^n $(\to \mathbb{RP}^n)$ and on $S^p \times S^q$ $(\to (S^P \times S^q)/\mathbb{Z}_2)$ (respectively). These two cases are related to *projective* and *conformal* differential geometry. Our main purpose is to introduce the projectively/conformally equivariant quantization and prove its uniqueness.

1.3 The crucial property common to these two cases is *maximality* in the following sense. The corresponding Lie algebras:

$$\mathfrak{g} = \mathfrak{sl}(n+1, \mathbb{R}) \quad \text{and} \quad \mathfrak{g} = \mathfrak{o}(p+1, q+1), \quad p+q \geq 3, \tag{1.3}$$

can both be viewed as maximal subalgebras of the Lie algebra $\mathrm{Vect}_{\mathrm{Pol}}(\mathbb{R}^n)$ of all polynomial vector fields on $\mathbb{R}^n$ (cf. [15, 2]). We believe (although this remains unproven) that the uniqueness of the equivariant quantization map is due to the maximality of the algebras of symmetry.

1.4 Consider the group of all diffeomorphisms $\mathrm{Diff}(M)$ and the Lie algebra of all smooth vector fields $\mathrm{Vect}(M)$. An important aspect of our approach is the study of $\mathrm{Diff}(M)$- and $\mathrm{Vect}(M)$-module structure on the space of differential operators .

Let $\mathcal{F}_\lambda$ be the space of *tensor densities* of degree λ on M:

$$\varphi = f(x^1, \dots, x^n) \left|dx^1 \wedge \dots \wedge dx^n\right|^\lambda \tag{1.4}$$

that is, of sections of the line bundle $\Delta_\lambda(M) = |\Lambda^n T^* M|^{\otimes\lambda}$ over M. Denote by $\mathcal{D}_{\lambda,\mu}(M)$ the space of differential operators

$$A : \mathcal{F}_\lambda \to \mathcal{F}_\mu. \tag{1.5}$$

This space is naturally a module over $\mathrm{Diff}(M)$ and $\mathrm{Vect}(M)$. Therefore, for a Lie subgroup $G \subset \mathrm{Diff}(M)$, one obtains a G-action on $\mathcal{D}_{\lambda,\mu}(M)$.

Remark 1.1. Assume that M is orientable and choose a volume form on M, one then naturally identifies $\mathcal{D}_{\lambda,\mu}(M)$ with $\mathcal{D}(M)$ (i.e., with $\mathcal{D}_{0,0}(M)$) as a vector space. One can, therefore, consider $\mathcal{D}_{\lambda,\mu}(M)$ as a two-parameter family of $\mathrm{Diff}(M)$- and $\mathrm{Vect}(M)$-actions on the same space, $\mathcal{D}(M)$.

It is worth noticing that many important examples of differential operators acting on tensor densities have already been studied by the classics. The most important case $\mathcal{D}_{\frac{1}{2},\frac{1}{2}}$ (of operators on *half*-densities) naturally arises in geometric quantization (see, e.g., [12]). We refer to recent articles [5, 14, 10, 16] (and references therein) for a systematic study and classification of the modules $\mathcal{D}_{\lambda,\mu}(M)$.

1.5 In many cases we have to assume that the G-action cannot be defined *globally* on M. An example is provided when M is not compact (e.g., $M = \mathbb{R}^n$) and one can only deal with a $\mathfrak{g}$-action. Moreover, there are many cases when even the $\mathfrak{g}$-action is only locally defined. The most general framework is provided by the notion of a flat G-*structure* (defined by a local action of G on M, compatible with a local identification of M with some homogeneous space G/H). Existence of such structures on a given smooth manifold is a difficult (and widely open) problem of topology (see, e.g., [17]).

We will assume that M is endowed either with a flat projective or a flat conformal structure.

1.6 In the particular case $n = 1$, both conformal and projective structures coincide. We refer to [3] (see also [18]) for a thorough study of $\mathfrak{sl}(2, \mathbb{R})$-equivariant quantization and of the corresponding invariant star product.

2. Existence results

In this section we formulate our most general existence theorems. We will treat the basic examples and give some explicit formulæ later.

2.1. Symbols with values in tensor densities

The space $\mathcal{S}(M)$ of contravariant symmetric tensor fields on M is naturally a $\mathrm{Diff}(M)$- and $\mathrm{Vect}(M)$-module. We will, however, need to consider a one-parameter family of Diff(M)- and Vect(M)-actions on this space. Put

$$\mathcal{S}_\delta = \mathcal{S}(M) \otimes_{C^\infty(M)} \mathcal{F}_\delta. \tag{2.1}$$

The space $\mathcal{S}_\delta$ is also, naturally, a Diff(M)- and Vect(M)-module.

2.2. Action of Vect(M) on the space of symbols

Space $\mathcal{S}(M)$ is a Poisson algebra isomorphic as a Vect(M)-module to the space of smooth functions on T^*M polynomial on the fibers. The action of $X \in \mathrm{Vect}(M)$ on $\mathcal{S}$ is given by the Hamiltonian vector field

$$L_X = \frac{\partial X}{\partial \xi_i}\frac{\partial}{\partial x^i} - \frac{\partial X}{\partial x^i}\frac{\partial}{\partial \xi_i}, \tag{2.2}$$

where (x^i, ξ_i) are local coordinates on T^*M (we identified vector fields on M with the first-order polynomials on T^*M, that is $X = X^i\xi_i$ in (2.2))*. The Poisson bracket on $\mathcal{S}(M)$ is usually called the (symmetric) Schouten bracket (see, e.g., [8]).

The Vect(M)-action on $\mathcal{S}_\delta$ is of the form:

$$L_X^\delta = L_X + \delta \mathrm{D}(X), \tag{2.3}$$

where

$$\mathrm{D}(X) = \partial_i X^i, \qquad \partial_i = \frac{\partial}{\partial x^i}. \tag{2.4}$$

Note that the formula (2.3) does not depend on the choice of local coordinates. The space $\mathcal{S}_\delta$ is naturally the space of symbols corresponding to the space of differential operators $\mathcal{D}_{\lambda,\mu}$ such that $\mu - \lambda = \delta$.

2.3. Isomorphism of $\mathfrak{g}$-modules

Let now M be a endowed with a flat projective or conformal structure (see, e.g., [17] and Section 3.1 for precise definition). The most general results of [15, 13, 6, 7] can be formulated as follows.

Theorem 2.1. *For generic values of δ there exists an isomorphism of $\mathfrak{sl}(n+1,\mathbb{R})$- or $\mathfrak{o}(p+1,q+1)$-modules (respectively):*

$$\widetilde{\mathcal{Q}}_{\lambda,\mu} : \mathcal{S}_\delta \xrightarrow{\cong} \mathcal{D}_{\lambda,\mu}, \qquad \delta = \mu - \lambda \tag{2.5}$$

which is unique provided the principal symbol be preserved at each order.

There exist values of δ such that the the modules $\mathcal{S}_\delta$ and $\mathcal{D}_{\lambda,\mu}$ with $\mu-\lambda=\delta$ are not isomorphic for generic λ. These particular values are called *resonant*.

*Throughout this paper the summation over repeated indices is understood.

Remark 2.2.

(a) In the projective case the resonance values of δ are $\delta = k/(n+1)$, where k is integer $\geq n+1$ (see [13]).

(b) In the conformal case the structure of resonances is much more complicated (see [7]).

(c) In the one-dimensional case, $M = S^1$ or $\mathbb{R}$, the above theorem still holds true but the resonances are simply $\delta = 1, \frac{3}{2}, 2, \frac{5}{2}, \ldots$ and appear in [3, 9]. (The projective or conformal structure is then replaced by the natural $\mathfrak{sl}(2, \mathbb{R})$ action.)

2.4. Definition of quantization map

Let us introduce a new operator on symbols that will eventually insure the symmetry of the corresponding differential operators. Define $\mathcal{I}_\hbar : \mathcal{S}_\delta \to \mathcal{S}_\delta[[i\hbar]]$ by

$$\mathcal{I}_\hbar(P)(\xi) = P(i\hbar\,\xi). \tag{2.6}$$

Note that we will understand $\hbar$ either as a formal parameter or as a fixed real number, depending upon the context. We will introduce the quantization map $\mathcal{Q}_{\lambda,\mu;\hbar} : \mathcal{S}_\delta \to \mathcal{D}_\lambda[[i\hbar]]$ depending on $\hbar$:

$$\mathcal{Q}_{\lambda,\mu;\hbar} = \widetilde{\mathcal{Q}}_{\lambda,\mu} \circ \mathcal{I}_\hbar \tag{2.7}$$

where $\widetilde{\mathcal{Q}}_{\lambda,\mu}$ is the isomorphism from Theorem 2.1.

2.5. Symmetric quantized operators

Let us recall that if $\lambda+\mu = 1$, there exists, for compactly-supported densities, a $\mathrm{Vect}(M)$-invariant pairing $\mathcal{F}_\lambda^{\mathbb{C}} \otimes \mathcal{F}_\mu^{\mathbb{C}} \to \mathbb{C}$ defined by

$$\varphi \otimes \psi \mapsto \int_M \overline{\varphi}\,\psi. \tag{2.8}$$

We can now formulate the important corollary:

Corollary 2.3. *For generic δ and $\lambda + \mu = 1$ the quantization*

$$\check{P} = \mathcal{Q}_{\frac{1-\delta}{2}, \frac{1+\delta}{2};\hbar}(P) \tag{2.9}$$

of any symbol $P \in \mathcal{S}_\delta$ is a symmetric (formally self-adjoint) operator.

Proof. Let us denote by A^* the adjoint of $A \in \mathcal{D}_{\lambda,1-\lambda}$ with respect to the pairing (2.8). Consider the symmetric operator

$$\mathrm{Symm}(\mathcal{Q}_{\lambda,\mu;\hbar}(P)) = \tfrac{1}{2}\Big(\mathcal{Q}_{\lambda,\mu;\hbar}(P) + (\mathcal{Q}_{\lambda,\mu;\hbar}(P))^*\Big),$$

which exists whenever $\lambda + \mu = 1$. Notice that it has the same principal symbol as $\mathcal{Q}_{\lambda,\mu;\hbar}(P)$. Now, the map $\mathrm{Symm}(\mathcal{Q}_{\lambda,\mu;\hbar})$ is obviously $\mathfrak{g}$-equivariant. The result follows then from the uniqueness of $\mathfrak{g}$-module isomorphism — see Theorem 2.1. □

2.6. Invariant star products

The special and most important value $\delta = 0$ is non-resonant. The existence and uniqueness of the isomorphism (2.5) in this case has been proven in [15, 7].

Define an associative bilinear operation (depending on λ)

$$*_{\lambda;\hbar} : \mathcal{S} \otimes \mathcal{S} \to \mathcal{S}[[i\hbar]] \tag{2.10}$$

such that

$$\mathcal{Q}_{\lambda;\hbar}(P *_{\lambda;\hbar} Q) = \mathcal{Q}_{\lambda;\hbar}(P) \circ \mathcal{Q}_{\lambda;\hbar}(Q). \tag{2.11}$$

Recall that an associative operation $*_\hbar : \mathcal{S} \otimes \mathcal{S} \to \mathcal{S}[[i\hbar]]$ is called a *star product* if it is of the form

$$P *_\hbar Q = PQ + \frac{i\hbar}{2}\{P, Q\} + O(\hbar^2), \tag{2.12}$$

where $\{\cdot,\cdot\}$ stands for the Poisson bracket on T^*M, and is given by bi-differential operators at each order in $\hbar$.

Theorem 2.4. *The associative, conformally invariant, operation $*_{\lambda;\hbar}$ defined by (2.11) is a star product if and only if $\lambda = \frac{1}{2}$.*

Let us emphasize that this theorem provides us precisely with the value of λ used in geometric quantization and, in some sense, links the latter to deformation quantization.

2.7. Quantum Hamiltonians in the (pseudo-)Riemannian case

Assume now that a (pseudo-)Riemannian metric g on M is fixed.

To recover the traditional Schrödinger picture of quantum mechanics, one needs to associate to the operator $\check{P}$ resulting from our quantization map (2.9) an operator

$$\widehat{P} : \mathcal{F}_0^{\mathbb{C}} \to \mathcal{F}_0^{\mathbb{C}} \tag{2.13}$$

on the space of complex-valued functions on a conformally flat manifold (M, g).

Definition 2.5. Using the natural identification $\mathcal{F}_0^{\mathbb{C}} \to \mathcal{F}_\lambda^{\mathbb{C}}$ between tensor densities and smooth functions given by

$$f \mapsto f\, |\mathrm{Vol}_{\mathrm{g}}|^{\lambda}, \tag{2.14}$$

one can define the differential operator $\widehat{P}$ by the commutative diagram

$$\begin{array}{ccc} \mathcal{F}_0^{\mathbb{C}} & \xrightarrow{\widehat{P}} & \mathcal{F}_0^{\mathbb{C}} \\ {\scriptstyle |\mathrm{Vol}_{\mathrm{g}}|^{\lambda}} \downarrow & & \downarrow {\scriptstyle |\mathrm{Vol}_{\mathrm{g}}|^{\mu}} \\ \mathcal{F}_\lambda^{\mathbb{C}} & \xrightarrow{\check{P}} & \mathcal{F}_\mu^{\mathbb{C}} \end{array} \tag{2.15}$$

where $\check{P}$ is given by (2.9) in the case $\lambda + \mu = 1$.

We call the operator $\widehat{P}$ corresponding to a polynomial P the quantum Hamiltonian.

Remark 2.6. An important feature of equivariant quantization in the conformal case is, of course, that the metric g can be chosen conformally flat (as a representative of the conformal structure on M).

3. Examples: quantizing quadratic Hamiltonians

Let us consider a homogeneous second order polynomial H on T^*M. In arbitrary local coordinates it is of the form:

$$H = h^{ij}(x)\xi_i\xi_j. \tag{3.1}$$

To illustrate the general Theorem 2.1 we will give explicit formulæ for the quantization in this case.

We will need special coordinate systems adopted to projective and conformal structures.

3.1. Local coordinates

The following description can be considered as a definition of flat projective and conformal structures.

Definition 3.1. A smooth manifold M is endowed with a projective (or conformal) structure if there exists an atlas U_i on M with local coordinates $(x^1, \ldots, x^n)_i$ in each chart U_i that are accorded on $U_i \cap U_j$ in the following way.

(1) *Projective case*: The linear span of the following $n(n+2)$ vector fields

$$\langle \partial_i\,, \quad x^i\partial_j\,, \quad x^i\mathcal{E} \rangle\,, \qquad \text{where} \quad \mathcal{E} = x^j\partial_j\,. \tag{3.2}$$

is independent of the choice of a chart (i.e., coincides on each $U_i \cap U_j$). The vector fields (3.2) obviously generate the Lie algebra isomorphic to $\mathfrak{sl}(n+1, \mathbb{R})$ and, therefore, one obtains a (locally defined) $\mathfrak{sl}(n+1, \mathbb{R})$-action on M.

It worth noticing that coordinate changes on $U_i \cap U_j$ in this case are nothing but linear-fractional transformations.

(2) *Conformal case*: The linear span of the following $(n+2)(n+1)/2$ vector fields

$$\langle \partial_i,\ x_i\partial_j - x_j\partial_i,\ \mathcal{E},\ x_jx^j\partial_i - 2x_i\mathcal{E} \rangle \tag{3.3}$$

coincide on each $U_i \cap U_j$ (we have used the notation $x_i = g_{ij}x^j$ where the flat metric $g = \mathrm{diag}(1, \ldots, 1, -1, \ldots, -1)$ is of signature (p, q)). Note that the vector fields (3.3) generate the Lie algebra isomorphic to $\mathfrak{o}(p+1, q+1)$.

The coordinate changes are then given by the conformal transformations.

3.2. The quantization map in the second-order case

We are now ready to give explicit expressions of our quantization maps. For the sake of simplicity we restrict ourself to the most interesting case $\lambda = \mu = \frac{1}{2}$ (see [15, 6, 7] for the general formulæ). We will use coordinates adopted to a projective or conformal structure on M.

(1) *Projective case*: The quantization map (2.7) associates to the polynomial (3.1) the following second-order differential operator

$$\mathcal{Q}_{\frac{1}{2};\hbar}(H) = -\hbar^2 \left(h^{ij}\partial_i\partial_j + \partial_j(h^{ij})\partial_i + \frac{(n+1)}{4(n+2)}\partial_i\partial_j(h^{ij}) \right) \tag{3.4}$$

on the space $\mathcal{F}^{\mathbb{C}}_{1/2}$ of half-densities.

(2) *Conformal case*: One has

$$\begin{aligned} \mathcal{Q}_{\frac{1}{2};\hbar}(H) \quad = \quad & -\hbar^2 \Big(h^{ij}\partial_i\partial_j + \partial_j(h^{ij})\partial_i + \frac{n}{4(n+1)}\partial_i\partial_j(h^{ij}) \\ & + \frac{n}{4(n+1)(n+2)}\partial_i\partial_i(h^{jj}) \Big) \end{aligned} \tag{3.5}$$

Let us also recall here for comparison the explicit expression of the most famous and standard quantization procedure, called the *Weyl* or *symmetric* quantization.

(3) *Weyl quantization*: In the case of second-order polynomials

$$\mathcal{Q}_{\text{Weyl}}(H) = -\hbar^2 \left(h^{ij}\partial_i\partial_j + \partial_j(h^{ij})\partial_i + \tfrac{1}{4}\partial_i\partial_j(h^{ij}) \right) \tag{3.6}$$

This formula makes sense globally only in the case $M = \mathbb{R}^n$ and it is then defined uniquely (up to normalization) by equivariance with respect to the standard $\mathfrak{sp}(2n, \mathbb{R})$-action on $T^*M = \mathbb{R}^{2n}$.

Remark 3.2. It can be checked that:

(a) the explicit formulæ (3.4,3.5) are independent of the choice of the adopted coordinate system;

(b) in both cases the differential operator $\mathcal{Q}_{\frac{1}{2};\hbar}(H)$ is symmetric.

3.3. Quantizing the geodesic flow

Assume now that there exists (a non-degenerate) metric

$$\mathrm{g} = \mathrm{g}_{ij} dx^i dx^j \tag{3.7}$$

corresponding to the polynomial (3.1), i.e., $H = \mathrm{g}^{-1}$, that is, $h^{ij} \equiv \mathrm{g}^{ij}$.

Let us apply the procedure of Section 2.7 to obtain the quantum Hamiltonian. It is easy to see that the fact that the operators (3.4–3.6) are symmetric implies in each case (1), (2), (3) that the corresponding quantum Hamiltonian has the general form:

$$\widehat{H} = -\hbar^2 \left(\Delta_{\mathrm{g}} + U\right) \tag{3.8}$$

where Δ_{g} is the Laplace–Beltrami operator and the function U (the potential) is different in each of the three cases.

The potential can be easily computed in each case (1), (2), (3). In particular, in the conformal case one can choose g *conformally flat.* It has been shown in [6] that in such a situation

$$\widehat{H} = -\hbar^2 \left(\Delta_{\mathrm{g}} - \frac{n^2}{4(n-1)(n+2)} R_{\mathrm{g}}\right), \tag{3.9}$$

where R_{g} stands for the scalar curvature of (M, g). It worth noticing that this result differs from other known versions of the quantization of the geodesic flow (cf. [4] and references therein).

4. Universal formula in the projective case

In the projective case there exists an explicit formula for the quantization map that is valid for polynomial functions on T^*M of an arbitrary order k (see [15, 13]). In this section we will give a new (but still equivalent) expression of the projectively equivariant quantization map. As above, we will consider only the case $\lambda = \mu = \frac{1}{2}$.

4.1. The Euler and divergence operators

Fix an arbitrary system of local coordinates $(x^1, \ldots, x^n)$ on M and introduce the following differential operators acting *locally* on the space of symbols (i.e., on the space of polynomials $\mathbb{C}[x^1, \dots, x^n, \xi_1, \dots, \xi_n]$):

$$\mathrm{E} = \xi_i \frac{\partial}{\partial \xi_i} + \frac{n}{2}, \qquad \mathrm{D} = \frac{\partial}{\partial \xi_i} \frac{\partial}{\partial x^i} \tag{4.1}$$

4.2. The confluent hypergeometric function

Recall that the confluent hypergeometric function is a series in z depending on two parameters $a, b \in \mathbb{R}$ (or $\mathbb{C}$):

$$F\left(\begin{array}{c} a \\ b \end{array} \middle| \, z \right) = \sum_{m=0}^{\infty} \frac{(a)_m}{(b)_m} \frac{z^m}{m!}$$

where

$$(a)_m := a(a+1) \cdots (a+m-1)$$

(see, e.g., [11]).

4.3. The explicit formula

Identify locally polynomials with differential operators by the normal ordering map

$$\sigma : A_k^{i_1 \dots i_k} \partial_{i_1} \cdots \partial_{i_k} \mapsto A_k^{i_1 \dots i_k} \xi_{i_1} \cdots \xi_{i_k} \tag{4.2}$$

Using this identification we will work with differential operators as with polynomials.

The following theorem provides an explicit expression for the projectively equivariant quantization map (2.7) in terms of a "noncommutative" confluent function.

Theorem 4.1. *The projectively equivariant quantization is of the form*

$$\mathcal{Q}_{\frac{1}{2};\hbar} = F\left(\begin{array}{c} 2\mathrm{E} \\ \mathrm{E} \end{array} \middle| \, \frac{i\hbar \mathrm{D}}{4} \right) \tag{4.3}$$

for every system of local coordinates adopted to the projective structure.

In other words, one has

$$\mathcal{Q}_{\frac{1}{2};\hbar} = \sum_{m=0}^{\infty} C_m(\mathrm{E}) \, (i\hbar \mathrm{D})^m \tag{4.4}$$

where the expression

$$C_m(\mathrm{E}) = \frac{1}{m!} \frac{(\mathrm{E} + \frac{1}{2})(\mathrm{E} + \frac{3}{2}) \cdots (\mathrm{E} + m - \frac{1}{2})}{(2\mathrm{E} + m)(2\mathrm{E} + m + 1) \cdots (2\mathrm{E} + 2m - 1)} \tag{4.5}$$

is also understood as an infinite (Taylor) series in E.

Remark 4.2.

(a) If the manifold M is endowed with a projective structure, then choosing an adapted coordinate system, the formula (4.3) is well defined globally on T^*M.

(b) One easily checks that for the second-order polynomials the formula (4.3) coincides with (3.4).

The expression (4.3) is a particular case (for $\lambda = 1/2$) of the quantization map given in [15] after restriction to the space of k-th order polynomials. Unfortunately, we do not know the analog of the formula (4.3) in the conformal case.

4.4. Comparison with the Weyl quantization

The Weyl quantization map, $\mathcal{Q}_{\mathrm{Weyl}}$, retains the very elegant form:

$$\begin{aligned}\mathcal{Q}_{\mathrm{Weyl}} &= \exp\left(\frac{i\hbar}{2}\mathrm{D}\right)\\ &= \mathrm{Id} + \frac{i\hbar}{2}\mathrm{D} - \frac{\hbar^2}{8}\mathrm{D}^2 + O(\hbar^3)\end{aligned} \tag{4.6}$$

(see [1]). Comparing (4.6) with (4.3) one notes that the exponential is, indeed, the most elementary hypergeometric function.

4.5. An open problem

It is well-known that the Weyl quantization can be extended for a much larger class of functions on $\mathbb{R}^{2n}$ than polynomials, namely to the space of pseudo-differential symbols.

We formulate here the following problem. Is it true that the formula (4.3) is also valid for pseudo-differential symbols? If the answer is positive it would be interesting to compute its integral expression.

References

[1] G.S. Argaval, E. Wolf, Calculus for functions of noncommuting operators and general phase space methods in quantum mechanics, I. Mapping theorems and ordering of functions on noncommuting operators, *Phys. Rev. D*, **2:10** (1970) 2161–2188.

[2] F. Boniver and P. B. A. Lecomte, A remark about the Lie algebra of infinitesimal conformal transformations of the Euclidian space, to appear in *J. of London Math. Soc.*

[3] P. Cohen, Yu. Manin and D. Zagier, Automorphic pseudodifferential operators, in *Algebraic aspects of integrable systems*, Progr. Nonlinear Dif-

ferential Equations Appl., 26, Birkhäuser Boston, Boston, MA, 1997, pp. 17–47.

[4] C. Duval, J. Elhadad and G.M. Tuynman, The BRS Method and Geometric Quantization: Some Examples, *Comm. Math. Phys.*, **126** (1990), 535–557.

[5] C. Duval and V. Ovsienko, Space of second order linear differential operators as a module over the Lie algebra of vector fields, *Adv. in Math.* **132**: 2 (1997), 316–333.

[6] C. Duval and V. Ovsienko, Conformally equivariant quantization, Preprint CPT-98/P.3610, math.DG/9801122.

[7] C. Duval, P. Lecomte and V. Ovsienko, Conformally equivariant quantization: existence and uniqueness, to appear in *Ann. Inst. Fourier.*

[8] D. B. Fuchs, *Cohomology of infinite-dimensional Lie algebras*, Consultants Bureau, New York, 1987.

[9] H. Gargoubi, Sur la géométrie des opérateurs différentiels linéaires sur $\mathbb{R}$, Preprint CPT, 1997, P.3472.

[10] H. Gargoubi and V. Ovsienko, Space of linear differential operators on the real line as a module over the Lie algebra of vector fields, *Int. Res. Math. Notes*, 1996, No. 5, 235–251.

[11] R.L. Graham, D.E. Knuth and O. Patashnik, *Concrete mathematics. A foundation for computer science*, Second edition, Addison-Wesley Publishing Company, Reading, MA, 1994.

[12] A. A. Kirillov, Geometric Quantization, in *Encyclopedia of Math. Sci.*, Vol. 4, Springer-Verlag, 1990.

[13] P. B. A. Lecomte, On the cohomology of $\mathfrak{sl}(m+1,\mathbb{R})$ acting on differential operators and $\mathfrak{sl}(m+1,\mathbb{R})$-equivariant symbol, Preprint Université de Liège, 1998.

[14] P. B. A. Lecomte, P. Mathonet and E. Tousset, Comparison of some modules of the Lie algebra of vector fields, *Indag. Math., N.S.*, **7**: 4 (1996), 461–471.

[15] P. B. A. Lecomte and V. Ovsienko, Projectively invariant symbol calculus, math.DG/9809061.

[16] P. Mathonet, Intertwining operators between some spaces of differential operators on a manifold, *Comm. in Algebra* (1999), to appear.

[17] W. P. Thurston, *Three-Dimensional Geometry and Topology*, Vol. 1, Princeton Univ. Press, 1997.

[18] E. J. Wilczynski, *Projective differential geometry of curves and ruled surfaces*, Leipzig, Teubner, 1906.

APPLICATION OF NONCOMMUTATIVE DIFFERENTIAL GEOMETRY ON LATTICE TO ANOMALY ANALYSIS IN ABELIAN LATTICE GAUGE THEORY

Takanori Fujiwara
Department of Mathematical Sciences
Ibaraki University, Mito 310-8512, JAPAN

Hiroshi Suzuki
Department of Mathematical Sciences
Ibaraki University, Mito 310-8512, JAPAN

Ke Wu*
Department of Mathematical Sciences
Ibaraki University, Mito 310-8512, JAPAN

Abstract The chiral anomaly in lattice abelian gauge theory is investigated by applying noncommutative differential geometry (NCDG). A new kind of double complex and descent equation is proposed on an infinite hypercubic lattice in arbitrary even-dimensional Euclidean space in the framework of NCDG. By using the general solutions of descent equation, we derive the chiral anomaly in Abelian lattice gauge theory. The topological origin of the anomaly is nothing but the Chern classes in NCDG.

Mathematics Subject Classification (2000): 81T27, 81T70, 81T75

* On leave of absence from Institute of Theoretical Physics, Academia Sinica, P.O.Box 2735, Beijing 100080, CHINA

Y. Maeda, et al. (eds.), Noncommutative Differential Geometry and Its Applications to Physics, 13–30.

Keywords: Abelian lattice gauge theory, axial anomaly, noncommutative geometry

1. Introduction

There is an important 'no go' theorem in lattice gauge theory proved by Nielson and Ninomiya (NN) [1]. The theorem states that all the naive properties for chiral fermions on a lattice can not hold simultaneously. Thus the construction of chiral field theories on lattice appears to be more difficult. Since then very many efforts have been made in order to get around the NN theorem. One method has been suggested by Ginsparg and Wilson [2] some years ago. They modified the chirality of Dirac operator D, one of the conditions in NN theorem, to the so called Ginsparg–Wilson (GW) relation as:

$$\gamma_5 D + D\gamma_5 = aD\gamma_5 D, \tag{1.1}$$

where a is the lattice spacing taken to be unity in the present paper.

Recently, there have been remarkable developments regarding this problem. [3, 4, 5, 6] It has been shown that the chiral symmetry may be preserved in lattice gauge theory at least to some extent, if the lattice Dirac operator takes a particular form. Although the proposed Dirac operators are very complicated, all of them satisfy the GW relation. It was soon recognized that the GW relation implies the index theorem [4, 5, 6] and the exact chiral symmetry[6] of the fermion action. The chiral or axial anomaly arises from the non-invariance of the fermion integration measure under the chiral transformations. In terms of the GW Dirac operator D, the chiral anomaly can be written as

$$q(x) = -\tfrac{1}{2}\mathrm{tr}\gamma_5 D(x,x). \tag{1.2}$$

From the property of the Dirac operator D satisfying GW relation, the anomaly is gauge invariant and local (in the sense that it locally depends on the gauge field). Moreover its variation under any local deformation of gauge field satisfies

$$\sum_x \delta q(x) \;=\; 0, \tag{1.3}$$

which reflects the topological nature of the anomaly.

The understanding of structure of the gauge anomaly on the lattice with finite spacing is quite important. Many papers [7] appeared recently, which study the properties of the chiral anomaly on the lattice, such as its perturbative property, continuum limit, etc. Among of them, Lüscher derived a theorem [8] to show the explicit form of the chiral anomaly $q(x)$ in abelian gauge theory on four-dimensional euclidean infinite hypercubic lattice,

$$q(x) = \alpha + \beta_{\mu\nu}F_{\mu\nu}(x) + \gamma\varepsilon_{\mu\nu\rho\sigma}F_{\mu\nu}(x)F_{\rho\sigma}(x+\widehat{\mu}+\widehat{\nu}) + \partial^*_\mu k_\mu(x). \tag{1.4}$$

The notation will be explained in the next section. Furthermore, Lüscher proved that there exists a gauge invariant lattice formulation of Abelian chiral gauge theory [9].

Lüscher's theorem has been extended by us to arbitrary even dimensional regular lattice using noncommutative geometry. Once the exterior derivative on a lattice with nilpotency and Leibniz' rule is obtained as in usual differential geometry, then BRST analysis, which was used to discuss the chiral anomaly in continuum theory, is extended to abelian lattice gauge theory in [10]. A double complex, descent equation and its general solution are used to derive the anomaly in [11]. In this paper we will continue our research on chiral anomaly of lattice gauge theory. The chiral anomaly will be re-derived in terms of a new kind of double complex and its descent equation in noncommutative differential geometry (NCDG).

The paper is organized as follows. First, some basic result of NCDG on lattice will be reviewed in Section 2. Then a new kind of double complex and descent equation are proposed, and the general solution to the descent equation will be studied in Section 3. The chiral anomalies are derived in Section 4. Finally, conclusion and remarks are included in Section 5.

2. Noncommutative differential geometry on the D-dimensional lattice Z^D

NCDG does not yet have a long history; however, it has obtained wide applications in physics, especially in quantum physics. In fact, one can trace back to Heisenberg and Dirac in the early part of this century, when quantum mechanics was discovered. They deformed the commutative classical phase space to quantum phase space which is noncommutative. In quantum mechanics the physical quantities have been quantized to be operators in Hilbert space. It is such representation that motivated von Neumann to study the operator algebras which, in turn, inspired the foundation of noncommutative geometry. The main method in NCDG by Connes [12] is to study the algebraic object like function algebra $\mathcal{A}(\mathrm{M})$ defined on manifold M rather than the geometric object such as manifold M itself. It can therefore be used to discuss some geometric objects which can not be treated in the usual way, such as the discrete set, or the quantum group. The NCDG on lattice used in this paper is one of the simple case of noncommutative differential calculus on a discrete set. Recently its application to lattice gauge theory has achieved some progress [8], although most of the mathematics involved has appeared before [12, 13]. Some basic results are reviewed in this section without proof.

The D-dimensional infinite hypercubic regular lattice Z^{D} could be understood as the discretization of continuous Euclidean space M^{D}. The discreteness can not be expressed properly by usual differential geometry. However it can be well described by noncommutative geometry. In this formalism the first remarkable property is that there are two types of vector field defined on the algebra $\mathcal{A}(\mathrm{Z^D})$, $\mathcal{A}(\mathrm{Z^D})$ being the algebra of function defined on Z^D: the forward difference operator ∂_μ and the backward difference operator ∂^*_μ, playing the role of the ordinary derivative ∂_x on $\mathcal{A}(\mathrm{M})$, as follows:

$$\begin{aligned} \partial_\mu f(x) &= f(x+\widehat{\mu}) - f(x), \\ \partial^*_\mu f(x) &= f(x) - f(x-\widehat{\mu}), \end{aligned} \tag{2.1}$$

where $f(x)$ is an element in $\mathcal{A}(\mathrm{Z^D})$, i.e., an arbitrary function on the lattice and as shown in 1.1, is taken as unity in the present paper.

In order to obtain the full differential calculus on $\mathcal{A}(\mathrm{Z^D})$ one needs to define differential forms on it. The basis of 1-form on it are abstract objects $dx_1, dx_2, \cdots, dx_D$, satisfying the Grassmann algebra relation:

$$dx_\mu dx_\nu = -dx_\nu dx_\mu. \tag{2.2}$$

Then we can define an exterior derivative operator d on $\mathcal{A}(\mathrm{Z^D})$ with forward difference operator as

$$df(x) = \partial_\mu f(x) dx_\mu, \tag{2.3}$$

where the summation on repeated indices is understood. For generic n-form,

$$f(x) = \frac{1}{n!} f_{\mu_1 \cdots \mu_n}(x)\, dx_{\mu_1} \cdots dx_{\mu_n}, \tag{2.4}$$

where the coefficient $f_{\mu_1 \cdots \mu_n}(x)$ is an antisymmetric tensor with rank n, the exterior derivative operator d maps an n-form to an $(n+1)$-form as

$$df(x) = \frac{1}{n!} \partial_\mu f_{\mu_1 \cdots \mu_n}(x)\, dx_\mu dx_{\mu_1} \cdots dx_{\mu_n}. \tag{2.5}$$

It is easy to notice that the difference operators do not satisfy the ordinary Leibniz' rule as a usual derivative operator. This is an important difference between commutative and noncommutative geometry. In order to obtain the correct differential calculus on $\mathcal{A}(\mathrm{Z^D})$ one should remedy this lack of a Leibniz' rule for the difference operators. Instead of a difference operator one can define the exterior derivative operator d to satisfy Leibniz' rule. There is an important lemma concerning the exterior derivative operator in order to obtain the correct Leibniz' rule for d.

Lemma 1: The exterior derivative operator d on $\mathcal{A}(\mathrm{Z^D})$ satisfies the nilpotency $d^2 = 0$ and Leibniz' rule

$$d[f(x)g(x)] = df(x)g(x) + (-1)^n f(x) dg(x), \tag{2.6}$$

where $f(x)$ is an n-form, if and only if the forms and functions satisfy the following conditions:

$$\begin{aligned} dx_\mu dx_\nu &= -dx_\nu dx_\mu, \\ dx_\mu f(x) &= f(x+\widehat{\mu})\, dx_\mu. \end{aligned} \tag{2.7}$$

The last relation shows that functions and 1-forms do not commute in order to have the right Leibniz' rule, such that noncommutativity appears. This is the essence of NCDG on a regular lattice and represents the discreteness of the lattice.

We can also introduce the gauge potential 1-form and the field strength 2-form on the lattice in terms of NCDG, once we get the whole form space and exterior derivative d acting on it, satisfying the nilpotency and the Leibniz' rule.

$$A(x) = A_\mu(x)\, dx_\mu, \qquad F(x) = dA(x) = \tfrac{1}{2} F_{\mu\nu}(x)\, dx_\mu dx_\nu, \tag{2.8}$$

where $F_{\mu\nu}(x) = \partial_\mu A_\nu(x) - \partial_\mu A_\nu(x)$ and d is the exterior derivative operator on $\mathcal{A}(\mathrm{Z^D})$. Note that the Bianchi identity $dF(x) = 0$ follows from the nilpotency of d. The field strength 2-form $F(x)$ is gauge invariant under a gauge transformation $A(x) \to A(x) - d\lambda(x)$. Although the gauge potential and the field strength have the same expressions as the ones in ordinary gauge theory, they have a different meaning because they are defined on lattice using NCDG. It is also called the noncommutative gauge theory on lattice, where the noncommutavity is widely used.

In lattice gauge theory the fundamental gauge field is the link variable the points x and $x + \widehat{\mu}$ and taking values in gauge group G. We restrict our discussion to the abelian case $G = U(1)$ in this paper. The gauge transformation acts on gauge fields as,

$$U_\mu(x) \to \Lambda(x) U_\mu(x) \Lambda(x+\widehat{\mu})^{-1}. \tag{2.9}$$

Where the parameter $\Lambda(x)$ of gauge transformation also takes values in $U(1)$. Unlike the noncommutative gauge theory on lattice, the gauge field strength is given by the plaquette variable

$$F_{\mu\nu}(x) = \frac{1}{i} \ln U_\mu(x) U_\nu(x+\widehat{\mu}) U_\mu(x+\widehat{\nu})^{-1} U_\nu(x)^{-1}. \tag{2.10}$$

We assume the 'admissibility' condition for the gauge field configuration [8], i.e.:

$$\sup_{x,\mu,\nu} |F_{\mu\nu}(x)| < \epsilon, \tag{2.11}$$

where ϵ is a fixed constant $0 < \epsilon < \pi/3$. Then we will obtain the relation of noncommutative gauge theory on lattice and lattice gauge theory by the lemma from Lüscher [8].

Lemma 2: Suppose $U_\mu(x)$ is admissible $U(1)$ gauge field. Then there exists a vector field $A_\mu(x)$ such that:

$$U_\mu(x) = e^{iA_\mu(x)}, \quad F_{\mu\nu}(x) = \partial_\mu A_\nu(x) - \partial_\mu A_\nu(x). \tag{2.12}$$

This lemma shows that the noncommutative gauge theory defined by NCDG on the lattice is equivalent to lattice gauge theory at least in the abelian case.

In noncommutative gauge theory on the lattice Z^D, the Chern classes of that gauge theory could be defined as:

$$c_k = F^k = FF\cdots F, \quad k = 1, 2, \cdots, \frac{D}{2}. \tag{2.13}$$

Although this looks like the Chern class in usual differential geometry, in fact, it is different from the usual Chern class.

$$\begin{aligned} c_k \quad = \quad & F_{\mu_1\nu_1}(x)F_{\mu_2\nu_2}(x + \widehat{\mu}_1 + \widehat{\nu}_1)\cdots \\ & \times F_{\mu_k\nu_k}(x + \widehat{\mu}_1 + \widehat{\nu}_1 + \cdots + \widehat{\mu}_{k-1} + \widehat{\nu}_{k-1}) \\ & \times dx_{\mu_1}dx_{\nu_1}dx_{\mu_2}dx_{\nu_2}\cdots dx_{\mu_k}dx_{\nu_k} \end{aligned} \tag{2.14}$$

One can easily prove that c_k is gauge invariant and closed $dc_k = 0$.

One needs the Poincaré lemma on the lattice, in order to understand the topological properties of the D-dimensional lattice Z^D, i.e., the cohomology properties of $\mathcal{A}(\mathrm{Z}^\mathrm{D})$ in terms of NCDG. For simplicity we will restrict our discussion to the k-forms with compact support or with exponentially decaying coefficients (this is corresponding to the locality in lattice gauge theory) and denote the linear space of all k-forms as Ω_k. The Poincaré lemma is as follows [8].

Lemma 3 (Poincaré Lemma for d): Let $f \in \Omega_k$ be a k-form satisfying: $df(x) = 0$ and $\sum_x f(x) = 0$ for $k = D$. Then there exists a form $g(x) \in \Omega_{k-1}$ such that $f(x) = dg(x)$.

In the proof of this lemma the locality plays a crucial role. If the reference point chosen in the proof is located in the compact support of $f(x)$, then $g(x)$ is supported on the same rectangular block of lattice as $f(x)$. The construction of the form $g(x)$ is explicit and its coefficients are just some particular linear combinations of the coefficients of $f(x)$.

In form space Ω_k the inner product, corresponding to the metric in D-dimensional lattice Z^D, is defined as:

$$\begin{aligned} \langle dx_\mu, dx_\nu \rangle &= \delta_{\mu,\nu}, \\ \langle dx_{\mu_1}\cdots dx_{\mu_k}, dx_{\nu_1}\cdots dx_{\nu_l} \rangle &= \delta_{k,l}\mathrm{det}(\langle dx_{\mu_i}, dx_{\nu_j}\rangle_{ij}). \end{aligned} \tag{2.15}$$

Then the divergence operator $d^* : \Omega_k \to \Omega_{k-1}$ defined by the backward difference operator

$$d^* f(x) = \frac{1}{(k-1)!} \partial^*_\mu f_{\mu\mu_2\cdots\mu_k}(x)\, dx_{\mu_2} \cdots dx_{\mu_n}, \tag{2.16}$$

is nothing but the dual operator of $-d$ in the Hilbert space $\Omega_\bullet = \sum_{k=0}^{k=D} \Omega_k$ with inner product $\sum_x \langle \cdot\,,\, \cdot \rangle$. It also is nilpotent $(d^*)^2 = 0$ and hence there is another Poincaré lemma for d^* [8].

Lemma 4 (Poincaré Lemma for d^*) Let $f \in \Omega_k$ be a k-form satisfying: $d^* f(x) = 0$ and $\sum_x f(x) = 0$ for $k = 0$. Then there exists a form $g(x) \in \Omega_{k+1}$ such that $f(x) = d^* g(x)$.

The proof of this lemma is almost the same as that of **Lemma 3**. Also the locality plays an important role.

One should notice that if one naively applies these lemmas to lattice gauge theory with a particular choice of the reference point, then the translation invariance will be lost. In order to overcome this difficulty, the 'bi-local' field $f(x,y)$ was used in Lüscher's paper [8], when one applies the Poincaré lemma to x then y is the reference point and vice versa. The 'bi-local' means that the composite field $f(x,y)$ decreases at least exponentially as $|x-y| \to \infty$. In this way, the locality is resumed, and the Poincaré Lemma can be applied to lattice gauge theory without problem.

Therefore we should work on the 'bi-local' fields. Let us introduce the noncommutative differential calculus on the algebra of all 'bi-local' fields, which is the functions of both x and y defined on $\mathrm{Z^D} \times \mathrm{Z^D}$. Besides two types of vectors of forward difference operator ∂_μ and backward difference operator ∂^*_μ for x, there is one more set of forward and backward difference operators acting on y, defined as,

$$\begin{aligned} \partial^y_\mu f(x,y) &= f(x,y)\overleftarrow{\partial}_\mu = f(x,y+\hat\mu) - f(x,y)\,, \\ \partial^{*y}_\mu f(x,y) &= f(x,y)\overleftarrow{\partial}^*_\mu = f(x,y) - f(x,y-\hat\mu)\,. \end{aligned} \tag{2.17}$$

Furthermore, we need another copy of the basis of exterior 1-forms $dy_\mu (\mu = 1, 2, \cdots, D)$ satisfying

$$dy_\mu dy_\nu = -dy_\nu dy_\mu, \quad dy_\mu f(x,y) = f(x,y+\hat\mu)\, dy_\mu. \tag{2.18}$$

In a similar way we will obtain the exterior derivative operator d_y, exterior divergence operator d^*_y and all the noncommutative differential calculus with respect to y.

A differential (k,l)-form on $\mathrm{Z}^D \times \mathrm{Z}^D$ is defined by

$$f = \frac{1}{k!l!} f_{\mu_1\cdots\mu_k;\nu_1\cdots\nu_l}(x,y)\mathrm{d}x_{\mu_1}\cdots \mathrm{d}x_{\mu_k}\mathrm{d}y_{\nu_1}\cdots \mathrm{d}y_{\nu_l}\,, \tag{2.19}$$

where $f_{\mu_1\cdots\mu_k;\nu_1\cdots\nu_l}(x,y)$ is completely antisymmetric in $\mu_1,\cdots,\mu_k$ and in $\nu_1,\cdots,\nu_l$, separately, also it is assumed to have compact support on $\mathrm{Z}^D\times\{y\}$ and $\{x\}\times\mathrm{Z}^D$ or to decrease at least exponentially as $|x-y|\to\infty$ in order to apply Poincaré lemma on it. The vector space of 'bi-local' field (k,l)-forms is denoted by $\Omega_{k,l}$.

The exterior differential with respect to x or y is denoted by d_x or d_y, acting on the (k,l)-form (2.19) as,

$$\begin{aligned} d_x f &= \frac{1}{k!l!}\partial_\mu f_{\mu_1\cdots\mu_k;\nu_1\cdots\nu_l}(x,y)dx_\mu dx_{\mu_1}\cdots dx_{\mu_k}dy_{\nu_1}\cdots dy_{\nu_l}\,, \\ d_y f &= \frac{(-1)^k}{k!l!}f_{\mu_1\cdots\mu_k;\nu_1\cdots\nu_l}(x,y)\overleftarrow{\partial}_\nu dx_{\mu_1}\cdots dx_{\mu_k}dy_\nu dy_{\nu_1}\cdots dy_{\nu_l}\,. \end{aligned} \tag{2.20}$$

Divergence operators d^*_x (2.16) and d^*_y may also be extended to (k,l)-forms:

$$\begin{aligned} d^*_x f &= \frac{1}{(k-1)!l!}\partial^*_\mu f_{\mu\mu_2\cdots\mu_k;\nu_1\cdots\nu_l}(x,y)dx_{\mu_2}\cdots dx_{\mu_k}dy_{\nu_1}\cdots dy_{\nu_l}\,, \\ d^*_y f &= \frac{(-1)^k}{k!(l-1)!}f_{\mu_1\cdots\mu_k;\nu\nu_2\cdots\nu_l}(x,y)\overleftarrow{\partial}^*_\nu dx_{\mu_1}\cdots dx_{\mu_k}dy_{\nu_2}\cdots dy_{\nu_l}\,. \end{aligned} \tag{2.21}$$

It is straightforward to show that these operators satisfy nilpotency and anticommutation relations between them, when they act on the 'bi-local' form space $\Omega_{k,l}$,

$$\begin{aligned} d_x^2 = d_y^2 = d_x^{*2} &= d_y^{*2} = 0\,, \\ (d_x+d_y)^2 = (d^*{}_x+d_y)^2 &= (d_x+d^*{}_y)^2 = (d^*_x+d^*_y)^2 = 0\,. \end{aligned} \tag{2.22}$$

There are four pairs of anticommuting nilpotent operators as $\{d_x,d_y\}$, $\{d_x,d^*_y\}$, $\{d^*_x,d_y\}$ and $\{d^*_x,d^*_y\}$, any one of them can be used to construct a double complex.

For example, given a (k,l) differential form $\omega^{k,l}(x,y)$ satisfying

$$d^*_x d^*_y\omega^{k,l}(x,y) = 0 \tag{2.23}$$

then by the Poincaré lemma there exists a form $\omega^{k-1,l+1}(x,y)$ satisfying

$$d^*_x\omega^{k,l}(x,y) + d^*_y\omega^{k-1,l+1}(x,y) = 0\,. \tag{2.24}$$

Since forms $\omega^{k-1,l+1}(x,y)$ also satisfy (2.23), they lead to new forms $\omega^{k-2,l+2}(x,y)$. Such a procedure can be continued until one ends up with $\omega^{0,k+l}(x,y)$ for $k+l\leq D$. In this way we have obtained a double complex for $\{d^*_x,d^*_y\}$. These formulas are also called descent equations, which were used [11] in analyzing the chiral anomaly in arbitrary even-dimensional lattice gauge theory. In this paper we want to use another double complex in terms of operators $\{d_x,d^*_y\}$ to discuss the chiral anomaly.

3. Double complex, descent equation, and its solutions

The technique of double complex and its descent equation are widely used in the anomaly analysis of continuous gauge field theory. It is our purpose to extend this technique to the anomaly analysis in lattice gauge theory in terms of NCDG on the lattice. As we did in [11] using $\{d_x^*, d_y^*\}$ double complex, here we will use $\{d_x, d_y^*\}$ double complex to perform anomaly analysis and to derive the chiral anomaly of abelian lattice gauge theory in arbitrary even dimensions.

We focus now on the 'bi-local' fields space $\bigcup_{k,l=0,1,\cdots,D} \Omega_{k,l}$ defined on $\mathrm{Z^D} \times \mathrm{Z^D}$ and start from an $(D,2)$-form $\omega^{D,2}(x,y)$ which satisfies

$$\sum_x d_y^* \omega^{D,2}(x,y) = 0. \tag{3.1}$$

Since $d_y^* \omega^{D,2}(x,y)$ is a D-form for x, in terms of Poincaré lemma there exist a $(D-1,1)$-form $\omega^{D-1,1}(x,y)$ such that

$$d_y^* \omega^{D,2}(x,y) + d_x \omega^{D-1,1}(x,y) = 0. \tag{3.2}$$

Acting d_y^* on the above equation we obtain, using Poincaré lemma for d_x again, there exists an $(D-2,0)$-form $\omega^{D-2,0}(x,y)$ such that

$$d_y^* \omega^{D-1,1}(x,y) + d_x \omega^{D-2,0}(x,y) = 0. \tag{3.3}$$

These are the descent equations for $\omega^{D,2}(x,y)$. For general $\omega^{D,l}(x,y)$ which satisfies

$$\sum_x d_y^* \omega^{D,l}(x,y) = 0, \tag{3.4}$$

one could obtain the following set of descent equations:

$$\begin{aligned} d_y^* \omega^{D,l}(x,y) \ &+ d_x \omega^{D-1,l-1}(x,y) &= 0, \\ d_y^* \omega^{D-1,l-1}(x,y) \ &+ d_x \omega^{D-2,l-2}(x,y) &= 0, \\ \vdots \qquad\qquad &\qquad\qquad \vdots & \\ d_y^* \omega^{D-l+1,1}(x,y) \ &+ d_x \omega^{D-l,0}(x,y) &= 0. \end{aligned} \tag{3.5}$$

It should be mentioned, according to the Poicaré Lemma [8, 11], that if $\omega^{D,l}(x,y)$ is gauge invariant and locally dependent on gauge field A_μ, then all $\omega^{D-m,l-m}(x,y)$ $(m = 1, 2, \cdots, l)$ appearing in (3.4) are gauge invariant and locally dependent on A_μ.

Now we are going to solve the descent equation. We start from $\omega^{D,l}(x,y)$, which is a D-form in x and an l-form in y, its dual version being 0-form in x

and $(D-l)$-form in y. We can use the dual version expression for simplicity to express the $\omega^{D,l}(x,y)$ as,

$$\omega^{D,l}(x,y) \\ = \frac{1}{l!}dx_{\mu_1}\cdots dx_{\mu_l}dy_{\mu_1}\cdots dy_{\mu_l}\omega_{\mu_{l+1},\cdots,\mu_D}(x,y)dx_{\mu_{l+1}}\cdots dx_{\mu_D}. \quad (3.6)$$

The position of the coefficient in the right side of (3.5) is of specific meaning. According to **Lemma 1**, moving the position of coefficients cross the forms, the value of the argument in the coefficients will shift according to the noncommutavity of forms and functions (2.6). So if two similar expressions of forms have different order in their coefficient and basis, then they absolutely different from each other. The position of coefficient $\omega_{\mu_{l+1},\cdots,\mu_D}(x,y)$ in the right side of (3.5) is chosen such that it will be more convenient for us in the following discussions.

In order to obtain the $\omega^{D-1,l-1}(x,y)$, first we introduce $\alpha^{D,l}(y)$ as,

$$\alpha^{D,l}(y) \\ = \sum_x \omega^{D,l}(x,y) \\ = \frac{1}{l!}dx_{\mu_1}\cdots dx_{\mu_l}dy_{\mu_1}\cdots dy_{\mu_l}\alpha_{\mu_{l+1},\cdots,\mu_D}(y)dx_{\mu_{l+1}}\cdots dx_{\mu_D}. \quad (3.7)$$

where $\alpha_{\mu_{l+1},\cdots,\mu_D}(y) = \sum_x \omega_{\mu_{l+1},\cdots,\mu_D}(x,y)$. Since the $\sum_x[\omega^{D,l}(x,y) - \delta_{x,y}\alpha^{D,l}(x)] = 0$ for a D-form for x, by Poicaré Lemma for d_x we obtain

$$\omega^{D,l}(x,y) = \delta_{x,y}\alpha^{D,l}(x) + d_x\vartheta^{D-1,l}(x,y). \quad (3.8)$$

Substituting this result into the first formula in (3.4) one obtains,

$$d_x\omega^{D-1,l-1}(x,y) = -d_y^*\delta_{x,y}\alpha^{D,l}(x) + d_x d_y^*\vartheta^{D-1,l}(x,y). \quad (3.9)$$

The final expression for $\omega^{D-1,l-1}(x,y)$ will be given soon after the following lemma.

Lemma 5: If the $\alpha^{D,l}(x)$ is given in (3.6), then there is a $(D-1,l-1)$-form $\alpha^{D-1,l-1}(x)$ satisfying,

$$-d_y^*[\delta_{x,y}\alpha^{D,l}(x)] = d_x[\delta_{x,y}\alpha^{D-1,l-1}(x)], \quad (3.10)$$

where $\alpha^{D-1,l-1}(x)$ is given by

$$\alpha^{D-1,l-1}(x) \\ = \frac{1}{(l-1)!}dx_{\mu_2}\cdots dx_{\mu_l}dy_{\mu_2}\cdots dy_{\mu_l}\alpha_{\mu_{l+1},\cdots,\mu_D}(x)dx_{\mu_{l+1}}\cdots dx_{\mu_D}. \quad (3.11)$$

Proof: As $\sum_x d_x f(x,y)$ vanishes for any local field $f(x,y)$, taking summation over x of the first equation in (3.4) one can obtain $\sum_x d_y^* \omega^{D,l}(x,y) = 0$. Changing the order of $\sum_x$ and d_y^* we get an important property of $\alpha^{D,l}(y)$,

$$d_y^* \alpha^{D,l}(y) = 0. \tag{3.12}$$

This means that

$$\frac{1}{l!} dx_{\mu_1} \cdots dx_{\mu_l} dy_{\mu_2} \cdots dy_{\mu_l} \times [\alpha_{\mu_{l+1},\cdots,\mu_D}(y+\widehat{\mu_1}) - \alpha_{\mu_{l+1},\cdots,\mu_D}(y)] dx_{\mu_{l+1}} \cdots dx_{\mu_D} = 0. \tag{3.13}$$

Expanding the both sides of (3.10) by definition of d_x and d_y^*,

$$\begin{aligned} & d_y^*[\delta_{x,y}\alpha^{D,l}(x)] \\ &= \frac{1}{(l-1)!}[\delta_{x,y} dx_{\mu_1} \cdots dx_{\mu_l} dy_{\mu_2} \cdots dy_{\mu_l} \alpha_{\mu_{l+1},\cdots,\mu_D}(x) dx_{\mu_{l+1}} \cdots dx_{\mu_D} \\ &\quad - \delta_{x,y-\widehat{\mu_1}} dx_{\mu_1} \cdots dx_{\mu_l} dy_{\mu_2} \cdots dy_{\mu_l} \alpha_{\mu_{l+1},\cdots,\mu_D}(x) dx_{\mu_{l+1}} \cdots dx_{\mu_D}] \\ & d_x[\delta_{x,y}\alpha^{D-1,l-1}(x)] \\ &= \frac{1}{(l-1)!}[\delta_{x+\widehat{\mu_1},y} dx_{\mu_1} \cdots dx_{\mu_l} dy_{\mu_2} \cdots dy_{\mu_l} \alpha_{\mu_{l+1},\cdots,\mu_D}(x) dx_{\mu_{l+1}} \cdots dx_{\mu_D} \\ &\quad - \delta_{x,y} dx_{\mu_1} \cdots dx_{\mu_l} dy_{\mu_2} \cdots dy_{\mu_l} \alpha_{\mu_{l+1},\cdots,\mu_D}(x-\widehat{\mu_1}) dx_{\mu_{l+1}} \cdots dx_{\mu_D}] \end{aligned} \tag{3.14}$$

One can easily prove this lemma in terms of (3.12) and the identity $\delta_{x+\widehat{\mu_1},y} = \delta_{x,y-\widehat{\mu_1}}$.

From this lemma and (3.9) it is easy to find

$$d_x[\omega^{D-1,l-1}(x,y) - \delta_{x,y}\alpha^{D-1,l-1}(x) - d_y^* \vartheta^{D-1,l}(x,y)] = 0. \tag{3.15}$$

Then we obtain the expression

$$\omega^{D-1,l-1}(x,y) = \delta_{x,y}\alpha^{D-1,l-1}(x) + d_y^* \vartheta^{D-1,l}(x,y) + d_x \vartheta^{D-2,l-1}(x,y). \tag{3.16}$$

In the same way we can find

$$\omega^{D-2,l-2}(x,y) = \delta_{x,y}\alpha^{D-2,l-2}(x) + d_y^* \vartheta^{D-2,l-1}(x,y) + d_x \vartheta^{D-3,l-2}(x,y), \tag{3.17}$$

where

$$\alpha^{D-2,l-2}(x) = \frac{1}{(l-2)!} dx_{\mu_3} \cdots dx_{\mu_l} dy_{\mu_3} \cdots dy_{\mu_l} \alpha_{\mu_{l+1},\cdots,\mu_D}(x) dx_{\mu_{l+1}} \cdots dx_{\mu_D}. \tag{3.18}$$

The general solution for descent equation is obtained similarly,

$$\omega^{D-k,l-k}(x,y) = \delta_{x,y}\alpha^{D-k,l-k}(x) + d_y^*\vartheta^{D-k,l-k+1}(x,y) + d_x\vartheta^{D-k-1,l-k}(x,y), \quad (3.19)$$

where

$$\alpha^{D-k,l-k}(x) = \frac{1}{(l-k)!}dx_{\mu_{k+1}}\cdots dx_{\mu_l}dy_{\mu_{k+1}}\cdots dy_{\mu_l} \times\alpha_{\mu_{l+1},\cdots,\mu_D}(x)dx_{\mu_{l+1}}\cdots dx_{\mu_D}, \quad (3.20)$$

for $k = 1, 2.\cdots, l$.

4. Axial anomaly of abelian lattice gauge theory

As mentioned in Section 1, the chiral anomaly (1.2) in abelian lattice gauge theory is gauge invariant, locally dependent on a gauge field A_μ, and its variation with respect to A_μ satisfies (1.3), which reflects the topological nature of the anomaly. Then its topological part will be fixed up to some constant coefficient. The result in the case of four dimensions is given by Lüscher's theorem (1.4), its extension to higher dimensions can be found in [10, 11]. Now we use the new double complex and descent equation proposed in the last section to derive the anomaly again. First we state the main result.

Theorem 1: In D-dimensional infinite hypercubic lattice, if the anomaly written in D-form $Q(x) = q(x)d^Dx$ is gauge invariant, locally depending on $U(1)$ gauge field A_μ and of topological nature

$$\sum_x \delta Q(x) = 0, \quad (4.1)$$

then,

$$Q(x) = \sum_{n=0}^{[D/2]} F^n B^{(D-2n)} + dK^{(D-1)}, \quad (4.2)$$

where F is the gauge field strength 2-form, $B^{(D-2n)}$ is constant $(D-2n)$-form and $K^{(D-1)}$ is gauge invariant $(D-1)$-form locally depending on A_μ.

One can easily find that the theorem is consistent with the result in [10, 11],

$$q(x) = \alpha + \sum_{n=1}^{[D/2]} \beta_{\mu_1\nu_1\mu_2\nu_2\cdots\mu_n\nu_n}F_{\mu_1\nu_1}(x)F_{\mu_2\nu_2}(x+\widehat{\mu}_1+\widehat{\nu}_1)\cdots \times F_{\mu_n\nu_n}(x+\widehat{\mu}_1+\widehat{\nu}_1+\cdots+\widehat{\mu}_{n-1}+\widehat{\nu}_{n-1}) + \partial_\mu^* k_\mu(x). \quad (4.3)$$

Obviously, one can recover Lüscher's theorem (1.4) exactly when $D = 4$.

The main theorem can be shown easily, once the following lemmas are obtained.

Lemma 6: On D-dimensional infinite hypercubic lattice, if D-form $Q(x) = q(x)d^D x$ is gauge invariant, locally depending on $U(1)$ gauge field A_μ and of topological nature $\sum_x \delta Q(x) = 0$, then

$$Q(x) \;=\; bd^D x + F\alpha^{D-2}(x) + d_x\vartheta^{D-1}(x), \tag{4.4}$$

where b is constant, $\alpha^{D-2}(x)$ and $\vartheta^{D-1}(x)$ are gauge invariant, locally depending on A_μ, and $d\alpha^{D-2}(x) = 0$.

Lemma 7: On D-dimensional infinite hypercubic lattice, if $2m$-form $\alpha^{2m}(x)$ $(2m < D)$ is gauge invariant, locally depending on A_μ, and $d\alpha^{2m}(x) = 0$, then

$$\alpha^{2m}(x) \;=\; B^{2m} + F\alpha^{2m-2}(x) + d_x\vartheta^{2m-1}(x), \tag{4.5}$$

where B^{2m} is constant $2m$ -form, $\alpha^{2m-2}(x)$ and $\vartheta^{2m-1}(x)$ are gauge invariant, locally depending on A_μ, and $d\alpha^{2m-2}(x) = 0$.

Substituting eq.(4.5) when $2m = D - 2$ into eq. (4.4), one can easily find the following result by the Bianchi identity $dF = 0$ and the Leibniz' rule for d in the NCDG,

$$\begin{aligned} Q(x) \;=\; & bd^D x + FB^{D-2} + F^2\alpha^{D-4}(x) \\ & + d_x(F\vartheta^{D-3}(x) + \vartheta^{D-1}(x)). \end{aligned} \tag{4.6}$$

Taking $2m = D - 4$ in **Lemma 7**, substituting $\alpha^{D-4}(x)$ in the above formula again and repeating this procedure up to $2m = 2$, we will arrive at the main **Theorem**.

Now we prove lemma 6 and 7.

Proof of Lemma 6: Following the discussion in [8, 11] some algebraic techniques can be used to separate $Q(x)$ into two parts, one is independent of the gauge field A_μ and other part depending on the gauge field A_μ. Also we can assume that the first part is a constant, since the translation invariance of lattice gauge theory forces all the dependence on x should be through the gauge field A_μ.

$$Q(x) \;=\; bd^D x + \int_0^1 dt \left(\frac{\partial Q(x)}{\partial t}\right)_{A\to tA}$$

$$= bd^D x + \sum_y \int_0^1 dt \left(\frac{\delta Q(x)}{\delta A_\nu(y)} \right)_{A \to tA} A_\nu(y)$$

$$= bd^D x + \sum_y \langle J^{D,1}(x,y), A(y) \rangle, \tag{4.7}$$

where $J^{D,1}(x,y)$ is a $(D,1)$-form in $\mathrm{Z^D} \times \mathrm{Z^D}$ as,

$$J^{D,1}(x,y) = \int_0^1 dt \left(\frac{\delta Q(x)}{\delta A_\nu(y)} \right)_{A \to tA} dy_\nu, \tag{4.8}$$

and $A(y)$ is gauge field 1-form.

According to the above assumption $Q(x)$ depends locally on the gauge field $A_\mu(x)$, where the locality means that the $J^{D,1}(x,y)$ decrease at least exponentially as $|x-y| \to \infty$. Then the summation over y in (4.6) is finite. Since the gauge group is abelian, the variation of the gauge invariant field with respect to gauge potential is gauge invariant, the 'bi-local' $J^{D,1}(x,y)$ is gauge invariant as well. Furthermore, we make an infinitesimal gauge transformation of $Q(x)$,

$$\delta_G Q(x) = \sum_y \langle J^{D,1}(x,y), d_y \lambda(y) \rangle, \tag{4.9}$$

from the gauge invariance of $Q(x)$, one obtains

$$d_y^* J^{D,1}(x,y) = 0. \tag{4.10}$$

This means

$$J^{D,1}(x,y) = d_y^* \omega^{D,2}(x,y), \tag{4.11}$$

where $\omega^{D,2}(x,y)$ is a 'bi-local' gauge invariant $(D,2)$-form. Then we have

$$Q(x) = bd^D x + \sum_y \langle \omega^{D,2}(x,y), F(y) \rangle, \tag{4.12}$$

where the dual property of $-d_y$ and d_y^* are used. The topological nature (1.3) of $Q(x)$ tells us

$$\begin{aligned} \sum_x J^{D,1}(x,y) &= \sum_x \int_0^1 dt \left(\frac{\delta Q(x)}{\delta A_\nu(y)} \right)_{A \to tA} dy_\nu \\ &= \int_0^1 dt \sum_x \left(\frac{\delta Q(x)}{\delta A_\nu(y)} \right)_{A \to tA} dy_\nu \\ &= 0. \end{aligned} \tag{4.13}$$

This shows that $\omega^{D,2}(x,y)$ satisfies the descent equation,

$$\sum_x d_y^* \omega^{D,2}(x,y) = 0. \tag{4.14}$$

Using the general solution of the descent equation (3.8) in the case of $l = 2$, i.e.,

$$\begin{aligned} \omega^{D,2}(x,y) &= \delta_{x,y}\alpha^{D,2}(x) + d_x\vartheta^{D-1,2}(x,y), \\ \alpha^{D,2}(x) &= \frac{1}{2!}dx_{\mu_1}dx_{\mu_2}dy_{\mu_1}dy_{\mu_2} \\ &\quad \times \alpha_{\mu_3,\cdots,\mu_D}(x)dx_{\mu_3}\cdots dx_{\mu_D}, \end{aligned} \tag{4.15}$$

and substituting them into (4.12), one can prove the lemma,

$$\begin{aligned} Q(x) &= bd^Dx + \sum_y \langle \delta_{x,y}\alpha^{D,2}(x) + d_x\vartheta^{D-1,2}(x,y), F(y)\rangle, \\ &= bd^Dx + \sum_y \langle \delta_{x,y}\frac{1}{2!}dx_{\mu_1}dx_{\mu_2}dy_{\mu_1}dy_{\mu_2}\alpha_{\mu_3,\cdots,\mu_D}(x) \\ &\qquad \times dx_{\mu_3}\cdots dx_{\mu_D}, F_{\rho_1\rho_2}dy_{\rho_1}dy_{\rho_2}\rangle \\ &\quad + d_x\sum_y \langle \vartheta^{D-1,2}(x,y), F(y)\rangle \\ &= bd^Dx + F\alpha^{D-2}(x) + d_x\vartheta^{D-1}(x), \end{aligned} \tag{4.16}$$

where $(D-2)$-form $\alpha^{D-2}(x) = (1/2!)\alpha_{\mu_3,\cdots,\mu_D}(x)dx_{\mu_3}\cdots dx_{\mu_D}$ and $(D-1)$-form $\vartheta^{D-1}(x) = \sum_y \langle \vartheta^{D-1,2}(x,y), F(y)\rangle$ are gauge invariant local fields. From the Leibniz' rule in NCDG one can easily prove that $(D-2)$-form $\alpha^{D-2}(x)$ is closed.

Proof of Lemma 7: The proof of this lemma is similar to that of **Lemma 6**. First we separate $\alpha^{2m}(x)$ into two parts, one is independent of the gauge field A_μ and the other part depends on the gauge field A_μ,

$$\begin{aligned} \alpha^{2m}(x) &= B^{2m} + \int_0^1 dt \left(\frac{\partial \alpha^{2m}(x)}{\partial t}\right)_{A\to tA} \\ &= B^{2m} + \sum_y \int_0^1 dt \left(\frac{\delta \alpha^{2m}(x)}{\delta A_\nu(y)}\right)_{A\to tA} A_\nu(y) \\ &= B^{2m} + \sum_y \langle J^{2m,1}(x,y), A(y)\rangle, \end{aligned} \tag{4.17}$$

where $J^{2m,1}(x,y)$ is a $(2m,1)$-form in $\mathrm{Z^D} \times \mathrm{Z^D}$ as,

$$J^{2m,1}(x,y) = \int_0^1 dt \left(\frac{\delta \alpha^{2m}(x)}{\delta A_\nu(y)} \right)_{A \to tA} dy_\nu . \tag{4.18}$$

From the locality of $\alpha^{2m}(x)$, we know that the $J^{2m,1}(x,y)$ is a 'bi-local' field, i.e., it decreases at least exponentially as $|x-y| \to \infty$ and the summation over y in (4.16) is finite. Since $\alpha^{2m}(x)$ is a gauge invariant field one obtains

$$d_y^* J^{2m,1}(x,y) \quad = \quad 0. \tag{4.19}$$

From Poincaré lemma it follows,

$$J^{2m,1}(x,y) \quad = \quad d_y^* \omega^{2m,2}(x,y), \tag{4.20}$$

where $\omega^{2m,2}(x,y)$ is a gauge invariant $(2m,2)$-form 'bi-local' field. We then have

$$\alpha^{2m}(x) \quad = \quad B^{2m} + \sum_y \langle \omega^{2m,2}(x,y), F(y) \rangle . \tag{4.21}$$

The closedness of $\alpha^{2m}(x)$ implies

$$d_x d_y^* \omega^{2m,2}(x,y) = d_x J^{2m,1}(x,y) = 0. \tag{4.22}$$

This shows that $\omega^{2m,2}(x,y)$ satisfies the descent equation. Using the general solution to the descent equation (3.18),

$$\begin{aligned} \omega^{2m,2}(x,y) &= \delta_{x,y} \alpha^{2m,2}(x) + d_x \vartheta^{2m-1,2}(x,y) \\ &\quad + d_y^* \vartheta^{2m,3}(x,y), \\ \alpha^{2m,2}(x) &= \frac{1}{2!} dx_{\mu_1} dx_{\mu_2} dy_{\mu_1} dy_{\mu_2} \\ &\quad \times \alpha_{\mu_3,\cdots,\mu_{2m}}(x) dx_{\mu_3} \cdots dx_{\mu_{2m}} . \end{aligned} \tag{4.23}$$

Substituting these equations into (4.21), we prove the lemma,

$$\begin{aligned} \alpha^{2m}(x) &= B^{2m} + \sum_y \langle \delta_{x,y} \alpha^{2m,2}(x) + d_x \vartheta^{2m-1,2}(x,y), F(y) \rangle , \\ &= B^{2m} + \sum_y \langle \delta_{x,y} \frac{1}{2!} dx_{\mu_1} dx_{\mu_2} dy_{\mu_1} dy_{\mu_2} \alpha_{\mu_3,\cdots,\mu_{2m}}(x) \\ &\qquad \times dx_{\mu_3} \cdots dx_{\mu_{2m}}, F_{\rho_1 \rho_2} dy_{\rho_1} dy_{\rho_2} \rangle \\ &\quad + d_x \sum_y \langle \vartheta^{2m-1,2}(x,y), F(y) \rangle \\ &= B^{2m} + F \alpha^{2m-2}(x) + d_x \vartheta^{2m-1}(x), \end{aligned} \tag{4.24}$$

where $(2m-2)$-form $\alpha^{2m-2}(x) = (1/2!)\alpha_{\mu_3,\cdots,\mu_{2m}}(x)dx_{\mu_3}\cdots dx_{\mu_{2m}}$ and $(2m-1)$-form $\vartheta^{2m-1}(x) = \sum_y \langle \vartheta^{2m-1,2}(x,y), F(y)\rangle$ are gauge invariant local fields. From the Leibniz' rule for NCDG one can easily prove that $(2m-2)$-form $\alpha^{2m-2}(x)$ is closed.

5. Conclusion

We would like to make a few remarks in this section.

We have extended the axial anomaly of ref. [8] in four dimensions to arbitrary even dimensions in abelian lattice gauge theory. In the derivation of the chiral anomaly in this paper, as well as in [8, 10, 11], the locality plays a very important role. The locality here means that taking the continuum limit one should obtain the correct continuous field theory, so we should keep this property in every step to make our derivation reliable.

We have given four different kinds of double complexes in section 2. Two of them have been used in the anomaly analysis in abelian lattice gauge theory, one is used in [11], and the second in this paper. The application of the other two double complexes is also very interesting, and is now investigated.

Lattice gauge theory is a kind of discretized field theory. The discreteness in this theory yields many specific characters of lattice gauge theory which differ from the usual continuum field theory. Such as the gauge field on lattice gauge theory is given by parallel transporter other than gauge vector field, the chirality as well as fermion doubling in lattice gauge theory are more serious problems and trouble us already many years even up to now. Discreteness could be represented by NCDG in lattice and very recently NCDG is used in lattice gauge theory to understand some quantum effect or topological effect such as the chiral anomaly in [8, 10, 11] and in this paper. However it is not hard to notice that in our discussion, we use some similar geometric concepts and tools as in ordinary commutative differential geometry. For example, the concept of double complex and descent equation lead us to the calculation of chiral anomaly in lattice gauge theory in the same spirit of anomaly analysis in continuum quantum field theory, although the exact meaning of them is essentially different. We hope that we can understand more about lattice theory along this way from point of view of NCDG in the near future.

Acknowledgments

One of us (K.W.) is very grateful to F. Sakata for the warm hospitality and to Faculty of Science of Ibaraki University for the financial support during his visiting at Ibaraki University. He would like to express his thanks also to the organizers of the workshop "Noncommutative Geometry and its Application to Physics" hold at Shonan-kokusaimura on May 31st to June 4th of 1999 for inviting him to join this workshop and the editor of this proceedings to carefully read and modify the draft of this paper.

References

[1] N.B. Nielson and M. Ninomiya, *Phys. Lett.* **B105** (1981) 219.

[2] P.H. Ginsparg and K.G. Wilson, *Phys. Rev.* **D25** (1982) 2649.

[3] H. Neuberger, *Phys. Lett.* **B417** (1998) 141; **B427** (1998) 353.
P. Hernández, K. Jansen and M. Lüscher, *Nucl. Phys.* **B552** (1999) 363.

[4] P. Hasenfratz, *Nucl. Phys. (Proc. Suppl.)* **63** (1998) 53; *Nucl. Phys.* **B525** (1998) 401.

[5] P. Hasenfratz, V. Laliena and F. Niedermayer, *Phys. Lett.* **B427** (1998) 125.

[6] M. Lüscher, *Phys. Lett.* **B428** (1998) 342.

[7] R. Narayanan, *Phys. Rev.* **D58** (1998) 097501.
F. Niedermayer, *Nucl. Phys. (Proc. Suppl.)* **73** (1999)105.
K. Fujikawa, *Nucl. Phys.* **B546** (1999) 480, hep-lat/9904007.
T. Reisz and H.J. Rothe, *Phys. Lett.* **B455** (1999)246.
H. Neuberger, *Phys. Rev.* **D59** (1999) 085006.
Y. Kikukawa and A. Yamada, *Phys. Lett.* **B448** (1999)265, *Nucl. Phys.* **547** (1999) 413. T. Aoyama and Y. Kikukawa, hep-lat/9905003.
D.H. Adams, hep-lat/9812003.
H. Suzuki, *Prog. Theo.Phys.* **102** (1999)1147, **102** (1999)141.
T.-W. Chiu, *Phys. Lett.* **B445** (1999) 371, hep-lat/9906007. T.-W. Chiu and T.-H. Hsieh, hep-lat/9901011.

[8] M. Lüscher, *Nucl. Phys.* **B538** (1999) 515.

[9] M. Lüscher, *Nucl. Phys.* **B549** (1999) 295.

[10] T. Fujiwara, H. Suzuki and K. Wu, Ibaraki University preprint IU-MSTP/35, hep-lat/9906015.

[11] T. Fujiwara, H. Suzuki and K. Wu, *Phys. Lett.* **B463** (1999) 63, hep-lat/9906016.

[12] A. Connes, *"Noncommutative Geometry"*, Academic Press, New York, 1994;

[13] A. Sitarz, *J. Geom. Phys.* **15** (1995) 123;
A. Dimakis and F. Müller-Hoissen, *Phys. Lett.* **B295** (1992) 242; A. Dimakis, F. Müller-Hoissen and T. Striker, *J. Phys. A: Math. Gen.* **26** (1993) 1927.
H.G. Ding, H.Y. Guo, J.M. Li and K. Wu, *Z. Phys.* **C64** (1994) 521;

GEOMETRICAL STRUCTURES ON NONCOMMUTATIVE SPACES

Olivier Grandjean
Harvard University
Cambridge MA 02138, USA

Abstract We review the connection between supersymmetric quantum mechanics and differential geometry. This gives a formulation of differential geometry that is well suited for generalization to the noncommutative setting. Two examples are presented. First, we show how the quantum target of the Wess–Zumino–Witten model on SU(2) can be encoded into '$N = 1$ spectral data'. Then, we explain how the Kähler structure on the noncommutative two-torus can be constructed starting with $N = 1$ spectral data.

Mathematics Subject Classification (2000): 81T40,81T70, 81T75

Keywords: noncommutative geometry, differential geometry, Wess–Zumino–Witten model

1. Introduction

Combining quantum theory with general relativity, heuristically, [10], [13], one obtains uncertainty relations for the spacial localization of an event along orthogonal directions. This fact forces us to give up the description of space-time as a smooth manifold, and calls for a (not yet known) generalization of geometry: quantum geometry.

Noncommutative geometry is a generalized version of (differential) geometry that allows one to describe a wide variety of spaces. It is very likely that some of its methods will be useful in the formulation of quantum geometry.

The most promising candidate for a unification of quantum theory with general relativity is certainly string theory. String vacua are described by certain

Y. Maeda, et al. (eds.), Noncommutative Differential Geometry and Its Applications to Physics, 31–48.

superconformal field theories. It was proposed in [4], [14] that one should be able to reconstruct space-time geometry from these superconformal field theories*. This proposal was extended to the idea that every conformal field theory is a σ-model with a possibly noncommutative target space. This idea can be checked in great details for the WZW model on SU(2) (see section 3), but the generalization to more complicated theories (like Gepner models, for example) is still missing. Trying to make progress on these matters lead to new developments in the opposite direction: the relation between superconformal field theories and classical manifolds leads to a enjoyable theory of noncommutative spaces endowed with additional structures, like Hermitian, Kähler or even hyper-Kähler structures. This formulation of noncommutative geometry will certainly be of great help to tackle the original problem.

Recently, noncommutative geometry has been a useful tool to study certain aspects of string theory [8], [23]. It appears that noncommutative spaces arise most naturally in the presence of D-branes [20]. In this context, noncommutativity is a consequence of inequivalent orderings of vertex operators on the world sheet's boundary [22], [23]. It has been shown that Kontsevich's star product on Poisson manifolds [19] arises in this way [3]. The general considerations mentioned above indicate that we do not have the full picture yet, and that many connections between noncommutative geometry and string theory are still to be discovered.

This short note presents some of the results obtained in [15] and [16], as well as a review of an argument given in [14]. None of the results covered here are claimed to be new. We refer the reader to the above sources for a more detailed presentation and other results.

Acknowledgments

I would like to warmly thank the organizers for the wonderful workshop. I also thank all participants for the enjoyable atmosphere they created during the workshop.

2. Spectral Triples

In order to introduce the concept of a noncommutative space, it is useful to recall the duality between topological spaces and Abelian C^*-algebras. This duality is the content of Gelfand's theorem (see [2] for example):

Theorem 1 *Let $\mathcal{A}$ be an Abelian C^*-algebra. Then, there is a unique (up to homeomorphism) locally compact Hausdorff space X such that $\mathcal{A}$ is isomorphic*

*Of course, this reconstruction is not, in general, unique, since different space-times sometimes lead to the same superconformal field theory

to the algebra $C_0(X)$ of continuous functions[†] *over X vanishing at infinity. Furthermore, X is compact if and only if $\mathcal{A}$ is unital.*

This duality means that the study of locally compact spaces is, in principle, equivalent to that of Abelian C^*-algebras. In particular, it implies that one can use tools from one field to study the other. One problem noncommutative geometry addresses is to extend notions and tools of topology and geometry to noncommutative algebras. One then think of noncommutative algebras as algebras of functions over 'noncommutative spaces'. The first step towards a generalization to the noncommutative setting is to build an algebraic formulation of geometry in terms of algebras and some additional structures. Let us sketch how this is done.

We consider a spin^c manifold M with Riemannian metric g. The spinor bundle S over M carries a spin connection ∇^S and a Clifford action,

$$c : T^*M \longrightarrow \mathrm{End} S \,,$$

satisfying the anticommutation relations,

$$\{c(\omega), c(\eta)\} = -2g(\omega, \eta),$$

for all $\omega, \eta \in T^*M$. The Dirac operator on M is defined by

$$\partial\!\!\!/ = c \circ \nabla^S : \Gamma(S) \longrightarrow \Gamma(S) \,.$$

In local coordinates x^μ on M, the Dirac operator reads

$$\partial\!\!\!/ = \gamma^\mu \nabla^S_\mu \,, \tag{1}$$

where $\gamma^\mu = c(dx^\mu)$ are the gamma matrices. We shall denote by $\mathcal{H}_e$ the space of square integrable sections of the spinor bundle, i.e.,

$$\mathcal{H}_e = L^2(S, \mathrm{dvol}_g) \,.$$

The main result is that [5],

> The triple $(\mathcal{H}_e, C^\infty(M), \partial\!\!\!/)$ completely encodes the topology and the geometry of M.

To interpret the above result in physical terms, we recall that Pauli's theory of non-relativistic electrons is the quantum mechanical system whose time evolution is governed by the Hamilton operator,

$$H = \frac{\hbar^2}{2m} \partial\!\!\!/^2 \,,$$

[†] In this note, all vector spaces are complex. In particular, functions are always complex valued

where m denotes the mass of the electron, acting on the Hilbert space $\mathcal{H}_e$. In this setting, the algebra $C^\infty(M)$ of smooth functions is the algebra of position measurements. When the manifold M is even dimensional the Clifford volume element γ defines a $\mathbb{Z}_2$-grading on the Hilbert space $\mathcal{H}_e$, and the self-adjoint operator,

$$Q = \frac{\hbar}{\sqrt{2m}} \partial\!\!\!/ \, ,$$

satisfies,

$$Q^2 = H \, , \quad \{Q, \gamma\} = 0 \, .$$

These equations mean that the system exhibits so called $N = 1$ *supersymmetry*. The above result shows that one can reconstruct physical space, M, from the theory of non-relativistic electrons on M.

It is also possible to recover the manifold from the theory of scalar particles on M, i.e., from the data $(L^2(M, dvol_g), C^\infty(M), \Delta)$, where Δ denotes the Laplace–Beltrami operator on M (see [14].) However, this formulation does not easily generalize to the noncommutative setting.

After these preliminaries, let us state the following:

Definition 1 ([5]) *A (compact)* noncommutative space *is described by an* $N = 1$ spectral triple $(\mathcal{H}, \mathcal{A}, D)$, *where*

- *$\mathcal{H}$ is a separable Hilbert space,*
- *$\mathcal{A}$ is a unital *-algebra of bounded operators on $\mathcal{H}$,*
- *D is a self-adjoint operator on $\mathcal{H}$ such that*
  1. *For each $a \in \mathcal{A}$, the commutator $[D, a]$ extends uniquely to a bounded operator on $\mathcal{H}$,*
  2. *The operator $e^{-\varepsilon D^2}$ is trace class for all $\varepsilon > 0$.*

The spectral triple is called even *if there is a $\mathbb{Z}_2$-grading γ on $\mathcal{H}$, i.e., a self-adjoint unitary, such that*

$$\{D, \gamma\} = 0 \, , \quad [\gamma, a] = 0 \, ,$$

for all $a \in \mathcal{A}$.

Under additional assumptions, it is possible to show that each spectral triple $(\mathcal{H}, \mathcal{A}, D)$, where $\mathcal{A}$ is an *Abelian* unital *-algebra, corresponds to a spin^c manifold (see [7], [21].)

Starting from a noncommutative space, one can develop the theory of differential forms, vector bundles, connections, etc.. We shall not enter that here, and refer the reader to [5] for a detailed presentation.

3. The target space of the SU(2) WZW model

The Wess–Zumino–Witten (WZW) model on $G = SU(2)$ is a quantum theory of maps,

$$g : \mathbb{R} \times S^1 = \Sigma \longrightarrow G .$$

The action functional of the theory is given by,

$$S[g]_k = \frac{k}{16\pi} \int_\Sigma d^2x \mathrm{Tr}(\partial^\mu g^{-1} \partial_\mu g) + k\Gamma[g] ,$$

where k is a positive integer and,

$$\Gamma[g] = \frac{1}{24\pi} \int_B d^3x \mathrm{Tr}((\widetilde{g}^{-1} d\widetilde{g})^{\wedge 3}) , \tag{1}$$

is the so called Wess–Zumino term. In eq.(1), B is a three dimensional manifold with boundary Σ, i.e., $\partial B = \Sigma$, and $\widetilde{g}$ denotes an extension of g from Σ to B. This theory is conformally invariant [24].

Let $\{t^A\}$ be a basis of $\mathfrak{su}(2)$ such that $\mathrm{Tr}(t^A t^B) = 2\delta^{AB}$, and let if^{ABC} denote the corresponding structure constants. We shall denote by V_j the spin j representation of $\mathfrak{su}(2)$ and by V_j^k the associated unitary highest weight representation of the Kac–Moody algebra $\widehat{\mathfrak{su}(2)}_k$,

$$[J_m^A, J_n^B] = if^{ABC} J_{m+n}^C + \frac{k}{2} m \delta^{AB} \delta_{m+n,0} ,$$

where $n, m \in \mathbb{Z}$. The Hilbert space of the theory is given by [24, 11, 12],

$$\mathcal{H} = \bigoplus_{j=0}^{k/2} V_j^k \otimes V_j^k . \tag{2}$$

This Hilbert space carries two commuting representations of $\widehat{\mathfrak{su}(2)}_k$ denoted by J_n^A and $\overline{J}_n^A$, acting on the left and right factors of eq.(2), respectively. The Sugawara construction,

$$L_n = \frac{1}{k+2} \sum_{m \in \mathbb{Z}} : J_{n-m}^A J_m^A : ,$$

produces two commuting representations of the Virasoro algebra with central charge

$$c = \frac{3k}{k+2} .$$

We shall denote by $\mathcal{H}_0$ the finite dimensional subspace of $\mathcal{H}$ annihilated by all positive modes of the $\widehat{\mathfrak{su}(2)}_k$ generators,

$$\mathcal{H}_0 \simeq \bigoplus_{j=0}^{k/2} V_j \otimes V_j \ .$$

Let $\{e^j_m\}$ denote an orthonormal basis of V_j, and $e^j_{m\overline{m}} = e^j_m \otimes e^j_{\overline{m}}$ the corresponding basis of $\mathcal{H}_0$. For each basis element $e^j_{m\overline{m}}$, there is an $\mathfrak{su}(2)$ primary field $V^j_{m\overline{m}}(z,\overline{z})$ and a function,

$$f^j_{m\overline{m}}(g) = \frac{1}{\sqrt{2j+1}}(e^j_m, \pi^j(g) e^j_{\overline{m}}) \ , \tag{3}$$

on $G = SU(2)$, where π^j denotes the representation of G on V_j. The three point functions of the primaries read [25],

$$\begin{aligned} &\langle e^{j_1}_{m_1\overline{m}_1} | V^{j_2}_{m_2\overline{m}_2}(1,1) | e^{j_3}_{m_3\overline{m}_3} \rangle \\ &\quad = C^k_{j_1 j_2 j_3} \int_G \overline{f^{j_1}_{m_1\overline{m}_1}(g)} f^{j_2}_{m_2\overline{m}_2}(g) f^{j_3}_{m_3\overline{m}_3}(g) d\mu(g) \ , \end{aligned} \tag{4}$$

where $d\mu$ denotes the normalized Haar measure on G. The constants $C^k_{j_1 j_2 j_3}$ satisfy

$$C^k_{j_1 j_2 j_3} = 0 \quad \text{unless} \quad \max(2j_1, 2j_2, 2j_3) \leq j_1 + j_2 + j_3 \leq k, \tag{5}$$

$$\lim_{k\to\infty} C^k_{j_1 j_2 j_3} = 1 \quad \text{if} \quad \max(2j_1, 2j_2, 2j_3) \leq j_1 + j_2 + j_3 \ . \tag{6}$$

We saw in section 2 that we can recover a manifold from the quantum theory of non-relativistic scalar particles or electrons on that manifold. Can we recover the group G from the WZW model? The answer given in [14] is that one can indeed reconstruct an effective target space, and that, at finite level k, it turns out to be a deformation of G. Let us sketch the argument.

The space $L^2(G, d\mu)$ of square integrable functions on G carries a unitary representation $\pi^{G\times G}$ of $G \times G$ given by,

$$(\pi^{G\times G}(g_1, g_2) f)(h) = f(g_1^{-1} h g_2) \ .$$

The decomposition of this representation into irreducibles reads,

$$L^2(G, d\mu) \simeq \overline{\bigoplus_{j=0}^{\infty} V_j \otimes V_j}^{\,d\mu} \ ,$$

where the isomorphism is given in eq.(3). Thus, we see that in the limit $k \to \infty$ the space of $\mathfrak{su}(2)$ primary fields is isomorphic to the space of functions on

G. Furthermore, in that limit, eqs.(4), (5), and (6) show that the three point functions are given by the structure constants of the algebra of functions on G. Finally, when restricted to the subspace $\mathcal{H}_0$, the operators L_0 and $\overline{L}_0$ are proportional to the Laplacian Δ_G on G. In summary, in the limit $k \to \infty$, we recover the data $(L^2(G, d\mu), C^\infty(G), \Delta_G)$ which, as mentioned in section 2, completely encode the geometry of the group G.

At finite level k the space of functions is cut at spin $k/2$, giving the finite-dimensional subspace $\mathcal{H}_0$, and the structure constants get deformed by the multiplicative constants $C^k_{j_1 j_2 j_3}$. The primary fields do not map the space $\mathcal{H}_0$ into itself. However, since we have a natural representation space, namely $\mathcal{H}_0$, it natural to define the deformed algebra of functions $\mathcal{A}_0$ as the algebra generated by the primaries acting on $\mathcal{H}_0$ followed by the orthogonal projection on $\mathcal{H}_0$. This construction yields an associative algebra, but it is noncommutative. The resulting algebra turns out to be the full matrix algebra on $\mathcal{H}_0$ [18][‡],

$$\mathcal{A}_0 = \mathrm{End}(\mathcal{H}_0) .$$

Thus, at finite level, the data describing the target space are

$$(\mathcal{H}_0, \mathcal{A}_0, L_0 + \overline{L}_0) .$$

As mentioned in section 2, these data are not appropriate for describing the deformed geometry of G since $\mathcal{A}_0$ is noncommutative. What is needed is a Dirac operator instead of the Laplacian $L_0 + \overline{L}_0$. Since we need spinors on G, it is natural to consider the supersymmetric WZW model.

The supersymmetric WZW model is equivalent to the tensor product of a bosonic WZW model with free massless fermions in the adjoint representation of G, [9], [17]. We denote the modes of the fermionic fields by ψ^A_r and $\overline{\psi}^A_r$, where $r \in \mathbb{Z} + a$, and $a = 0, 1$ in the Ramond and Neveu–Schwarz sectors, respectively. The fermionic Fock space carries its own $\widehat{\mathfrak{su}(2)}_k$ and Virasoro representations, and the full Hilbert space is endowed with two commuting representations of the super-Virasoro algebra. The supersymmetry generators read,

$$Q_r = \sqrt{\frac{2}{k}} \Big(\sum_s \psi^A_s J^A_{r-s} - \frac{i}{6} f^{ABC} \sum_{t,u} : \psi^A_t \psi^B_u \psi^C_{r-t-u} : \Big) ,$$

and the zero mode satisfy,

$$Q_0^2 = 2L_0^{\mathrm{tot}} - \frac{c^{\mathrm{tot}}}{12} ,$$

[‡] It has also been shown that $\mathcal{A}_0$ is a full matrix algebra for all simply laced, compact, connected and simply connected groups

where $L_0^{\rm tot}$ denotes the sum of the bosonic and fermionic contributions to the zero mode of the energy-momentum tensor. Being an odd square root of the Laplacian $L_0 + \overline{L}_0 = 2L_0$ on $\mathcal{H}_0$, the operator Q_0 is the Dirac operator we are looking for. Thus, the target space is described by the $N = 1$ spectral triple,

$$(\mathcal{H}_0 \otimes W, \mathcal{A}_0, Q_0) \,,$$

where W is the subspace of the Ramond Fock space spanned by all vectors annihilated by the positive fermionic modes,

$$\psi_n^A = \overline{\psi}_n^A = 0 \quad \text{on } W, \qquad \text{for all } n > 0 \,.$$

The resulting noncommutative space was studied in great details in [18], [16], and we shall not enter that issue here.

4. Supersymmetry and Geometry

Classical differential geometry allows one to determine whether a given manifold can be endowed with additional geometrical structures like symplectic, complex, or Kähler structures, for example. It turns out that $N = 1$ spectral triples are not quite appropriate to describe these higher geometrical structures, even in the classical setting. As we shall see, the solution arises naturally in the context of supersymmetric quantum mechanics. It is important to mention that there are well known links between supersymmetric σ-models and the geometry of manifolds (see [1] for an early result.) In fact, the notations used below are strongly inspired by superconformal field theories and their realizations as σ-models.

To begin with, we reformulate Riemannian geometry in a slightly different way than we did in section 2. Recall that Pauli's theory of non-relativistic electrons was formulated on a spin^c manifold M, that we shall take to be even dimensional for simplicity. We denote by S, c, and ∇^S, the spinor bundle, Clifford action, and the spin connection on S, respectively. Let $\overline{S}$ denote the complex (or charge) conjugate spinor bundle to S, $\overline{c}$ its Clifford action and $\nabla^{\overline{S}}$ its spin connection. The tensor product bundle $S \otimes \overline{S}$ is endowed with the tensor product connection,

$$\nabla^{S\otimes\overline{S}} = \nabla^S \otimes 1 + 1 \otimes \nabla^{\overline{S}} \,,$$

and two anticommuting Clifford actions,

$$\mathbf{c} = c \otimes 1 \,, \quad \overline{\mathbf{c}} = \gamma \otimes \overline{c} \,, \tag{1}$$

where γ denotes the Clifford chirality element. The total $\mathbb{Z}_2$-grading is,

$$\Gamma = \gamma \otimes \overline{\gamma} \,.$$

Each Clifford action gives a self-adjoint Dirac operator,

$$\mathcal{D} = \mathbf{c} \circ \nabla^{S\otimes\overline{S}}, \quad \overline{\mathcal{D}} = \overline{\mathbf{c}} \circ \nabla^{S\otimes\overline{S}},$$

that anticommutes with the $\mathbb{Z}_2$-grading,

$$\{\mathcal{D}, \Gamma\} = \{\overline{\mathcal{D}}, \Gamma\} = 0 ,$$

and satisfy the so called $N = (1,1)$ *algebra,*

$$\mathcal{D}^2 = \overline{\mathcal{D}}^2 , \quad \{\mathcal{D}, \overline{\mathcal{D}}\} = 0 .$$

In physical terms, sections of S and $\overline{S}$ describe electrons and positrons, respectively. Also, sections of $S\otimes\overline{S}$ describe the center of mass of an electron-positron bound state, i.e., positronium. More precisely, positronium is the quantum mechanical system whose time evolution is governed by the Hamiltonian,

$$H_{ep} = \mathcal{D}^2 = \overline{\mathcal{D}}^2 ,$$

(up to physical constants), acting on the Hilbert space

$$\mathcal{H}_{ep} = L^2(S \otimes \overline{S}, \mathrm{dvol}_g) \simeq \overline{\mathcal{H}_e \otimes_{C^\infty(M)} \mathcal{H}_p} .$$

The resulting data,

$$(\mathcal{H}_{ep}, C^\infty(M), \mathcal{D}, \overline{\mathcal{D}}, \Gamma) ,$$

are called $N = (1,1)$ *data* and they encode M as a Riemannian manifold (see [15] for details.) Of course, in the case at hand, the Hilbert space is simply the space of square integrable differential forms,

$$\mathcal{H}_{ep} \simeq L^2(\Lambda^\bullet T^* M, \mathrm{dvol}_g) ,$$

and the exterior derivative can be written as,

$$d = \tfrac{1}{2}(\mathcal{D} - i\overline{\mathcal{D}}) .$$

Furthermore, the $\mathbb{Z}_2$-grading Γ is the mod 2 reduction of a $\mathbb{Z}$-grading,

$$T = p \quad \text{on} \quad \Lambda^p T^* M ,$$

generating a $\mathfrak{u}(1)$ symmetry and satisfying,

$$\begin{aligned} [T, \mathcal{D}] &= -i\overline{\mathcal{D}} , \qquad [T, \overline{\mathcal{D}}] = i\mathcal{D} , \\ [T, f] &= 0 \quad \text{for all} \quad f \in C^\infty(M) . \end{aligned}$$

This additional symmetry T generating rotations in the $(\mathcal{D} - \overline{\mathcal{D}})$-plane makes the $N = (1,1)$ data into $N = 2$ *data*,

$$(\mathcal{H}_{ep}, C^\infty(M), \mathcal{D}, \overline{\mathcal{D}}, T) \; .$$

This extension from $N = (1,1)$ to $N = 2$ does not always hold in the noncommutative setting. From the physical point of view, the $\mathfrak{u}(1)$ symmetry generated by T is a global gauge symmetry, and it is natural to ask if we can extend it to a bigger symmetry group.

Let us try to embed $\mathfrak{u}(1)$ into $\mathfrak{su}(2)$. We are looking for operators L^3, $L^\pm$ on $\mathcal{H}_{ep}$ satisfying,

$$[L^3, L^\pm] = \pm 2L^\pm \; , \quad [L^+, L^-] = L^3 \; ,$$

such that,

(A1) $L^3 = T - \frac{1}{2}\mathrm{dim}M$,

(A2) $[L^\pm, f] = [L^3, f] = 0 \quad$ for all $f \in C^\infty(M)$,

(A3) d transforms under a spin $\frac{1}{2}$ representation, i.e.,

$$[L^+, d] = 0 \; , \quad [L^-, [L^-, d]] = 0 \; .$$

Property (A1) simply makes the spectrum of L^3 symmetric around zero. Property (A2) states that the symmetries should be local, and property (A3) ensures that we do not generate too many new supersymmetries at once.

It turns out that the above conditions cannot be fulfilled unless the manifold, M, admits a symplectic 2-form, ω. When M is symplectic we have

$$L^+ = \omega\wedge \; , \quad L^- = \omega\llcorner \; ,$$

where $\wedge$ and $\llcorner$ denote the wedge and interior products, respectively. We might then ask if the representation of $\mathfrak{su}(2)$ on the space of differential forms is unitary (unitarity is defined with respect to the scalar product induced by the Riemannian metric g on M.) To answer this question we notice that since both g and ω are non-degenerate, we can define an endomorphism $J \in \Gamma(\mathrm{End}(TM))$ by setting,

$$\omega(X, Y) = g(JX, Y) \; , \quad \text{for all } X, Y \in \Gamma(TM) \; .$$

The following statements are then equivalent,

- $\mathfrak{su}(2)$ acts unitarily, i.e., $(L^+)^* = L^-$
- M is almost Kähler, i.e., $J^2 = -1$.

This simply means that the compatibility of the symplectic structure with the Riemannian metric is equivalent to the unitarity if the $\mathfrak{su}(2)$ representation induced by the symplectic form.

For later convenience, let us define,

$$\widetilde{d}^* = [L^-, d] \ .$$

When the manifold is symplectic, the doublet $(d, \widetilde{d}^*)$ transforms under the spin $\frac{1}{2}$ representation of $\mathfrak{su}(2)$, and when M is almost Kähler, so does the conjugate doublet $(d^*, \widetilde{d})$. In the almost Kähler case, we have,

$$d^2 = \widetilde{d}^2 = 0 \ , \quad \{d, \widetilde{d}^*\} = 0 \ , \quad \{d, d^*\} = \Delta \ ,$$

where Δ denotes the Laplace operator on M. However, it is not true, in general, that d anticommutes with $\widetilde{d}$. In fact, the following statements are equivalent,

- $\{d, \widetilde{d}\} = 0$
- M is Kähler, i.e., the almost complex structure J is integrable.

Furthermore, when M is Kähler, we have the relations[§],

$$\{\widetilde{d}, \widetilde{d}^*\} = \Delta \ , \quad \partial = \tfrac{1}{2}(d - i\widetilde{d}) \ , \quad \overline{\partial} = \tfrac{1}{2}(d + i\widetilde{d}) \ ,$$

and Δ is central in the supersymmetry algebra, i.e., it commutes with all differential operators d, d^*, $\widetilde{d}$, $\widetilde{d}^*$, and with the $\mathfrak{su}(2)$ generators.

In the Kähler case, there is a further $\mathfrak{u}(1)$ symmetry $\mathcal{T} - \overline{\mathcal{T}}$, where,

$$\mathcal{T} = p \ , \quad \overline{\mathcal{T}} = q \quad \text{on } \Lambda^{p,q} M \ ,$$

are the holomorphic and antiholomorphic degree of differential forms. In summary, Kähler manifolds are encoded in $N = 4^+$ *data*,

$$(\mathcal{H}_{ep}, C^\infty(M), \partial, \partial^*, \overline{\partial}, \overline{\partial}^*, \mathfrak{su}(2) \times \mathfrak{u}(1)) \ ,$$

where the '+' superscript indicates the presence of an additional $\mathfrak{u}(1)$ symmetry, compared to the usual $N = 4$ algebra. This additional Cartan generator appears because Kähler manifolds are already described by $N = (2,2)$ *data*,

$$(\mathcal{H}_{ep}, C^\infty(M), \partial, \partial^*, \overline{\partial}, \overline{\partial}^*, \mathfrak{u}(1) \times \mathfrak{u}(1)) \ ,$$

obtained by forgetting about the $L^\pm$ generators.

Of course, one might continue the process and try to extend $N = (2,2)$ data to $N = (4,4)$ data with $\mathfrak{su}(2) \times \mathfrak{su}(2)$ symmetry. This leads to hyper-Kähler

---

[§] The symbols ∂ and $\overline{\partial}$ denote the holomorphic and antiholomorphic exterior differentials, respectively.

manifolds, for which one automatically obtains the $N = 8$ supersymmetry algebra with $Sp(4)$ symmetry group. We refer the interested reader to [15] for a detailed presentation.

In summary, we see that the geometry of a Riemannian manifold can be recovered from the quantum theory of positronium. Additional structures are described in terms of supersymmetry algebras acting on the space of differential forms, and we may view the passage from general geometries to more special ones as a symmetry enhancement. Conversely, going from special geometries to more general ones, appears as a symmetry breaking. It is important to notice that all generators of the supersymmetry algebra have a geometrical interpretation. The nice feature of this formulation of differential geometry is that it can be extended to the noncommutative setting (see [16].)

5. The Noncommutative 2-Torus

We explain how the step by step extension of spectral data described in the previous section is implemented in the case of the noncommutative 2-torus at irrational deformation parameter.

First, we introduce the $N = 1$ spectral data describing the classical 2-torus T^2. Let g be a constant (flat) metric on T^2. The associated Clifford algebra reads,

$$\{\gamma^\mu, \gamma^\nu\} = -2g^{\mu\nu} .$$

Since the torus is parallelizable we may take the spinor bundle, S, to be a trivial bundle of rank 2. Under Fourier transformation, we have,

$$\begin{aligned} C^\infty(T^2) &\simeq \mathcal{A}_0 := \mathcal{S}(\mathbb{Z}^2) , \\ L^2(S) &\simeq \mathcal{H} := \ell^2(\mathbb{Z}^2) \oplus \ell^2(\mathbb{Z}^2) . \end{aligned}$$

The pointwise product in $C^\infty(T^2)$ goes over to the convolution product in $\mathcal{S}(\mathbb{Z}^2)$ and the Dirac operator D acts on vectors in $\mathcal{H}$ as follows:

$$(D\xi)(p) = ip_\mu \gamma^\mu \xi(p) , \quad \xi \in \mathcal{H} , \quad p \in \mathbb{Z}^2 .$$

There is a $\mathbb{Z}_2$-grading on $\mathcal{H}$ given by,

$$\sigma = \frac{i}{2}\sqrt{g}\varepsilon_{\mu\nu}\gamma^\mu\gamma^\nu ,$$

where ε denotes the antisymmetric symbol,

$$\varepsilon_{12} = -\varepsilon_{21} = 1 .$$

The 2-torus is encoded in the even $N = 1$ spectral data,

$$(\mathcal{H}, \mathcal{A}_0, D, \sigma) .$$

Let ω denote the integer-valued antisymmetric bilinear form on $\mathbb{Z}^2$ defined by,

$$\omega(p,q) = p_1 q_2 - p_2 q_1 \,, \quad p,q \in \mathbb{Z}^2 \,.$$

The noncommutative 2-torus T^2_θ at irrational deformation parameter θ is the noncommutative space described by the $N = 1$ spectral data,

$$(\mathcal{H}, \mathcal{A}_\theta, D, \sigma) \,.$$

The algebra $\mathcal{A}_\theta$ is the algebra $\mathcal{A}_0$ endowed with the ω-twisted convolution product,

$$(f \cdot g)(p) = \sum_{q \in \mathbb{Z}^2} f(q) g(p-q) e^{i\pi\theta\omega(p,q)} .$$

Our first task is to extend the $N = 1$ data to $N = (1,1)$ data. Recall how we made this extension in the previous section:

(A) We replaced $\Gamma(S)$ by $\Gamma(S \otimes \overline{S}) \simeq \Gamma(S) \otimes_{C^\infty(M)} \Gamma(\overline{S})$.

(B) We constructed two Dirac operators $\mathcal{D}$ and $\overline{\mathcal{D}}$ out of the tensor product connection on $S \otimes \overline{S}$ and the two anticommuting Clifford actions.

Let us first take care of (A). We want to construct $\widetilde{\mathcal{H}} = \mathcal{H} \otimes_{\mathcal{A}_\theta} \mathcal{H}$. The first problem is that the Hilbert space $\mathcal{H}$ does not have an $\mathcal{A}_\theta$-action on the right. This point is solved by introducing a real structure J (see [6].) We define an antiunitary map,

$$\kappa : \mathcal{H} \longrightarrow \mathcal{H}, \quad (\kappa\xi)(p) = \overline{\xi(p)},$$

and a matrix $C \in \mathbb{M}_2(\mathbb{C})$ such that,

$$C = C^* = C^{-1} \,, \quad C\gamma^\mu = -\overline{\gamma}^\mu C \,.$$

Then, the operator $J = C\kappa$ is a *real structure* (see [6]), i.e., it is an antilinear operator on $\mathcal{H}$ such that,

$$J^2 = 1 \,, \quad J\sigma = -\sigma J \,, \quad JD = DJ \,,$$

and,

$$[JaJ^*, b] = [JaJ^*, [D,b]] = 0 \,, \quad \text{for all } a, b \in \mathcal{A}_\theta \,. \tag{1}$$

The equation (1) means, in particular, that conjugation by J maps the algebra $\mathcal{A}$ into its commutant. We might therefore think of J as a kind of Tomita–Takesaki modular involution (see [2] for example.) The real structure allows us to make $\mathcal{H}$ into an $\mathcal{A}_\theta$-bimodule by setting,

$$\xi \cdot a = JaJ^*\xi \,, \quad \text{for } a \in \mathcal{A}_\theta \,, \ \xi \in \mathcal{H} \,.$$

Let us denote by $\mathring{\mathcal{H}}$ the dense subspace $\mathcal{S}(\mathbb{Z}^2) \oplus \mathcal{S}(\mathbb{Z}^2)$ of $\mathcal{H}$. There is a natural scalar product on the space $\mathring{\mathcal{H}} \otimes_{\mathcal{A}_\theta} \mathring{\mathcal{H}}$ (see [16]) and we denote the Hilbert space completion by $\widetilde{\mathcal{H}}$.

We then need to construct the Dirac operators acting on $\widetilde{\mathcal{H}}$ as stated in (B). Since we do not have any spin connection at our disposal, we momentarily choose an arbitrary connection¶ (see the lectures of M. Dubois-Violette for the definition and properties of connections)

$$\nabla : \mathring{\mathcal{H}} \longrightarrow \Omega^1_D(\mathcal{A}_\theta) \otimes_{\mathcal{A}_\theta} \mathring{\mathcal{H}} \, .$$

The antilinear map,

$$\Psi : \Omega^1_D(\mathcal{A}_\theta) \otimes_{\mathcal{A}_\theta} \mathring{\mathcal{H}} \longrightarrow \mathring{\mathcal{H}} \otimes_{\mathcal{A}_\theta} \Omega^1_D(\mathcal{A}_\theta) \, ,$$

defined by,

$$\Psi(\omega \otimes \xi) = J\xi \otimes \omega^* \, ,$$

allows us to define a right connection,

$$\begin{aligned} \overline{\nabla} : \mathring{\mathcal{H}} &\longrightarrow \mathring{\mathcal{H}} \otimes_{\mathcal{A}_\theta} \Omega^1_D(\mathcal{A}_\theta) \, , \\ \overline{\nabla}\xi &= -\Psi(\nabla J^* \xi) \, . \end{aligned}$$

This right connection satisfies the Leibniz rule on the right,

$$\overline{\nabla}(\xi a) = \overline{\nabla}\xi a + \xi \otimes [D, a] \, .$$

We may then introduce a 'tensor product connection',

$$\widetilde{\nabla} = \overline{\nabla} \otimes 1 + 1 \otimes \nabla : \mathring{\mathcal{H}} \otimes_{\mathcal{A}_\theta} \mathring{\mathcal{H}} \longrightarrow \mathring{\mathcal{H}} \otimes_{\mathcal{A}_\theta} \Omega^1_D(\mathcal{A}_\theta) \otimes_{\mathcal{A}_\theta} \mathring{\mathcal{H}} \, ,$$

and two linear 'Clifford actions',

$$\mathbf{c}, \overline{\mathbf{c}} : \mathring{\mathcal{H}} \otimes_{\mathcal{A}_\theta} \Omega^1_D(\mathcal{A}_\theta) \otimes_{\mathcal{A}_\theta} \mathring{\mathcal{H}} \longrightarrow \mathring{\mathcal{H}} \otimes_{\mathcal{A}_\theta} \mathring{\mathcal{H}} \, ,$$

¶The space of 1-forms is defined by,

$$\Omega^1_D(\mathcal{A}_\theta) = \{ \sum_{j=1}^{n} a_0^j [D, a_1^j] \mid a_i^j \in \mathcal{A}_0 \, , \ n \in \mathbb{N} \} \, .$$

The exterior derivative

$$d : \mathcal{A}_0 \longrightarrow \Omega^1_D(\mathcal{A}_\theta)$$

is defined by

$$da = [D, a] \, , \quad \text{for all } a \in \mathcal{A}_0 \, .$$

defined by

$$\begin{aligned} \mathbf{c}(\xi_1 \otimes \omega \otimes \xi_2) &= \xi_1 \otimes \omega\xi_2 , \\ \overline{\mathbf{c}}(\xi_1 \otimes \omega \otimes \xi_2) &= \xi_1\omega \otimes \sigma\xi_2 . \end{aligned} \tag{2}$$

Notice that the $\mathbb{Z}_2$-grading σ enters the definition of $\overline{\mathbf{c}}$ in eq.(2) in the same way as in eq.(1). Finally, we define the Dirac operators on $\widetilde{\mathcal{H}}$ by setting,

$$\mathcal{D} = \mathbf{c} \circ \widetilde{\nabla} , \quad \overline{\mathcal{D}} = \overline{\mathbf{c}} \circ \overline{\nabla} .$$

Since the connection ∇ on $\mathring{\mathcal{H}}$ was arbitrary, our Dirac operators might not satisfy the $N = (1,1)$ algebra,

$$\mathcal{D}^* = \mathcal{D} , \quad \overline{\mathcal{D}}^* = \overline{\mathcal{D}} , \quad \mathcal{D}^2 = \overline{\mathcal{D}}^2 , \quad \{\mathcal{D}, \overline{\mathcal{D}}\} = 0 .$$

However, there is a unique tensor product connection $\widetilde{\nabla}$ on $\mathring{\mathcal{H}} \otimes_{\mathcal{A}_\theta} \mathring{\mathcal{H}}$, of the above form, for which the $N = (1,1)$ algebra is satisfied. This connection is given by

$$\widetilde{\nabla} e_i \otimes e_j = 0 , \quad i, j = 1, 2 ,$$

where the vectors $e_i \in \mathcal{H}$ are given by

$$e_1(p) = \delta_{p,0} \begin{pmatrix} 1 \\ 0 \end{pmatrix} , \quad e_2(p) = \delta_{p,0} \begin{pmatrix} 0 \\ 1 \end{pmatrix} .$$

Using this connection we obtain $N = (1,1)$ data,

$$(\widetilde{\mathcal{H}}, \mathcal{A}_\theta, \mathcal{D}, \overline{\mathcal{D}}) .$$

Up to the choice of a particular tensor product connection on $\widetilde{\mathcal{H}}$, we did not use the fact that we were dealing with the noncommutative torus. It is thus reasonable that this procedure gives $N = (1,1)$ data out of $N = 1$ data in more general situations. The natural $\mathbb{Z}_2$-grading on $\widetilde{\mathcal{H}}$ is,

$$\Gamma = \sigma \otimes \sigma ,$$

and, as in the classical case, it lifts uniquely to a $\mathbb{Z}$-grading,

$$T = \frac{1}{2i} g_{\mu\nu} \gamma^\mu \otimes \gamma^\nu \sigma ,$$

commuting with the action of $\mathcal{A}_\theta$ and such that

$$[T, d] = d ,$$

where the exterior derivative is given by

$$d = \tfrac{1}{2}(\mathcal{D} - i\overline{\mathcal{D}}) \, .$$

This implies that the $N = (1,1)$ data extend to $N = 2$ data. In order to extend these data further to $N = (2,2)$, we need to split the exterior derivative and the $\mathrm{u}(1)$ generator as

$$d = \partial + \overline{\partial} \, , \quad T = \mathcal{T} + \overline{\mathcal{T}} \, ,$$

in such a way that ∂, $\overline{\partial}$ together with their adjoints and $\mathcal{T}$, $\overline{\mathcal{T}}$ satisfy the $N = (2,2)$ algebra. In other words, we are, looking for operators

$$\widetilde{d} = i(\partial - \overline{\partial}) \, , \quad I = i(\mathcal{T} - \overline{\mathcal{T}}) \, ,$$

satisfying the relations,

$$\begin{gathered} [I, a] = 0 \quad \text{for all } a \in \mathcal{A}_\theta \, , \\ [I, \Gamma] = 0 \, , \quad \widetilde{d}^2 = 0 \, , \quad [I, \widetilde{d}] = \widetilde{d} \, , \\ [I, \sigma \otimes 1] = 0 \, . \end{gathered} \tag{3}$$

The last condition, eq.(3), is the requirement that the operator I commutes with the Hodge *-operator. This is explained in details in [16]. Solving all the above conditions leaves only one interesting solution for I, namely,

$$I = \frac{i}{2}(\sigma \otimes 1 + 1 \otimes \sigma) \, .$$

It turns out that the complex structure we obtain on the noncommutative 2-torus is the same as that found by A. Connes using another approach that relies on the equivalence of complex and conformal structures in two dimensions. The advantage of our notion of complex structure is that it is independent of the space dimension.

References

[1] Alvarez-Gaumé, L., Freedman, D.Z., Geometrical Structure and Ultraviolet Finiteness in the Supersymmetric σ-Model, *Commun. Math. Phys.* **80**, (1981) 443–451.

[2] Bratteli, O., and Robinson, D.W., *Operator Algebras and Quantum Statistical Mechanics 1,* 2nd ed. New York, Berlin, Heidelberg, London, Paris, Tokyo: Springer Verlag, 1987.

[3] Cattaneo, A.S., Felder, G., A Path Integral Approach to the Kontsevich Quantization Formula, `math.QA/9902090`, 1999.

[4] Chamseddine, A.H., Fröhlich, J., Some Elements of Connes' Non-Commutative Geometry, and Space-Time Geometry, In: *Chen Ning Yang, a great physicist of the twentieth century,* C.S. Liu and S.-T. Yau (eds.), Cambridge, MA: International Press (1995), 10–34.

[5] Connes, A., *Noncommutative Geometry,* London, New York: Academic Press, 1994.

[6] Connes, A., Noncommutative Geometry and Reality, *J. Math. Phys.* **36** (1995), 6194–6231.

[7] Connes, A., Gravity Coupled with Matter and Foundation of Noncommutative Geometry, *Commun. Math. Phys.* **186** (1996), 731–750.

[8] Connes, A., Douglas, M.R., Schwarz, A., Noncommutative Geometry and Matrix Theory: Compactification on Tori, *JHEP* 9802:003, 1998.

[9] Di Vecchia, P., Knizhnik, V.G., Petersen, J.L., Rossi, P., A Supersymmetric Wess–Zumino Lagrangian in Two Dimensions, *Nucl. Phys.* **B253** (1985), 701–726.

[10] Doplicher, S., Fredenhagen, K., Roberts, J.E., The Quantum Structure of Space-Time at the Planck Scale and Quantum Fields, *Commun. Math. Phys.* **172** (1995), 187–220.

[11] Felder, G., Gawędzki, K., Kupiainen, A., The Spectrum of Wess–Zumino–Witten Models, *Nucl. Phys.* **B299** (1988), 355–366.

[12] Felder, G., Gawędzki, K., Kupiainen, A., Spectra of Wess–Zumino–Witten Models with Arbitrary Simple Groups, *Commun. Math. Phys.* **117** (1988), 127–158.

[13] Fröhlich, J., The Non-Commutative Geometry of Two-Dimensional Supersymmetric Conformal Field Theory, In: *PASCOS, Proc. of the Fourth Intl. Symp. on Particles, Strings, and Cosmology.* K.C. Wali (ed.), Singapore: World Scientific 1995.

[14] Fröhlich, J., Gawędzki, K., Conformal Field Theory and the Geometry of Strings, *CRM Proceedings and Lecture Notes*, Vol.**7** (1994), 57–97.

[15] Fröhlich, J., Grandjean, O., Recknagel, A., Supersymmetric Quantum Theory and Differential Geometry, *Commun. Math. Phys.* **193** (1998), 527–594.

[16] Fröhlich, J., Grandjean, O., Recknagel, A., Supersymmetric Quantum Theory and Non-Commutative Geometry, *Commun. Math. Phys.* **203** (1999), 119–184.

[17] Fuchs, J., More on the Super WZW Theory, *Nucl. Phys.* **B318** (1989), 631–654.

[18] Grandjean, O., *Non-Commutative Differential Geometry,* Ph.D. Thesis, ETH Zürich, 1997. .

[19] Kontsevich, M., Deformation Quantization of Poisson Manifolds, `q-alg/9709040`, 1997.

[20] Polchinski, J., Dirichlet Branes and Ramond-Ramond Charges, *Phys. Rev. Lett.* **75** (1995), 4724–4727.

[21] Rennie, A., Commutative Geometries are Spin Manifolds, `math-ph/9903021`, 1999.

[22] Schomerus, V., D-Branes and Deformation Quantization, *JHEP* 9906:030, 1999.

[23] Seiberg, N., Witten, E., String Theory and Noncommutative Geometry, `hep-th/9908142`, 1999.

[24] Witten, E., Non-Abelian Bosonization in Two Dimensions, *Commun. Math. Phys.* **92** (1984), 455–472.

[25] Zamolodchikov, A.B., Fateev, V.A., Operator Algebra and Correlation Functions in the Two-Dimensional Wess–Zumino SU(2)$\times$SU(2) Chiral Model, *Sov. J. Nucl. Phys.* **43** (1986), 657–664.

A RELATION BETWEEN COMMUTATIVE AND NONCOMMUTATIVE DESCRIPTIONS OF D-BRANES

Nobuyuki Ishibashi
KEK Theory Group, Tsukuba, Ibaraki 305, JAPAN

Abstract In string theory D-branes can be expressed as a configuration of infinitely many lower-dimensional D-branes. Using this relation, the worldvolume theory of D-branes can be regarded as the worldvolume theory of the infinitely many lower dimensional branes. In the description in terms of the lower dimensional branes, some of the worldvolume coordinates become noncommutative. Actually this noncommutative theory can be regarded as noncommutative Yang–Mills theory. Therefore the worldvolume theory of D-branes have two equivalent descriptions, namely the usual static gauge description using ordinary Yang–Mills theory and the noncommutative description using noncommutative Yang–Mills theory. It will be shown that these two descriptions correspond to two different ways of gauge fixing of the reparametrization invariance and its generalization. We will give an explicit relation between commutative gauge field and noncommutative gauge field in semiclassical approximation, when the gauge group is $U(1)$.

Mathematics Subject Classification (2000): 81T40, 81T75

Keywords: String theory, D-brane, Noncommutative Yang–Mills

1. Introduction

Many physicists are interested in noncommutative geometry, because they expect that it captures some features of quantum gravity. It is intriguing to see that in string theory and M theory, which is considered to be the most promising model of quantum gravity, we come across several occasions in which a space-

Y. Maeda, et al. (eds.), Noncommutative Differential Geometry and Its Applications to Physics, 49–61.

time coordinate becomes noncommutative[1]–[15]. In this note we would like to discuss an example in which noncommutativity of space-time coordinates appears in string theory. The example we study here is the worldvolume theory of D-branes. As was pointed out in [16, 17], Dp-branes can be represented as a configuration of infinitely many D$(p-2r)$-branes. If such a relation hold, the worldvolume theory of the Dp-branes can also be regarded as the worldvolume theory of infinitely many D$(p-2r)$-branes. We will show that some of the coordinates on the worldvolume of the Dp-branes become noncommutative if one consider it as the worldvolume theory of D$(p-2r)$-branes. Actually the noncommutative theory we have is noncommutative Yang–Mills theory. Such a noncommutative description of the Dp-branes should be equivalent to the usual commutative descriptions. We will pursue this equivalence and show that these two descriptions correspond to two different way to fix the reparametrization invariance and its generalization on the worldvolume. Therefore we have here an example where a noncommutative theory is equivalent to a commutative theory. Such an equivalence in a similar context was studied in a recent paper [18]. We will comment on the relation between our results and theirs.

In section 2 we explain how Dp-branes can be expressed as a configuration of infinitely many D$(p-2r)$-branes. In section 3 we study the worldvolume theory of the Dp-branes regarding it as a configuration of the D$(p-2r)$-branes. Section 4 is devoted to discussions.

This note is based on a talk presented at the "Workshop on Noncommutative Differential Geometry and its Application to Physics", Shonan-Kokusaimura, Japan, May31–June 4, 1999.

After I had completed this note I was informed that there are papers [19] [20] whose results have considerable overlap with ours. Especially, in the first paper, they realized that the commutative and noncommutative descriptions of D-branes correspond to two different ways of gauge fixing, and the explicit relation between the commutative and noncommutative gauge fields, which coincides with ours were given in [20].

2. Dp-branes from D$(p-2r)$-branes

In this section we will explain how Dp-branes can be expressed as a configuration of infinitely many D$(p-2r)$-branes. For simplicity, the special case of expressing one Dp-brane by D-instantons, namely the $p = 2r + 1$ case, will be treated here. We will comment on more general cases at the end of this section.

We study this problem in the Euclidean space $\mathbf{R}^D$. * D should be 26 for a bosonic string and 10 for a superstring. We will first deal with bosonic string

*Construction of Dp-branes from D$(p-2r)$-branes was done on torus in [21]. Things discussed here are partially generalized to the space compactified on a torus in [22].

theory in which the whole of the manipulations are simpler. Later we will explain how one can generalize the arguments to the superstring case. The configuration of infinitely many D-instantons can be expressed by the $\infty \times \infty$ Hermitian matrices X^M $(M = 1, \cdots, D)$. The one we consider is

$$\begin{aligned} X^i &= \widehat{P}^i, \qquad (i = 1, \cdots, p+1), \\ X^M &= 0 \qquad (M = p+2, \cdots, D), \end{aligned} \tag{1}$$

where $\widehat{P}^i$ $(i = 1 \cdots, p+1)$ satisfy

$$[\widehat{P}^i, \widehat{P}^j] = i\theta^{ij}. \tag{2}$$

In this note, the $(p+1) \times (p+1)$ matrix θ is assumed to be invertible.

Let us show that this configuration of D-instantons is equivalent to a Dp-brane. In order to show that two different configurations of D-branes are equivalent, one should prove that the open string theory corresponding to these configurations are equivalent.

A quick way to see the equivalence is to look at the boundary states. The boundary state $|B\rangle$ corresponding to the configuration eq.(1) can be written as follows:

$$|B\rangle = \mathrm{TrP}e^{-i\int_0^{2\pi} d\sigma p_i \widehat{P}^i}|B\rangle_{-1}. \tag{3}$$

Here $p_i(\sigma)$ is the variable conjugate to the string coordinate $x^i(\sigma)$ and is equal to $(1/2\pi\alpha')\dot{x}_i$ in the usual flat background. $|B\rangle_{-1}$ denotes the boundary state for a D-instanton at the origin and satisfies $x^i(\sigma)|B\rangle_{-1} = 0$. $|B\rangle_{-1}$ includes also the ghost part which is not relevant to the discussion here. The factor in front of $|B\rangle_{-1}$ is an analogue of the Wilson loop and corresponds to the background eq.(1). Eq.(3) can be rewritten by using path integral as

$$|B\rangle = \int [dP] e^{\frac{i}{2}\int d\sigma P^i \partial_\sigma P^j \omega_{ij} - i\int d\sigma p_i P^i}|B\rangle_{-1}, \tag{4}$$

where $\omega_{ij} = (\theta^{-1})_{ij}$.

It is straightforward to perform the path integral in eq.(4). Using the Fock space representation of $|B\rangle_{-1}$, the path integral is Gaussian. The determinant factor can be regularized in the usual way [23], and one can show that $|B\rangle$ coincides with the boundary state for a Dp-brane with the $U(1)$ gauge field strength $F_{ij} = \omega_{ij}$ on the worldvolume. Knowing that the path integration is Gaussian, it is easy to confirm this fact. Indeed, one can show that the following identity holds:

$$\begin{aligned} 0 &= \int [dP] \frac{\delta}{\delta P^i(\sigma)} e^{\frac{i}{2}\int d\sigma P^i \partial_\sigma P^j \omega_{ij} - i\int d\sigma p_i P^i}|B\rangle_{-1} \\ &= [i\omega_{ij}\partial_\sigma x^j - ip_i(\sigma)]|B\rangle. \end{aligned} \tag{5}$$

Therefore $|B\rangle$ should coincide with $\exp[\frac{i}{2}\int d\sigma x^i \partial_\sigma x^j \omega_{ij}]|B\rangle_p$ up to normalization. Here $|B\rangle_p$ denotes the boundary state for a Dp-brane satisfying $p_i(\sigma)|B\rangle_p = 0$. Hence $|B\rangle$ is equivalent to the boundary state for a Dp-brane with the $U(1)$ gauge field strength $F_{ij} = \omega_{ij}$ on the worldvolume.

In the above analysis using the path integral expression eq.(4), the overall normalization of the boundary state is ambiguous. Actually one can prove the equivalence including the normalization by showing that the open string theory corresponding to the configuration eq.(1) is equivalent to the one corresponding to a Dp-brane. We refer to [24] for more details.

It is easy to supersymmetrize the above arguments. In the NSR formalism the boundary state for D-instanton can be written as a sum of four states $|B;\pm\rangle_{-1,I}$ where $I = NS, R$ indicates the sector it belongs to and

$$
\begin{aligned}
x^M(\sigma)|B;\pm\rangle_{-1,I} &= 0, \\
(\psi^M(\sigma) \pm i\tilde{\psi}^M(\sigma))|B;\pm\rangle_{-1,I} &= 0.
\end{aligned} \tag{6}
$$

Supersymmetric generalization of eq.(4) can be given for each $|B;\pm\rangle_{-1,I}$ as

$$
|B;\pm\rangle_I = \int [dPd\chi] e^{\frac{1}{2}\int d\sigma (iP^i\partial_\sigma P^j + \chi^i\chi^j)\omega_{ij} - \int d\sigma (ip_i P^i - \pi_i \chi^i)} |B;\pm\rangle_{-1,I}, \tag{7}
$$

where $\pi^M(\sigma) = \frac{1}{2}(\psi^M(\sigma) \mp i\tilde{\psi}^M(\sigma))$. We can show following the same arguments as above that this boundary state coincides with the boundary state for a Dp-brane up to normalization. It is also possible to prove the equivalence of the open string theories [25].

Since the arguments in this section are essentially about the variables x^i, ψ^i $(i = 1, \cdots, p+1)$ on the worldsheet, it is quite straightforward to apply the arguments here to prove that a Dp-brane can be expressed as a configuration of infinitely many D$(p-2r)$-branes. It is also easy to generalize the argument to the case of N Dp-branes. In such a case we should consider the block diagonal background

$$
X^i = \widehat{P}^i \otimes I_N, \tag{8}
$$

where I_N is the $N \times N$ identity matrix which is an element of $U(N)$ Lie algebra. The expression of the boundary state in eq.(4) should be modified to

$$
|B\rangle = \int [dP] \mathrm{TrP} e^{\frac{i}{2}\int d\sigma P^i \partial_\sigma P^j \omega_{ij} - i\int d\sigma p_i P^i} |B\rangle_{-1}, \tag{9}
$$

where TrP here means the trace of the path-ordered product with respect to the $U(N)$ indices. It is easy to see that this configuration corresponds to N Dp-branes following the same arguments as above.

3. Worldvolume Theory

In the previous section, we explained that the open string theory corresponding to the configuration of D-instantons in eq.(1) is equivalent to the one for a Dp-brane with $F = \omega$. This means that the worldvolume theory of one Dp-brane can also be described as the worldvolume theory of D-instantons. In this section, we investigate the worldvolume theory from two different points of view, i.e., either as the worldvolume theory of one Dp-brane or D-instantons. We call them Dp-brane picture and D-instanton picture respectively. We will show how these two are related with each other. The argument in this section will be done for bosonic string case for simplicity. For superstring case, similar results can be derived starting from the expression eq.(7).

Our argument in this section can be applied to study the worldvolume theory of Dp-branes by regarding it as the worldvolume theory of infinitely many D$(p-2r)$-branes for $2r < p+1$. It is also straightforward to deal with the case when the number of the Dp-brane is more than one. We will comment on these generalizations at the end of this section.

3.1. Dp-brane Picture

Let us start from the following expression of the boundary state for the Dp-brane:

$$|B\rangle = \int [dP] e^{\frac{i}{2}\int d\sigma P^i \partial_\sigma P^j \omega_{ij} - i\int d\sigma p_i P^i} |B\rangle_{-1}. \tag{10}$$

This corresponds to a Dp-brane longitudinal to x^i ($i = 1, \cdots, p+1$) directions with the $U(1)$ gauge field strength $F = \omega$. The worldvolume theory of a Dp-brane consists of a gauge field A_i and scalar fields ϕ^M ($M = p+2, \cdots, D$). ϕ^M correspond to the shape of the worldvolume which can be expressed by the equation $x^M = \phi^M(x^1, \cdots, x^{p+1})$. We are considering here the field configurations in the static gauge, so that the coordinates on the worldvolume are taken to be $x^1, \cdots, x^{p+1}$. The boundary state corresponding to a configuration of A_i, ϕ^M is easily guessed to be

$$\begin{aligned} |B\rangle = \int [dP] \exp[i \int d\sigma A_i(P)\partial_\sigma P^i \\ - i\int d\sigma (p_i P^i + \sum_{M=p+2}^{D} p_M \phi^M(P))]|B\rangle_{-1}. \end{aligned} \tag{11}$$

Indeed this coincides with eq.(10) when $F = \omega, \phi^M = 0$. For small deformations $\delta A_i, \delta\phi^M$ from the background $F = \omega, \phi^M = 0$, we expect that the boundary state $|B\rangle$ in eq.(10) is modified by the vertex operator as

$$(1 + i\int d\sigma(\delta A_i(x)\partial_\sigma x^i - \delta\phi^M(x) p_M))$$

$$\times \int [dP] e^{\frac{i}{2}\int d\sigma P^i \partial_\sigma P^j \omega_{ij} - i\int d\sigma p_i P^i} |B\rangle_{-1}, \tag{12}$$

which is consistent with eq.(11). Moreover since $x^M(\sigma)|B\rangle_{-1} = 0$ this boundary state exactly describes the emission of a closed string from the worldvolume $x^M = \phi^M(x^1, \cdots, x^{p+1})$. The contribution of the gauge field is in the form of the Wilson loop. This state will be BRS invariant for only those A_i, ϕ^M satisfying the equations of motion. $Q_B|B\rangle = 0$ implies that the path integral measure is invariant under the reparametrization $\sigma \to \sigma'(\sigma)$. Imposing such a condition, one may be able to deduce the equations of motion after calculations similar to [23].

Thus in this picture, the worldvolume theory is a $U(1)$ gauge theory with scalars ϕ^M. In this note we always assume that Pauli–Villars regularization on the worldsheet is taken so that the noncommutativity because of the regularization discussed in [18] does not occur.

3.2. D-instanton Picture

Now let us consider the worldvolume theory as the worldvolume theory of D-instantons. The boundary state eq.(10) corresponds to the configuration eq.(1) of D-instantons. General configuration of D-instantons can be described as

$$X^M = \phi^M(\widehat{P}) \; (M = 1, \cdots, D), \tag{13}$$

if one assumes that the operators $\widehat{P}^i$ $(i = 1, \cdots, p)$ generate all the possible operators acting on the Chan–Paton indices of D-instantons. Here in defining the functions ϕ^M, we need to specify the ordering of the operators $\widehat{P}^i$, which will be given shortly. What will be the form of the boundary state corresponding to the configuration eq.(13)? A natural guess is

$$|B\rangle = \int [dP] e^{\frac{i}{2}\int d\sigma P^i \partial_\sigma P^j \omega_{ij} - i\int d\sigma p_M \phi^M(P)} |B\rangle_{-1}. \tag{14}$$

For small deformations $\delta\phi^M$ from the background eq.(1), we expect that the boundary state in eq.(10) is modified as

$$\int [dP](1 - i\int d\sigma \delta\phi^M(P) p_M) e^{\frac{i}{2}\int d\sigma P^i \partial_\sigma P^j \omega_{ij} - i\int d\sigma p_i P^i} |B\rangle_{-1}. \tag{15}$$

Since the vertex operators corresponding to the transverse deformations ϕ^M $(M = p+2, \cdots, D)$ coincides with those in eq.(12), we expect that the transverse ϕ^M appear in the same way as in eq.(11). Eq.(14) is unique choice satisfying this condition and the D-dimensional rotational invariance. The boundary state eq.(14) describes the emission of a closed string from the hypersurface $X^M = \phi^M(P)$. Therefore the fields $\phi^M(P)$ correspond to the shape of the D-branes and P^i play the role of the coordinates of the p-brane.

In order to be consistent with the path integral form eq.(14), the ordering in eq.(13) should be chosen to be Weyl ordering [26]. To be more explicit, for each c-number function $f(P)$, let us define the Weyl-ordered function $f(\widehat{P})$ to be

$$f(\widehat{P}) = \int d^{p+1}k e^{ik_i \widehat{P}^i} \tilde{f}(k), \tag{16}$$

where

$$\tilde{f}(k) = \int \frac{d^{p+1}P}{(2\pi)^{p+1}} e^{-ik_i P^i} f(P). \tag{17}$$

Then $\phi^M(\widehat{P})$ on the right hand side of eq.(13) should be understood to be the Weyl-ordered function corresponding to the c-number function $\phi^M(P)$ in eq.(14).

The action and other physical quantities in the worldvolume theory of D-instantons are written as a trace of a function of the Weyl-ordered operators in eq.(16). However it is more convenient for us to rewrite everything in terms of the c-number functions $\phi^M(P)$ in eq.(17). We can do so by using the following formula[27]:

$$\begin{aligned} &Tr(f_1(\widehat{P}) f_2(\widehat{P}) \cdots f_n(\widehat{P})) \\ &= \int \frac{d^{p+1}P}{(2\pi)^{(p+1)/2}\sqrt{|\det\theta|}} f_1(P) * f_2(P) \cdots f_n(P). \end{aligned} \tag{18}$$

Here the $*$-product is defined as

$$f(P) * g(P) = e^{\frac{i}{2}\theta^{ij}\frac{\partial}{\partial\xi^i}\frac{\partial}{\partial\zeta^j}} f(P+\xi) g(P+\zeta)|_{\xi=\zeta=0}. \tag{19}$$

Hence, a trace of a function of Weyl-ordered operators can be rewritten in terms of the corresponding c-number functions by replacing product of operators by $*$-product of the corresponding c-number functions and trace by integral.

Thus if one regards the worldvolume theory as a theory of D-instantons, the description should be noncommutative. P^i can be considered as the coordinates on the p-brane and they are noncommutative under the $*$-product reflecting the commutation relation eq.(2). Now let us discuss what kind of theory this noncommutative field theory is. The Lagrangian of the worldvolume theory of D-instantons can be written in terms of the commutators of the matrices $\phi^M(\widehat{P})$. Since we started from the background in eq.(1), let us express ϕ^i $(i = 1, \cdots, p+1)$ in the form of the background and the fluctuations around it:

$$\phi^i(\widehat{P}) = \widehat{P}^i + \theta^{ij} a_j(\widehat{P}). \tag{20}$$

The c-number expression corresponding to the commutators of $\phi^i(\widehat{P})$ are easily calculated to be

$$[\phi^i(\widehat{P}), \phi^j(\widehat{P})] \to i\theta^{ij} - i(\theta f \theta)^{ij}, \tag{21}$$

where

$$f_{ij} = \partial_i a_j(P) - \partial_j a_i(P) - i a_i * a_j(P) + i a_j * a_i(P). \tag{22}$$

f_{ij} can be considered as the field strength of a noncommutative Yang–Mills field a_i. $\phi^i(\widehat{P})$ essentially corresponds to the covariant derivative $\partial + ia$. Thus the commutators of ϕ^i with other fields give the covariant derivative of these fields. Other commutators are interpreted as the gauge covariant commutators in the noncommutative Yang–Mills theory. Since the Lagrangian is written in terms of these commutators, the noncommutative theory we have here can be considered as noncommutative Yang–Mills theory [28]. The gauge invariance of the theory stems from the transformation

$$\delta\phi^M(\widehat{P}) = i[\epsilon, \phi^M(\widehat{P})]. \tag{23}$$

As a theory of D-instantons this is the $U(\infty)$ transformation under which the theory should is invariant. In the c-number formulation such a transformation corresponds to the coordinate transformation

$$\delta P = \theta^{ij}\partial_j \epsilon(P). \tag{24}$$

This is the coordinate transformation which preserves $\omega = \theta^{-1}$. If one regards ω as a symplectic form, such transformations are the canonical transformations. The invariance under the canonical transformation will be discussed in the next subsection.

3.3. Relation between the Two Pictures

In the previous subsections we obtain two points of view about the worldvolume theory. Since they are supposed to describe the same thing, there should be correspondence between the two. In the Dp-brane picture, the worldvolume fields are the gauge field A_i $(i = 1, \cdots, p+1)$ and scalars ϕ^M $(M = p+2, \cdots, D)$, where the coordinates on the worldvolume is taken to be x^i $(i = 1, \cdots, p+1)$. On the other hand, the worldvolume fields in the D-instanton picture are $\phi^M(P)$ $(M = 1, \cdots, D)$ and P^i are the coordinates on the worldvolume.

As we noticed in the previous section, the fields ϕ^M $(M = p+2, \cdots, D)$ common to both correspond to each other. Therefore we should find how the fields A_i and ϕ^i are related. Let us first consider small deformations $\delta A_i, \delta\phi^i$ from the background eq.(1). From eq.(12), one can see that δA_i changes the boundary state as

$$\delta|B\rangle = i\int [dP] \int d\sigma \delta A_i(P)\partial_\sigma P^i e^{\frac{i}{2}\int d\sigma P^i \partial_\sigma P^j \omega_{ij} - i\int d\sigma p_i P^i}|B\rangle_{-1}, \tag{25}$$

which should be compared with the variation corresponding to $\delta\phi^i$ from eq.(15):

$$\delta|B\rangle = -i\int[dP]\int d\sigma\delta\phi^i(P)p_i e^{\frac{i}{2}\int d\sigma P^i\partial_\sigma P^j\omega_{ij}-i\int d\sigma p_i P^i}|B\rangle_{-1}. \quad (26)$$

The relation between these two variations can be derived from the following identity

$$\begin{aligned} 0 &= \int[dP]\frac{\delta}{\delta P^i(\sigma)}e^{\frac{i}{2}\int d\sigma P^i\partial_\sigma P^j\omega_{ij}-i\int d\sigma p_i P^i}|B\rangle_{-1} \\ &= \int[dP][i\omega_{ij}\partial_\sigma P^j - ip_i]e^{\frac{i}{2}\int d\sigma P^i\partial_\sigma P^j\omega_{ij}-i\int d\sigma p_i P^i}|B\rangle_{-1}, \quad (27)\end{aligned}$$

which implies that $\delta|B\rangle$ in eqs.(25)(26) coincide with each other when $\delta A_i = \omega_{ij}\delta\phi^j$. Such a relation was given in [29].

In order to see the relation between the two pictures for finite deformations, the most convenient way is to consider a description involving fields A_i and ϕ^M $(M = 1,\cdots,D)$. From eqs.(11)(14), the boundary state involving all these fields should be

$$|B\rangle = \int[dP]e^{i\int d\sigma A_i(P)\partial_\sigma P^i - i\int d\sigma p_M\phi^M(P)}|B\rangle_{-1}. \quad (28)$$

However, there are two many fields in such a description and there should be symmetry to reduce the number of them. The boundary state eq.(28) is invariant under gauge transformation $\delta A_i = \partial_i\lambda$. Moreover, in the D$p$-brane picture, considering A_i and ϕ^M just means considering the theory before the gauge fixing of reparametrization invariance. Therefore the theory should have reparametrization invariance.

Indeed we can argue that the boundary state eq.(28) is invariant under the following transformation

$$\begin{aligned} \delta A_i(P) &= -\epsilon^j(P)F_{ji}(P), \\ \delta\phi^M(P) &= -\epsilon^i(P)\partial_i\phi^M(P), \quad (29)\end{aligned}$$

because the variation is proportional to a sum of the equations of motion for P^i. This transformation coincides with the coordinate transformation up to field-dependent gauge transformation because

$$\delta A_i(P) = -\epsilon^j\partial_j A_i - \partial_i\epsilon^j A_j + \partial_i(\epsilon^j A_j). \quad (30)$$

Since the boundary state eq.(28) is gauge invariant, it is reparametrization invariant.

Therefore the description involving A_i and ϕ^M $(M = 1,\cdots,D)$ is invariant under the reparametrization on the worldvolume. The Dp-brane picture obviously corresponds to the static gauge. On the other hand, one can see from eq.(14) that the D-instanton picture apparently corresponds to the gauge $F_{ij} = \omega_{ij}$. We are not sure if such a gauge can be taken for arbitrary configuration of the gauge field F, but at least when we are thinking about the fluctuation from the background $F = \omega$ perturbatively, it seems all right. Such a gauge does not fix the whole reparametrization invariance on the worldvolume. The residual invariance consists of the coordinate transformation preserving ω_{ij}, i.e., the canonical transformation. We saw such an invariance as the c-number counterpart of the $U(\infty)$ invariance in the previous subsection.

Since the difference is from the gauge choice, we can give the explicit relation between the static gauge variables $A_i(x), \phi^M_{st}(x)$ $(M = p+2, \cdots, D)$ and the variables $\phi^M_{nc}(P)$ $(M = 1, \cdots, D)$ † in the noncommutative description at least classically, i.e., for small θ,:

$$\phi^M_{nc}(P) = \phi^M_{st}(\phi^1_{nc}(P), \cdots, \phi^{p+1}_{nc}(P)), \tag{31}$$

$$\omega_{ij} = F_{kl}(\phi^1_{nc}(P), \cdots, \phi^{p+1}_{nc}(P)) \frac{\partial \phi^k_{nc}}{\partial P^i} \frac{\partial \phi^l_{nc}}{\partial P^j}. \tag{32}$$

Eq.(32) can be rewritten in terms of the noncommutative Yang–Mills field a_i using eq.(20) as

$$(\theta^{-1})_{ij} = (M^T F(P + \theta a) M)_{ij}, \tag{33}$$

where

$$M^i{}_j = \delta^i{}_j + \theta^{ik} \partial_j a_k(P). \tag{34}$$

This gives an explicit relation between commutative gauge field $A_i(x)$ and noncommutative gauge field $a_i(P)$.

Now let us comment on two generalizations of our results in this section. First one is to consider more than one Dp-branes. Starting from the background eq.(8), we can follow the arguments of $N = 1$ case. This time all the fields A_i and ϕ^M are in the adjoint representation of $U(N)$. We should put TrP in front of the right hand sides of eqs.(10), (11), (14), (28) and then we can follow the same arguments as $N = 1$ case. The reparametrization invariance of the boundary state in this case can be derived as follows. The path-ordered trace version of eq.(28) can be rewritten by introducing fermions ψ in the fundamental representation $U(N)$ as

$$|B\rangle = \int [dPd\psi] \exp\left[\int d\sigma \psi^\dagger \partial_\sigma \psi - i \int d\sigma A^a_i(P) \partial_\sigma P^i \psi^\dagger t^a \psi \right.$$
$$\left. -i \int d\sigma p_M \phi^{M,a}(P) \psi^\dagger t^a \psi \right] |B\rangle_{-1}. \tag{35}$$

†Here the subscript st and nc are for distinguishing ϕ^M in different gauges.

Here t^a are the generators of $U(N)$ in the fundamental representation. In this form, we can see that the boundary state is invariant under the transformation

$$\begin{aligned}\delta A_i(P) &= -\epsilon^j(P)F_{ji}(P),\\ \delta\phi^M(P) &= -\epsilon^i(P)D_i\phi^M(P),\end{aligned} \tag{36}$$

because the variation is proportional to the equations of motion for P^i, ψ. This transformation is again equivalent to the coordinate transformation up to field-dependent gauge transformation. Moreover we can argue that the boundary state is invariant under

$$\begin{aligned}\delta A_i(P) &= -\epsilon^j(P)t^aF_{ji}(P),\\ \delta\phi^M(P) &= -\epsilon^i(P)t^a\partial_i\phi^M(P),\end{aligned} \tag{37}$$

because the variation is proportional to

$$\lim_{\sigma'\to\sigma}\psi^\dagger t^a\psi(\sigma')\times(\text{equations of motion})(\sigma).$$

This transformation can be considered to be a non-abelian generalization of coordinate transformation up to field-dependent gauge transformation. Fixing such invariance by taking static gauge or $F=\omega$ and we get commutative or noncommutative descriptions respectively.

Secondly it is also possible to study the worldvolume theory of Dp-branes regarding it as the worldvolume theory of infinitely many D$(p-2r)$-branes. We obtain a description in which some of the coordinates on the worldvolume become noncommutative.

4. Discussions

In this note, we have shown that Dp-branes with constant field strength F_{ij} can be represented as a configuration of infinitely many D$(p-2r)$-branes. The worldvolume theory of the Dp-branes can be analyzed by regarding it as the worldvolume theory of the D$(p-2r)$-branes and we obtain a noncommutative description of the worldvolume theory. The system we studied here is gauge equivalent to the one studied in [18]. In that paper D-branes in constant B_{ij} background was considered and the authors get commutative and noncommutative descriptions of the worldvolume theory depending on the regularization. Moreover there exist descriptions with continuously varying θ which connect the commutative and noncommutative descriptions [30]. Actually we can realize such descriptions also in our formalism by considering our system in constant B_{ij} background [31]. Therefore we suspect that our noncommutative description is equivalent to a choice of θ in [18]. Since we have an explicit relation between the commutative and noncommutative descriptions which is valid for small θ, it may be interesting to compare our relation and theirs. Here we just comment on one crucial difference. In [18] the gauge transformations of the commutative and noncommutative descriptions are related, but in the relation which we have obtained the gauge invariance of noncommutative theory is the residual reparametrization invariance which is not related to the commutative gauge invariance. Therefore the relation we obtained in eq.(32) is for gauge invariant quantities.

Since we have an example in which there is a relation between commutative and noncommutative theory, it may be generalized and be used in studying other noncommutative theories. One example is the noncommutative geometric formulation of open string field theory[32]. We think that the relation we studied in this note may be relevant in revealing symmetries hidden in the string field theory. We hope that we can come back to this problem in the future.

Acknowledgments

We would like to thank the organizers of the workshop for the wonderful workshop. I am grateful to H. Aoki, S. Iso, H. Kawai, Y. Kitazawa and T. Tada for collaborations and K. Okuyama for discussions. This work was supported by the Grant-in-Aid for Scientific Research from the Ministry of Education, Science and Culture of Japan.

References

[1] E. Witten, *Nucl. Phys.* **B460**, 335 (1996) hep-th/9510135.

[2] A. Connes, M.R. Douglas and A. Schwarz, *JHEP* **02**, 003 (1998) hep-th/9711162.

[3] T. Banks, W. Fischler, S. H. Shenker and L. Susskind, *Phys. Rev.* **D55** 5112 (1997) hep-th/9610043.

[4] N. Ishibashi, H. Kawai, Y. Kitazawa and A. Tsuchiya, *Nucl. Phys.* **B498** 467 (1997) hep-th/9612115.

[5] T. Banks, N. Seiberg and S. Shenker, *Nucl. Phys.* **B490**, 91 (1997) hep-th/9612157.

[6] P. Ho and Y. Wu, *Phys. Lett.* **B398**, 52 (1997) hep-th/9611233.

[7] D.B. Fairlie, *Mod. Phys. Lett.* **A13**, 263 (1998) hep-th/9707190.

[8] M.R. Douglas and C. Hull, *JHEP* **02**, 008 (1998) hep-th/9711165.

[9] Y.E. Cheung and M. Krogh, *Nucl. Phys.* **B528**, 185 (1998) hep-th/9803031.

[10] T. Kawano and K. Okuyama, *Phys. Lett.* **B433**, 29 (1998) hep-th/9803044.

[11] F. Ardalan, H. Arfaei and M.M. Sheikh-Jabbari, hep-th/9803067; F. Ardalan, H. Arfaei and M.M. Sheikh-Jabbari, *JHEP* **02**, 016 (1999) hep-th/9810072; F. Ardalan, H. Arfaei and M.M. Sheikh-Jabbari, hep-th/9906161.

[12] C. Chu and P. Ho, *Nucl. Phys.* **B550**, 151 (1999) hep-th/9812219; C. Chu and P. Ho, hep-th/9906192.

[13] V. Schomerus, *JHEP* **06**, 030 (1999) hep-th/9903205.

[14] J. Frohlich, O. Grandjean and A. Recknagel, hep-th/9706132; J. Frohlich, O. Grandjean and A. Recknagel, *Commun. Math. Phys.* **203**, 119 (1999) math-ph/9807006.

[15] F. Lizzi and R.J. Szabo, *Chaos Solitons Fractals* **10**, 445 (1999) hep-th/9712206; G. Landi, F. Lizzi and R.J. Szabo, hep-th/9806099; F. Lizzi and R.J. Szabo, hep-th/9904064.

[16] P. K. Townsend, *Phys. Lett.* **B373** 68 (1996) hep-th/9512062.

[17] B. de Wit, J. Hoppe and H. Nicolai, *Nucl. Phys.* **B305**[FS23] 545 (1988).

[18] N. Seiberg and E. Witten, hep-th/9908142.

[19] L. Cornalba and R. Schiappa, hep-th/9907211.

[20] L. Cornalba, hep-th/9909081.

[21] O.J. Ganor, S. Ramgoolam and W.I. Taylor, *Nucl. Phys.* **B492**, 191 (1997) hep-th/9611202; W.I. Taylor, hep-th/9801182.

[22] M. Kato and T. Kuroki, *JHEP* **03**, 012 (1999) hep-th/9902004.

[23] C.G. Callan, C. Lovelace, C.R. Nappi and S.A. Yost, *Nucl. Phys.* **B308**, 221 (1988).

[24] N. Ishibashi, *Nucl. Phys.* **B539**, 107 (1999) hep-th/9804163.

[25] N. Ishibashi, to appear.

[26] T.D. Lee, *Particle physics and introduction to field theory*, Harwood Academic Publishers (1981).

[27] H. Aoki, N. Ishibashi, S. Iso, H. Kawai, Y. Kitazawa and T. Tada, hep-th/9908141.

[28] A. Connes and M. Rieffel, Yang–Mills for noncommutative two-tori, in *Operator Algebras and Mathematical Physics* (Iowa City, Iowa, 1985), pp.237, *Contemp. Math. Oper. Algebra. Math. Phys.* 62, AMS 1987.

[29] M. Li, *Nucl. Phys.* **B499** 149 (1997) hep-th/9612222.

[30] B. Pioline and A. Schwarz, *JHEP* **08**, 021 (1999) hep-th/9908019.

[31] N. Ishibashi, S. Iso, H. Kawai and Y. Kitazawa, to appear.

[32] E. Witten, *Nucl. Phys.* **B268**, 253 (1986).

INTERSECTION NUMBERS ON THE MODULI SPACES OF STABLE MAPS IN GENUS 0

Alexandre Kabanov *
Institut für Mathematik, Universität Zürich–Irchel
Winterthurerstr. 190, CH-8057 Zürich, SWITZERLAND
kabanov@math.unizh.ch

Takashi Kimura †
Department of Mathematics, 111 Cummington Street
Boston University, Boston, MA 02215, USA
kimura@math.bu.edu

Abstract Let V be a smooth, projective, convex variety. We define tautological ψ and κ classes on the moduli space of stable maps $\overline{\mathcal{M}}_{0,n}(V)$, give a (graphical) presentation for these classes in terms of boundary strata, derive differential equations for the generating functions of the Gromov–Witten invariants of V twisted by these tautological classes, and prove that these intersection numbers are completely determined by the Gromov–Witten invariants of V. This results in families of genus zero cohomological field theory structures on the cohomology ring of V which includes the quantum cohomology as a special case.

Mathematics Subject Classification (2000): 14H10, 32G15, 81T40

Keywords: Cohomological field theory, Gromov–Witten invariants, tautological cohomology classes, moduli spaces of stable curves, moduli spaces of stable maps, large phase space

* Research of the first author was partially supported by NSF grant number DMS-9803553.
† Research of the second author was partially supported by NSF grant number DMS-9803427.

Y. Maeda, et al. (eds.), Noncommutative Differential Geometry and Its Applications to Physics, 63–98.

1. Introduction

There has recently been a great deal of interest in $\overline{\mathcal{M}}_{g,n}(V)$, the moduli space of stable maps of genus g with n marked points into a smooth, projective variety V, an object whose construction was envisioned by Kontsevich as the proper algebro-geometric setting for Gromov–Witten invariants and quantum cohomology. These spaces may be regarded as the building blocks of a rigorous formulation of the physical theory of topological gravity coupled to the topological sigma model with target variety V [54] in the language of algebraic geometry. The state space of this theory is $H^\bullet(V)$ together with its Poincaré pairing η. The correlators of the theory (multilinear operations on the $H^\bullet(V)$) can be constructed from intersection numbers on $\overline{\mathcal{M}}_{g,n}(V)$. Furthermore, the factorization properties of the correlators endows $(H^\bullet(V), \eta)$ with the structure of a cohomological field theory (CohFT) in the sense of Kontsevich–Manin [36]. CohFTs have a simple description in the language of operads — they are nothing more than algebras over the modular operad $\{ H_\bullet(\overline{\mathcal{M}}_{g,n}) \}$ where $\overline{\mathcal{M}}_{g,n}$ is the moduli space of stable curves of genus g with n marked points. Let us briefly review these constructions.

The space $\overline{\mathcal{M}}_{g,n}(V)$ is a compactification of the moduli space of holomorphic maps from a genus g Riemann surface with n marked points to V by allowing the surfaces to degenerate forming double points away from the marked points provided that a stability condition is satisfied. The moduli space of stable maps $\overline{\mathcal{M}}_{g,n}(V)$ should be regarded as a generalization of the moduli space of stable curves $\overline{\mathcal{M}}_{g,n}$, due to Deligne–Mumford and Knudson [9, 33], as it reduces to the latter in the special case where V is a point. However, unlike the moduli space of curves, $\overline{\mathcal{M}}_{g,n}(V)$ has worse than orbifold singularities (it is a singular Deligne–Mumford stack) when the relevant obstruction bundle fails to vanish. Intersection theory on stacks requires the usage of the virtual fundamental class [39, 4, 5] in lieu of the topological one. However, when V is a smooth, projective, convex variety $\overline{\mathcal{M}}_{0,n}(V)$ are complex orbifolds [17]. We shall henceforth assume V satisfies these conditions and we restrict ourselves to genus zero.

The spaces $\overline{\mathcal{M}}_{0,n}(V)$ are equipped with evaluation morphisms

$$\mathrm{ev}_i : \overline{\mathcal{M}}_{0,n}(V) \to V$$

for $i = 1, \ldots, n$ which evaluates the stable map on the ith puncture. The Gromov–Witten invariants of V are the intersection numbers of the pull back of cohomology classes on V via these evaluation morphisms. The Gromov–Witten invariants often solve enumerative problems in algebraic geometry, *e.g.*, the number of rational curves in $\mathbb{CP}^2$ counted with suitable multiplicity [36] (see also [10, 17]). The Gromov–Witten invariants of V endow $H^\bullet(V)$ with a deformed cup product resulting in the structure of quantum cohomology (see also [48]) which makes $H^\bullet(V)$ into a cohomological field theory (CohFT)

[36, 43] (or equivalently, a (formal) Frobenius manifold [12, 25, 43]). The Gromov–Witten invariants assemble into a generating function (called the potential) on $H^\bullet(V)$ which must obey the WDVV (Witten–Dijkgraaf–Verlinde–Verlinde) equations. The WDVV equations allows the recursive computation of the Gromov–Witten invariants of certain homogeneous spaces, *e.g.*, projective spaces, Grassmannians, etc. [36, 17, 10]

Another set of cohomology classes on $\overline{\mathcal{M}}_{0,n}(V)$ are those associated to its universal curve. The tautological ψ classes are the first Chern classes of tautological line bundles on $\overline{\mathcal{M}}_{0,n}(V)$. Certain intersection numbers of the ψ classes and the pull back of cohomology classes on V via the evaluation morphisms are the called Gromov–Witten invariants twisted by the ψ classes. These twisted invariants are of great interest as the associated potential is conjectured to obey a highest weight condition for the Virasoro Lie algebra [13, 14, 29], a conjecture which has some intriguing evidence behind it [2, 19, 23, 46]. This conjecture has been proven in the case where V is a point and is a consequence of Kontsevich's proof of the Witten conjecture [42, 35, 11]. There, the highest weight condition gives rise to recursion relations which completely determine these intersection numbers in all genera. The conjecture has also been proven in genus 0 [41, 20].

In this paper, we begin by reviewing the notion of an operad and algebras over them. We focus our attention on the operad $\{\overline{\mathcal{M}}_{0,n}\}$, thereafter, and, by applying the homology functor, the operad $\{H_\bullet(\overline{\mathcal{M}}_{0,n})\}$ which is the main operad relevant to this paper. The latter can be described purely in terms of spaces of decorated graphs modulo some relations. Algebras over the operad $\{H_\bullet(\overline{\mathcal{M}}_{0,n})\}$ (together with a cyclicity condition) are genus zero CohFTs and can be characterized by Feynman diagrams associated to these decorated graphs.

Next, we review $\overline{\mathcal{M}}_{0,n}(V)$, exploiting the canonical stratification of $\overline{\mathcal{M}}_{0,n}(V)$ by complex orbifolds which generalize the stratification of $\overline{\mathcal{M}}_{0,n}$. These strata are indexed by decorated trees which can be used to develop a graphical language describing cohomology classes on $\overline{\mathcal{M}}_{0,n}(V)$. This graphical language is particularly useful in describing the behavior of these classes under restriction to the strata and under the push forward and pull back with respect to the forgetful map $\overline{\mathcal{M}}_{0,n+1}(V) \to \overline{\mathcal{M}}_{0,n}(V)$.

We then introduce the tautological ψ and κ classes on $\overline{\mathcal{M}}_{0,n}(V)$ generalizing those classes from $\overline{\mathcal{M}}_{0,n}$. The definition of the κ classes is new and has the novel feature that their proper definition involves the pull backs via the evaluation morphisms of the cohomology classes on V. We prove restriction properties for these classes to the boundary strata and derive a (graphical) presentation of the ψ and κ classes in terms of boundary strata generalizing the presentation for the same classes in [26] on the moduli space of curves.

We prove that the Gromov–Witten classes twisted by the ψ and κ classes endow $H^\bullet(V)$ with a formal family of CohFT structures by pushing them forward to the moduli space of curves. These CohFT structures are deformations of the cup product containing, as a special case, the product in quantum cohomology.

Finally, we prove recursion relations for the intersection numbers of the Gromov–Witten invariants twisted by the ψ and κ classes by using these presentations and the restriction properties of these cohomology classes. These recursion relations can be encoded in terms of a potential for these intersection numbers satisfying a system of differential equations called the topological recursion relations. We also prove two other equations, the analogs of the puncture and dilaton equations, which do not use the presentation of the ψ and κ classes and show that all of these intersection numbers are completely determined by the Gromov–Witten invariants of V.

Consider the special case where there are no κ classes. These topological recursion relations were originally written down by Witten [54] in the context of topological gravity coupled to topological matter before the moduli space of stable maps were even defined! A proof of Witten's equations in the algebro-geometric setting appeared in [19]. In the setting of symplectic geometry, these equations were proven in [48]. The proof of the puncture and dilaton equations for the ψ classes appeared in [54, 19, 37, 43].

Our work is a generalization of these results to the case where the κ classes are included. The κ classes are important here because their restriction properties manifestly yield families of CohFT structures on $H^\bullet(V)$ and provide the necessary cohomology classes (with values in the tensor algebra of $H^\bullet(V)$) on $\overline{\mathcal{M}}_{0,n}(V)$ which define the CohFT. This is in marked contrast to the case involving only the ψ classes where the tensor algebra-valued forms are not manifest (see [26, 38, 30] for a special case).

Furthermore, both the Gromov–Witten invariants twisted by only the κ classes and the same twisted by only the ψ classes give rise to families of genus zero CohFT structures on $H^\bullet(V)$. The families corresponding to the κ classes and those associated to the ψ classes are, in some sense, dual to each other and this correspondence extends to higher genera [27].

The paper is structured as follows. In the first section, we review operads, algebras over them, the moduli space of stable curves, and genus zero cohomological field theories. In the second section, we review the geometry of the moduli space of genus 0 stable maps $\overline{\mathcal{M}}_{0,n}(V)$ with n marked points associated to a smooth, projective, convex variety V. In the third section, we develop a graphical calculus to describe cohomology classes on $\overline{\mathcal{M}}_{0,n}(V)$. In the fourth section, we introduce the tautological ψ and κ classes on $\overline{\mathcal{M}}_{0,n}(V)$. We prove restriction properties and lifting properties of these classes and obtain a graphical presentation of these classes in terms of boundary strata. In the fifth section, we show that these intersection numbers endow the cohomology ring $H^\bullet(V)$ with formal families of CohFT structures. In the last section, we derive differential equations satisfied by the potentials for these intersection numbers and prove that they are completely determined by the Gromov–Witten invariants.

2. Operads, stable curves, and cohomological field theories

In this section, we review operads and algebras over them. An operad is an object which parametrizes multilinear operations on a vector space. While operads have their origins in homotopy theory, they appear naturally in mathematical physics in the context of topological field theories since the correlators in such theories are typically multilinear operations on the state space of the theory which are parametrized by cycles (or cohomology classes) on various moduli spaces.

While we give an overview of operads in this section, the main constructions in this paper primarily utilizes only the third description of cohomological field theories.

Operads

Definition 2.1. Let $\mathcal{P} := \{\mathcal{P}(n)\}_{n\geq 1}$ be objects in a symmetric monoidal category, e.g., the category of topological spaces, vector spaces, differential graded vector spaces (or chain complexes), and so forth. Suppose each $\mathcal{P}(n)$ is equipped with the action of S_n, the symmetric group on n letters. Furthermore, suppose there are *composition morphisms*

$$\circ_i : \mathcal{P}(n) \times \mathcal{P}(n') \to \mathcal{P}(n+n'-1)$$

taking

$$(\Sigma, \Sigma') \mapsto \Sigma \circ_i \Sigma'$$

for all $i = 1\ldots, n$, Σ in $\mathcal{P}(n)$, and Σ' in $\mathcal{P}(n')$ satisfying the following properties:

1. (Associativity) For all $i = 1, \ldots, n$, $j = 1 \ldots, n'$, $\Sigma \in \mathcal{P}(n)$, $\Sigma' \in \mathcal{P}(n')$, and $\Sigma'' \in \mathcal{P}(n'')$, the following condition is satisfied:

$$\Sigma \circ_i (\Sigma' \circ_j \Sigma'') = (\Sigma \circ_i \Sigma') \circ_{i+j-1} \Sigma''$$

2. (S-equivariance) The morphisms $\circ_i : \mathcal{P}(n) \times \mathcal{P}(n') \to \mathcal{P}(n+n'-1)$ for all $i = 1 \ldots, n$ are equivariant under the action of the symmetric groups S_n, $S_{n'}$, and $S_{n+n'-1}$ in the natural fashion.

3. (Unit element) There exists a *unit element* $\mathbf{1} \in \mathcal{P}(1)$ such that

$$\Sigma \circ_i \mathbf{1} = \mathbf{1} \circ_1 \Sigma$$

for all $i = 1 \ldots, n$ and Σ in $\mathcal{P}(n)$.

It is possible to have an object satisfying all of the above axioms except for existence of the unit. In such cases, one may always adjoin a unit element.

Example 2.2. Let $\mathcal{H}$ be a vector space. Let $\mathcal{E}nd_{\mathcal{H}} := \{\, \mathcal{E}nd_{\mathcal{H}}(n) \,\}$ where $\mathcal{E}nd_{\mathcal{H}}(n) := \mathrm{Hom}(\mathcal{H}^n, \mathcal{H})$. There is an action of S_n on $\mathcal{E}nd_{\mathcal{H}}(n)$ which permutes the tensor factors. The composition morphisms for all f in $\mathcal{E}nd_{\mathcal{H}}(n)$ and g in $\mathcal{E}nd_{\mathcal{H}}(n')$ taking $(f, g) \mapsto f \circ_i g \in \mathcal{E}nd_{\mathcal{H}}(n)$ are defined by

$$
\begin{aligned}
(f \circ_i g)(v_1, \ldots, v_{n+n'-1}) \\
&:= f(v_1, \ldots, v_{i-1}, g(v_i, \ldots, v_{i+n'-1}), v_{i+n'}, \ldots, v_{n+n'-1})
\end{aligned}
$$

for all v_j in $\mathcal{H}$ where $j = 1 \ldots, n + n' - 1$. The resulting operad in the category of vector spaces, $\mathcal{E}nd_{\mathcal{H}}$, is the *endomorphism operad of* $\mathcal{H}$. More generally, if $\mathcal{H}$ is a chain complex then the operad $\mathcal{E}nd_{\mathcal{H}}$ in the category of chain complexes is defined in the exactly the same fashion.

Example 2.3. The *little disks operad,* $\mathcal{D} := \{\, \mathcal{D}(n) \,\}$, is an operad in the category of topological spaces where $\mathcal{D}(n)$ is the space of configurations of n ordered little disks embedded in the unit disk whose interiors do not intersect pairwise. $\mathcal{D}(n)$ is an open submanifold of $(\mathbb{R}^2 \times \mathbb{R}^+)^n$. S_n acts on $\mathcal{D}(n)$ by permuting the ordering of the little disks. For Σ in $\mathcal{D}(n)$ and Σ' in $\mathcal{D}(n')$, $\Sigma \circ_i \Sigma'$ in $\mathcal{D}(n+n'-1)$ is obtained by identifying the unit disk of $\mathcal{D}(n')$ with the i-th little disk of $\mathcal{D}(n)$ after the appropriate dilation and translation.

If $\mathcal{P}$ is an operad of topological spaces then one can define an operad of chain complexes $C_\bullet(\mathcal{P}) := \{\, C_\bullet(\mathcal{P}(n)) \,\}$ where $C_\bullet(\mathcal{P}(n))$ are, say, singular chains on $\mathcal{P}(n)$. Applying the homology functor one obtains the operad of homology groups $H_\bullet(\mathcal{P}) := \{\, H_\bullet(\mathcal{P}(n)) \,\}$. Of course, an operad in the category of vector spaces may be regarded as an operad in the category of complexes with the differential equal to zero.

Given an operad, one can construct 'representations' of them as follows.

Definition 2.4. Let $\mathcal{P}$ be an operad and $\mathcal{H}$ be a complex. *$\mathcal{H}$ is said to be a $\mathcal{P}$-algebra* given a morphism of operads $\mathcal{P} \to \mathcal{E}nd_{\mathcal{H}}$.

Example 2.5. Let $\mathcal{H}$ be a $\mathbb{Z}$-graded vector space vector space with two binary operations. The first operation, multiplication taking $a \otimes b \mapsto a \cdot b$ is commutative, associative and of degree zero. The second operation, the bracket, is of degree 1 taking $a \otimes b \mapsto [a, b]$ satisfying $[a, b] = -(-1)^{(|a|+1)(|b|+1)}[b, a]$ and $[a, [b, c]] = [[a, b], c] + (-1)^{(|a|+1)(|b|+1)}[b, [a, c]]$. Finally, a graded Leibnitz rule is satisfied, namely, $[a, b \cdot c] = [a, b] \cdot c + (-1)^{|b|(|a|+1)} b \cdot [a, c]$. $\mathcal{H}$ together with these operations is called a *Gerstenhaber algebra* a variant of a Poisson superalgebra where the bracket has degree 1. An $H_\bullet(\mathcal{D})$-algebra structure on $\mathcal{H}$ is [6, 7] equivalent to $\mathcal{H}$ being endowed with the structure of a Gerstenhaber algebra.

The operad $H_\bullet(\mathcal{D})$ has two interesting suboperads. The suboperad $\mathcal{C} := \{ H_0(\mathcal{D}(n)) \}$ is the operad which governs commutative, associative algebras. The suboperad $\{ H_{n-1}(\mathcal{D}(n)) \}$ is the operad which governs Lie superalgebras with a bracket of degree 1.

An interesting operad of chain complexes which arose directly from homotopy theory is the A_∞ operad due to Stasheff. We will realize this operad as a real analog of the moduli space of stable curves of genus zero.

Example 2.6. Let $\mathcal{A} := \{ \mathcal{A}(n) \}$ be an operad of cell complexes defined as follows. Let $\mathcal{M}_n^{\mathbb{R}} := \{ [\Sigma; x_1, \dots, x_n] \}$ be the moduli space of oriented circles $\mathbb{RP}^1$, Σ, with n distinct, ordered marked points $x_1, \dots, x_n$ where any two such configurations are regarded as equivalent should they be related by an action of $PSL(2, \mathbb{R})$. This space admits a compactification $\overline{\mathcal{M}_n^{\mathbb{R}}}$ by allowing circles to undergo mitosis to become a tree of circles attached at nodes. $\overline{\mathcal{M}_n^{\mathbb{R}}}$ is the disjoint union of $n!$ identical connected components each of which is an oriented convex polytope (and hence a cell complex) of dimension $n-3$. The union of the interiors of these polytopes is $\mathcal{M}_n^{\mathbb{R}}$. Let $\mathcal{A}(n) := C_\bullet(\overline{\mathcal{M}_{n+1}^{\mathbb{R}}})$ denote the natural cell complex associated to the polytopes. The composition maps $\mathcal{A}(n) \times \mathcal{A}(n') \to \mathcal{A}(n+n'-1)$ taking $(\Sigma, \Sigma') \mapsto \Sigma \circ_i \Sigma'$ by attaching the i-th puncture of Σ to the $(n'+1)$-th puncture of Σ' is a morphism of cell complexes. Furthermore, S_n acts on $\mathcal{A}(n)$ by permuting the ordering of the first n marked points — this just permutes the $n!$ connected components of $\mathcal{A}(n)$. The resulting operad $\mathcal{A}$ is called the A_∞ *(or homotopy associative) operad.* Algebras over $\mathcal{A}$ are called A_∞ *algebras* [52, 53] and consist of a $\mathbb{Z}$-graded vector space $\mathcal{H}$ with a differential d of degree -1 together with operations m_n of degree $n-2$ in $\mathrm{Hom}(\mathcal{H}^{\otimes n}, \mathcal{H})$ for $n \geq 2$ satisfying the identities

$$dm_n(v_1, \dots, v_n) + \sum_{i=1}^{n} \epsilon(i) m_n(v_1, \dots, dv_i, \dots, v_n)$$

$$= \sum_{\substack{k+l=n+1 \\ k,l \geq 2}} \sum_{i=1}^{l-1} \epsilon(k,i)\, m_l(v_1, \dots, v_i, m_k(v_{i+1}, \dots, v_{i+k}), v_{i+k+1}, \dots, v_n),$$

where $\epsilon(i) = (-1)^{|v_1|+\cdots+|v_{i-1}|}$ is the sign picked up by passing d through $v_1, \dots, v_{i-1}$ and $\epsilon(k,i) = (-1)^{k(|v_1|+\cdots+|v_i|)}$ is the sign picked up by passing m_k through $v_1, \dots, v_i$.

Let us mention another closely related operad which is of great interest to mathematical physicists.

Example 2.7. Let $\mathcal{T} := \{\mathcal{T}(n)\}$ be the operad where $\mathcal{T}(n)$ is the moduli space of configurations of $(n+1)$ ordered, holomorphically embedded unit disks (with nonintersecting interiors) in the Riemann sphere $\mathbb{CP}^1$. The S_n action on $\mathcal{T}(n)$ permutes the order of the disks while the composition maps taking $(\Sigma, \Sigma') \mapsto \Sigma \circ_i \Sigma'$ are obtained cutting out the $(n+1)$st disk of Σ' and the ith disk of Σ and then sewing them together along the boundary. A smooth morphism of operads $\mathcal{T} \to \mathcal{E}nd_{\mathcal{H}}$ is essentially a genus zero $c = 0$ conformal field theory in the sense of Segal [49, 50, 51]. The space $\mathcal{H}$ is the state space of the theory. Similarly, an algebra over the operad of chains $C_\bullet(\mathcal{T})$ yields a definition of a *genus zero* $c = 0$ *topological conformal field theory (TCFT).*

Relaxing the $c = 0$ condition above corresponds to constructing a determinant line bundle over $\mathcal{T}(n)$. We refer the interested reader to [49].

There is a morphism of operads $\mathcal{D} \to \mathcal{T}$ taking a unit disk with n little disks to the Riemann sphere with $(n+1)$ holomorphically embedded disks by identifying the unit disk with the upper hemisphere of $\mathbb{CP}^1$ and then endowing the lower hemisphere with the standard chart to obtain the $(n+1)$st holomorphically embedded disk. It follows that if $\mathcal{H}$ is the state space of a genus zero $c = 0$ TCFT then $H_\bullet(\mathcal{H})$ is a Gerstenhaber algebra. It can be shown that $H_\bullet(\mathcal{H})$ has the structure of a Batalin–Vilkovisky algebra [18] which is a Gerstenhaber algebra together with an additional unary operation Δ of degree 1 such that $\Delta^2 = 0$ and

$$[a, b] = (-1)^{|a|}(\Delta(a \cdot b) - (\Delta(a) \cdot b + (-1)^{|a|}\, a \cdot \Delta(b))$$

for all a, b in $\mathcal{H}$. Δ appears because $\mathcal{T}$ is homotopy equivalent to the framed little disks operad, an operad in which $\mathcal{D}$ embeds as a suboperad, such that the center of each little disk is equipped with a tangent direction. Algebras over the operad of homology groups of the framed little disks operad are nothing more than Batalin–Vilkovisky algebras [18]. This algebraic structure has also been constructed from the purely algebraic point of view in terms of chiral algebras [40].

In [31] it was proved that there is the structure of an L_∞-algebra (a Lie version of the A_∞ algebra) on a subspace of a genus zero $c = 0$ TCFT provided that a mild condition is satisfied.

Throughout the remainder of this paper, we shall focus upon one particular operad of great physical interest, namely the operad of homology groups associated to the moduli space of stable curves of genus 0. This is an important operad because of the crucial role that it plays in the theory of Gromov–Witten invariants.

The moduli space of stable curves

Let $\mathcal{M}_{g,n} := \{[\Sigma; x_1, \ldots, x_n]\}$ denote the moduli space of Riemann surfaces of genus g, Σ, together with n distinct marked points $x_1, \ldots, x_n$ where any two such data are identified if they are related by a conformal automorphism of Σ. We also require that $[\Sigma; x_1, \ldots, x_n]$ has no infinitesimal automorphisms. This is the case when $2g - 2 + n > 0$. These spaces are generally not compact, but they admit a natural compactification $\overline{\mathcal{M}}_{g,n}$ (due to Deligne, Knudsen, and Mumford [9, 33]) in which $\mathcal{M}_{g,n}$ sits as an open dense subset. The space $\overline{\mathcal{M}}_{g,n}$ is the moduli space of stable curves of genus g with n marked points. This means that a connected complex curve Σ is allowed to have, at worst, nodal singularities (double points) away from the marked points subject to the stability condition that each connected component of the curve minus its marked points and nodal singularities must have negative Euler characteristic. The spaces $\overline{\mathcal{M}}_{g,n}$ are connected, compact, complex orbifolds with complex dimension $3g - 3 + n$. Let us now restrict ourselves to $g = 0$.

Associated with any stable curve $[\Sigma; x_1, \ldots, x_n]$ in $\overline{\mathcal{M}}_{0,n}$, one obtains its associated *dual graph* by collapsing each irreducible component to a point to obtain a vertex, connecting any two vertices with an edge if their corresponding components share a node and attaching a tail to a vertex for each marked point on that component. The set of dual graphs associated to points in $\overline{\mathcal{M}}_{0,n}$, denoted by $\mathcal{G}_{0,n}$, is the set of all trees whose edges are labeled (in a 1–1 fashion) with elements in the set $\{1, \ldots, n\}$ and with vertices each of which has a valency greater than or equal or 3. (The latter comes from the stability condition.) The graphs in $\mathcal{G}_{0,n}$ are also called *stable graphs of genus g with n tails (or legs).*

The space $\overline{\mathcal{M}}_{0,n}$ are stratified spaces where each stratum is indexed by an element of $\mathcal{G}_{0,n}$ as follows. For each Γ in $\mathcal{G}_{0,n}$, let $\overline{\mathcal{M}}_\Gamma$ denote the closure of the locus of points in $\overline{\mathcal{M}}_{0,n}$ whose dual graph is equal to Γ. It is a closed irreducible subvariety of (complex) codimension equal to the number of edges of Γ. There is a canonical isomorphism

$$\chi_\Gamma : \prod_{v \in V(\Gamma)} \overline{\mathcal{M}}_{0,n(v)} \to \overline{\mathcal{M}}_\Gamma,$$

where $V(\Gamma)$ denotes the set of vertices of Γ and $n(v)$ is the valence of the vertex v. This morphism may be viewed as taking a collection of Riemann spheres with marked points, one for each vertex v in $V(\Gamma)$, and then attaching them together along the marked points to obtain a node in the manner indicated by the stable graph Γ. This gives rise to a morphism

$$\rho_\Gamma : \prod_{v \in V(\Gamma)} \overline{\mathcal{M}}_{0,n(v)} \to \overline{\mathcal{M}}_\Gamma \to \overline{\mathcal{M}}_{0,n}$$

which is a composition of χ_Γ with the inclusion map.

Each $\overline{\mathcal{M}}_{0,n+1}$ has an action of S_n which permutes the ordering of the first n marked points. Furthermore, let Γ_i, $1 \leq i \leq n(v_1) - 1$, be a graph in $\mathcal{G}_{0,n+1}$ with one edge and two vertices v_1, v_2 which is obtained as follows. Take two vertices: v_1 with $n(v_1)$ tails labeled by the integers from 1 to $n(v_1)$ and v_2 with $n(v_2)$ tails labeled by the integers from 1 to $n(v_2)$. Select the tail numbered i from v_1 and the tail numbered $n(v_2)$ from v_2 and join them. Then relabel the remaining $n+1 = (n(v_1)-1)+(n(v_2)-1)$ tails in the following order: first take $i-1$ tails from v_1, then the tails from v_2, and then the remaining tails from v_1 so that the order on v_1 and v_2 is preserved. The associated morphisms

$$\rho_{\Gamma_i} : \overline{\mathcal{M}}_{0,(n(v_1)-1)+1} \times \overline{\mathcal{M}}_{0,(n(v_2)-1)+1} \to \overline{\mathcal{M}}_{0,n+1}$$

endow the collection $\{\,\overline{\mathcal{M}}_{0,n+1}\,\}$ with the structure of an operad.

There is clearly something artificial about treating the $(n+1)$st marked point of curves in $\overline{\mathcal{M}}_{0,n+1}$ differently from the rest. This asymmetry can be removed through the notion of a *cyclic operad* of which this is an example. We refer the interested reader to [22].

Applying the homology functor one obtains an induced morphism

$$\rho_{\Gamma *} : \otimes_{v \in V(\Gamma)} H_\bullet(\overline{\mathcal{M}}_{0,n(v)}) \to H_\bullet(\overline{\mathcal{M}}_{0,n}).$$

There is a nice description of $H_\bullet(\overline{\mathcal{M}}_{0,n})$ purely in terms of stable graphs. Let $\mathbb{C}[\mathcal{G}_{0,n}]$ denote the vector space with basis $\mathcal{G}_{0,n}$. There is a map $\mu_{0,n} : \mathbb{C}[\mathcal{G}_{0,n}] \to H_\bullet(\overline{\mathcal{M}}_{0,n})$ taking a stable graph $\Gamma \mapsto [\overline{\mathcal{M}}_\Gamma]$. This map is not injective as two cycles associated to different graphs can be homologous. Let $\mathcal{I}_{0,n}$ denote the kernel of $\mu_{0,n}$. The map $\mu_{0,n}$ is known to be surjective. This is nothing more than the fact that $H_\bullet(\overline{\mathcal{M}}_{0,n})$ is generated by the boundary classes [32]. Therefore, the quotient $\mathcal{Q}_{0,n} := \mathbb{C}[\mathcal{G}_{0,n}]/\mathcal{I}_{0,n}$ is isomorphic to $H_\bullet(\overline{\mathcal{M}}_{0,n})$ thereby yielding a nice graphical description of these homology groups.

It may seem curious that this operad can be written as decorated graphs modulo relations but this is, in fact, a general feature of operads in the category of finite dimensional vector spaces and is due to Ginzburg–Kapranov [24].

Cohomological field theories and Feynman diagrams

Let $(\mathcal{H}, \eta)$ be a vector space $\mathcal{H}$ with a metric η. Let T^n denote the nth tensor power. Associate to every graph Γ in $\mathcal{G}_{0,n}$ an element, I_Γ, in $T^n\mathcal{H}^*$ via the following *Feynman rules*. Associate to each vertex of valence r, an element $I_{0,r}$ in $S^r\mathcal{H}^*$ where S^r denotes the rth symmetric product. Associate to each edge of the graph the inverse metric η^{-1}. Now contract the $I_{0,n(v)}$ for each vertex v in Γ with the inverse metrics associated to the each edge in the manner indicated by the graph Γ.

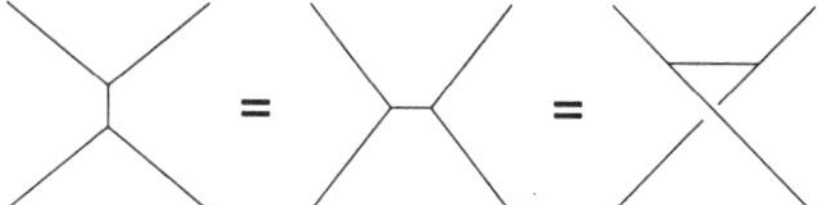

The generator of all relations between trivalent graphs.

Definition 2.8. If the map $\mathbb{C}[\mathcal{G}_{0,n}] \to T^n\mathcal{H}^*$ taking $\Gamma \mapsto I_\Gamma$ induces a map $H_\bullet(\overline{\mathcal{M}}_{0,n}) \cong \mathcal{Q}_{0,n} \to T^n\mathcal{H}^*$ then the induced map endows $(\mathcal{H}, \eta)$ with the structure of a *genus zero cohomological field theory (CohFT)*. The maps I_Γ are the *correlators* of the theory.

Notice that if $(\mathcal{H}, \eta)$ were $\mathbb{Z}_2$-graded then one could clearly assign Feynman rules which associate a graded commutative tensor to each vertex. We refer to [36] for details.

Suppose we restrict ourselves to the subspaces, $\mathcal{F}_{0,n}$, of $\mathcal{Q}_{0,n}$ which are represented by trivalent graphs, graphs whose vertices all have valence 3. The Feynman rules associate an element $I_{0,3} \in S^3\mathcal{H}^*$ to each vertex. Using the metric, one can view $I_{0,3}$ as a commutative, binary operation $\mathcal{H} \otimes \mathcal{H} \to \mathcal{H}$. The relations between the trivalent graphs are easy to characterize — they are generated by the relations in the figure below. Consequently, it is easy to see that $(\mathcal{H}, \eta)$ inherits the structure of a Frobenius algebra. That is, $\mathcal{H}$ is a commutative, associative algebra with an invariant metric η. This should not be surprising since $\mathcal{F}_{0,n}$ is isomorphic to $H_0(\overline{\mathcal{M}}_{0,n}) \cong \mathbb{C}$, *i.e.,* the operad $\{ H_0(\overline{\mathcal{M}}_{0,n+1}) \}$ is precisely the commutative operad $\mathcal{C}$ and the additional cyclic structure insures that the multiplication is invariant with respect to the inner product η.

A genus zero CohFT has, in addition, higher order products associated to the higher order vertices which must satisfy additional relations. The relations between all of these operations can most easily characterized in terms of a potential function. Indeed, it yields an alternative way to describe a genus zero CohFT.

Definition 2.9. The *potential* Φ of a genus zero CohFT is given by a linear map $I : H_\bullet(\overline{\mathcal{M}}_{0,\bullet}, \mathbb{C}) \to T^n\mathcal{H}^*$ of rank $r+1$ is defined by choosing a basis $\{ e_0, \dots, e_r \}$ for $\mathcal{H}$ and letting $\mathbf{x} = \sum_{\alpha=0}^{r} x^\alpha e_\alpha$ represent an arbitrary cohomology class in $\mathcal{H}$. Let $I_{0,n}(\mathbf{x}, \mathbf{x}, \cdots, \mathbf{x})$ denote the element in $\mathbb{C}[[x^0, x^1 \dots, x^r]]$, the ring of formal power series with $\mathbb{C}$ coefficients in $x^0, \dots, x^r$ in the graded sense, obtained by using h to pair $\mathcal{I}_{0,n}([\overline{\mathcal{M}}_{0,n}])$ with $\mathbf{x} \otimes \mathbf{x} \otimes \dots \otimes \mathbf{x}$ and

$$\Phi(\mathbf{x}) := \sum_{n=3}^{\infty} I_{0,n}(\mathbf{x}, \mathbf{x}, \dots, \mathbf{x})\frac{1}{n!}$$

which is regarded as an element in $\mathbb{C}[[x^0, \dots, x^r]]$.

The potential completely characterizes the cohomological field theory in genus zero by encoding the relations between the $I_{0,n}$ in terms of a differential equation satisfied by the potential function as can be seen in the following theorem.

Theorem 2.10. *An element Φ in $\mathbb{C}[[x^0, \dots, x^r]]$ is the potential of a rank r, genus zero CohFT $(\mathcal{H}, \eta)$ if and only if [36, 43] it contains only terms which are cubic and higher order in the variables $x^0, \dots, x^r$ and it satisfies the WDVV equation*

$$(\partial_a \partial_b \partial_e \Phi)\, \eta^{ef}\, (\partial_f \partial_c \partial_d \Phi) \;=\; (-1)^{|x_a|(|x_b|+|x_c|)}\, (\partial_b \partial_c \partial_e \Phi)\, \eta^{ef}\, (\partial_f \partial_a \partial_d \Phi),$$

where $\eta_{ab} := \eta(e_a, e_b)$, η^{ab} is in inverse matrix to η_{ab}, ∂_a is derivative with respect to x^a, and the summation convention has been used.

The theorem is a consequence of the work of Keel [32] who proved that all relations between boundary divisors in $H_\bullet(\overline{\mathcal{M}}_{0,n})$ arise from lifting the basic codimension one relation on $\overline{\mathcal{M}}_{0,4}$ which is precisely the equality depicted in the figure above.

There is a third description of a CohFT in terms of cohomology classes. A CohFT is a pair $(\mathcal{H}, \eta)$ together with a collection $\Omega := \{\, \Omega_{0,n} \,\}$ where $\Omega_{0,n}$ is an even element in $H^\bullet(\overline{\mathcal{M}}_{0,n}) \otimes T^n\mathcal{H}^*$ defined for stable pairs $(0, n)$ satisfying the following (where the summation convention has been used):

i $\Omega_{0,n}$ is invariant under the action of the symmetric group S_n.

ii For each partition of $[n] = S_1 \sqcup S_2$ such that $|S_1| = n_1$ and $|S_2| = n_2$, consider the inclusion map $\rho : \overline{\mathcal{M}}_{0,S_1\sqcup *} \times \overline{\mathcal{M}}_{0,S_2\sqcup *} \to \overline{\mathcal{M}}_{0,n}$ where $*$ denotes the two marked points that are attached under the inclusion map. The forms satisfy the restriction property

$$\begin{aligned} &\rho^*\, \Omega_{0,n}(\gamma_1, \gamma_2\, \dots, \gamma_n) \\ &\quad = \pm\, \Omega_{0,n_1}((\bigotimes_{\alpha\in S_1} \gamma_\alpha) \otimes e_\mu)\, \eta^{\mu\nu} \otimes \Omega_{0,n_2}(e_\nu \otimes \bigotimes_{\alpha\in S_2} \gamma_\alpha) \end{aligned}$$

where the sign $\pm$ is the usual one obtained by applying the permutation induced by S to $(\gamma_1, \gamma_2, \dots \gamma_n)$ taking into account the grading of $\{\, \gamma_i \,\}$ and where $\{\, e_\alpha \,\}$ is a homogeneous basis for $\mathcal{H}$.

Let Γ be a stable graph then the canonical morphism ρ_Γ can be constructed from morphisms in (ii) above and the $\Omega_{0,n}$ satisfies a restriction property of the form

$$\rho_\Gamma^* \Omega_{0,n} = \eta_\Gamma^{-1} \left(\bigotimes_{v\in V(\Gamma)} \Omega_{0,n(v)} \right) \tag{1}$$

where

$$\eta_\Gamma^{-1} : \bigotimes_{v\in V(\Gamma)} H^\bullet(\overline{\mathcal{M}}_{0,n(v)}) \otimes T^{n(v)}\mathcal{H}^* \to H^\bullet(\overline{\mathcal{M}}_{0,n}) \otimes T^n\mathcal{H}^*$$

is the linear map contracting tensor factors of $\mathcal{H}$ using the metric η induced from successive application of equations (ii) above.

The collection Ω forms a genus zero CohFT $H_\bullet(\overline{\mathcal{M}}_{0,n}) \to T^n\mathcal{H}^*$ via $[c] \mapsto \int_{[c]} \Omega_{0,n}$. In particular, the correlators are given by $I_{0,n} = \int_{[\overline{\mathcal{M}}_\Gamma]} \Omega_{0,n}$. It is this formulation of CohFTs which arises most naturally in algebraic geometry and which we shall use in the sequel. Notice that the definition of a cohomological field theory and potentials is valid even when enlarges the ground ring from $\mathbb{C}$ to another ring $\mathcal{K}$.

All of these objects can be extended to higher genera $\{\overline{\mathcal{M}}_{g,n}\}$ where the collection forms a modular operad, a higher genus generalization of the notion of an operad where the stable graphs are now allowed to have loops and the vertices are labeled by a nonnegative integer (the genus). This is beyond the scope of this paper, however.

We now show that given a convex, smooth, projective variety V, tautological κ classes and the Gromov–Witten classes on the moduli space of stable maps, one can construct formal families of CohFTs structures where $\mathcal{H} := H^\bullet(V)$ and η is the Poincaré pairing. These families can be viewed as a further deformation of the cup product and contains the quantum cohomology of V as a special case. In order to perform the construction, however, we need to review the moduli space of stable maps into a variety V, a generalization of the moduli space of stable curves.

3. The Moduli Spaces of Stable Maps

In this section we recall properties of the moduli space of stable maps whose intersection numbers are the main object of study in this paper. These spaces can be viewed as a generalization of the moduli spaces of stable curves due to Deligne–Mumford, a canonical compactification of the moduli space of configurations of n labeled, points on Riemann surfaces of genus g.

To describe the moduli space of stable maps, one needs to generalize the notion of a stable curve.

Definition 3.1. A *quasi-stable curve of genus g with n marked points*, denoted by $(C; x_1, x_2, \ldots, x_n)$, is a connected, complex projective, reduced, curve with (at most) double points of arithmetic genus g together with distinct points $x_1, x_2, \ldots, x_n$ in C away from the double points. A *special point* of C is either a marked point or a double point.

Definition 3.2. Let V be a smooth, projective variety and β in $H_2(V, \mathbb{Z})$. The *moduli space of stable maps* $\overline{\mathcal{M}}_{g,n}(V, \beta) := \{[C, f; x_1, x_2, \ldots, x_n]\}$

consists of isomorphism classes of the following data: a quasi-stable curve of genus g with n marked points $(C; x_1, x_2, \dots, x_n)$ together with a morphism $f : C \to V$ such that $\beta = [\,f(C)\,]$ satisfying the stability condition that each irreducible component which is mapped to a point in V musts be stable in the usual sense, *i.e.* each irreducible component minus its special points must have negative Euler characteristic.

Not all $\beta \in H_2(V, \mathbb{Z})$ are represented by such curves in V. Let $B(V)$ be the semi-subgroup of $H_2(V, \mathbb{Z})$ generated by the images of the curves in V.

Definition 3.3. The *evaluation morphism* associated to the ith marked point $ev_i : \overline{\mathcal{M}}_{g,n}(V, \beta) \to V$ takes the point $[C, f; x_1, x_2, \dots, x_n]$ to $f(x_i)$ for all $i = 1, \dots, n$. The *stabilization morphism* $st : \overline{\mathcal{M}}_{g,n}(V, \beta) \to \overline{\mathcal{M}}_{g,n}$, defined when $2g - 2 + n > 0$, takes the point $[C, f; x_1, x_2, \dots, x_n]$ to $[C; x_1, x_2, \dots, x_n]^{\text{st}}$. The latter means that one contracts all components that became unstable after forgetting f. Let $\pi : \overline{\mathcal{M}}_{g,n+1}(V, \beta) \to \overline{\mathcal{M}}_{g,n}(V, \beta)$ be the morphism which forgets the ith point followed by stabilization, *i.e.* $[C, f; x_1, x_2, \dots, x_{n+1}] \mapsto [C, f; x_1, \dots, \widehat{x_i}, \dots, x_{n+1}]^{\text{st}}$. This morphism is the universal curve over $\overline{\mathcal{M}}_{g,n}(V, \beta)$ [3, 5].

Theorem 3.4. *[4],[17] Let V be a smooth, projective variety and β in $H_2(V, \mathbb{Z})$. The moduli spaces $\overline{\mathcal{M}}_{g,n}(V, \beta)$ exist and are complete, Deligne-Mumford stacks.*

The moduli space of stable maps becomes the moduli space of curves by choosing the variety V to be a point. The moduli space of curves $\overline{\mathcal{M}}_{g,n}$ is a smooth stack (orbifolds in the algebraic category) for all genera. In general, $\overline{\mathcal{M}}_{g,n}(V, \beta)$ is highly singular. A useful criterion by which one can obtain nice moduli spaces, at least in genus zero, is that V be convex.

Definition 3.5. A smooth, projective variety V is said to be *convex* if for all morphisms $f : \mathbb{CP}^1 \to V$, the cohomology $H^1(\mathbb{CP}^1, f^* T_V)$ vanishes where T_V is the (holomorphic) tangent bundle.

Theorem 3.6. *[5],[17] Let V be a smooth, projective, convex variety of dimension D then the space $\overline{\mathcal{M}}_{0,n}(V, \beta)$ is a locally normal, projective variety of pure (complex) dimension $D + n - 3 + \int_\beta c_1(T_V)$ and can have, at worst, orbifold singularities. In addition, the divisor corresponding to the locus of the singular curves is a normal crossing divisor up to a quotient by a finite group.*

Definition 3.7. Let V be a smooth, projective, convex variety then the *(genus zero) Gromov–Witten invariants of V* are maps $(H^\bullet(V))^{\otimes n} \to \mathbb{C}$ given by

$$\gamma_1 \otimes \gamma_2 \otimes \dots \otimes \gamma_n \mapsto \langle\, \mathrm{ev}_1^* \gamma_1\, \mathrm{ev}_2^* \gamma_2 \dots \mathrm{ev}_n^* \gamma_n \,\rangle_\beta$$

where

$$\langle\, \mathrm{ev}_1^* \gamma_1\, \mathrm{ev}_2^* \gamma_2 \dots \mathrm{ev}_n^* \gamma_n \,\rangle_\beta := \int_{\overline{\mathcal{M}}_{0,n}(V,\beta)} \mathrm{ev}_1^* \gamma_1\, \mathrm{ev}_2^* \gamma_2 \dots \mathrm{ev}_n^* \gamma_n$$

where $\gamma_1, \gamma_2, \ldots, \gamma_n$ are elements in $H^\bullet(V)$, n is the number of marked points, β belongs to $B(V)$, and the product is understood to be the cup product.

Gromov–Witten invariants can also be defined in higher genera and for certain non-convex varieties provided that one uses the proper notion of fundamental class (the so called *virtual fundamental class* [39, 4, 5]) of the moduli spaces of stable maps.

The moduli space $\overline{\mathcal{M}}_{0,n}(V, \beta)$ for V a smooth, projective, convex variety is endowed with a canonical stratification by strata which are indexed by certain decorated trees called stable trees.

Let us introduce some notation regarding graphs (and trees). All graphs we shall consider are connected and have vertices with a valence greater than or equal to zero. Given such a tree Γ, let $V(\Gamma)$ be its set of vertices, $E(\Gamma)$ be its set of edges, and $S(\Gamma)$ be its set of tails. Each edge has two endpoints belonging to $V(\Gamma)$ which are allowed to be the same. Each tail has only one endpoint. For all $v \in V(\Gamma)$, let $n(v)$ be the number of half-edges emanating from v. Each edge gives rise to two half-edges, and each tail to one half-edge.

The graphs associated to $\overline{\mathcal{M}}_{0,n}(V)$ are generalizations of the stable graphs associated to $\overline{\mathcal{M}}_{0,n}$ and shall also be called a stable graph. A *stable graph* consists of a quadruple (Γ, g, β, μ), where Γ is a connected graph as above, $g : V(\Gamma) \to \mathbb{Z}_{\geq 0}$, $\beta : V(\Gamma) \to B(V)$, and μ is a bijection between $S(\Gamma)$ and a given set I. We also require that for each vertex v the stability condition be satisfied: if $\beta(v) = 0$, then $2g(v) - 2 + n(v) > 0$. We define the *genus,* $g(\Gamma)$, *of* Γ to be $b_1(\Gamma) + \sum_{v \in V(\Gamma)} g(v)$, where b_1 is the first Betti number. We also define the *degree,* $\beta(\Gamma)$, *of* Γ to be $\sum_{v \in V(\Gamma)} \beta(v)$. A *stable tree* is a stable graph of genus zero.

To each stable map $[C, f; x_1, \ldots, x_n] \in \overline{\mathcal{M}}_{0,n}(V, \beta)$ one can associate a stable tree, called its *dual tree*, by collapsing each irreducible component to a point forming a vertex, drawing a half-edge emanating from this vertex for each special point on that irreducible component and connecting any two half-edges if the the two components share a double point, and then decorating each vertex v with the arithmetic genus $g(v)$ of its corresponding irreducible component and with $\beta(v)$ which is the image of fundamental class of that irreducible component in $B(V)$. Finally, the tails are decorated with integers $\{1, \ldots, n\}$ corresponding to the marked points on C.

In this way each stable tree Γ with n tails of genus 0 and degree β determines a closed sub-orbifold $\overline{\mathcal{M}}_\Gamma$ of $\overline{\mathcal{M}}_{0,n}(V, \beta)$ by taking the closure of the locus of stable maps whose dual tree is Γ [17]. Each $\overline{\mathcal{M}}_\Gamma$ is a smooth stack (orbifold) of complex codimension $|E(\Gamma)|$ in $\overline{\mathcal{M}}_{0,n}(V, \beta)$. We denote the orbifold fundamental class of $\overline{\mathcal{M}}_\Gamma$ by $[\overline{\mathcal{M}}_\Gamma]$. According to [34] the stratification defined above is finite.

The stable trees Γ play a further role since it indicates how each stratum $\overline{\mathcal{M}}_\Gamma$ can be expressed as a fibered product of fundamental classes of the moduli spaces quotiented by the action of the automorphism group of the stable tree. More precisely, if $v \in V(\Gamma)$ then we denote by $\overline{\mathcal{M}}(v)$ the moduli space $\overline{\mathcal{M}}_{0,n(v)}(V, \beta(v))$. Each edge of Γ joining vertices v_1 and v_2 determines evaluation morphisms $\overline{\mathcal{M}}(v_1) \to V$ and $\overline{\mathcal{M}}(v_2) \to V$. The stratum $\overline{\mathcal{M}}_\Gamma$ is isomorphic as a stack (orbifold) to the fibered product of $\overline{\mathcal{M}}(v)$, $v \in V(\Gamma)$ over $V^{|E(\Gamma)|}$ with respect to the evaluation morphisms corresponding to each edge quotiented by the obvious action of $\mathrm{Aut}(\Gamma)$, the automorphism group of Γ. We denote by $\widehat{\prod}_{v\in V(\Gamma)}\overline{\mathcal{M}}(v)$ the fibered product defined above, by ρ_Γ the composition

$$\widehat{\prod}_{v\in V(\Gamma)}\overline{\mathcal{M}}(v) \to \overline{\mathcal{M}}_\Gamma \to \overline{\mathcal{M}}_{0,n}(V, \beta),$$

and by j_Γ the inclusion

$$\widehat{\prod}_{v\in V(\Gamma)}\overline{\mathcal{M}}(v) \to \prod_{v\in V(\Gamma)} \overline{\mathcal{M}}(v).$$

4. Presentation of Cohomology Classes via Graphs

In this section we introduce a graphical notation for certain cohomology classes on the moduli spaces of stable maps which uses the graphical presentation of the boundary strata. Here we generalize the notation introduced in [26, 2.2].

Throughout the rest of this paper we assume that $g = 0$ and that V is a convex, smooth, projective variety.

We start with explaining how to push forward and pull back the cohomology classes represented by graphs under the universal curve morphism. Let $\pi : \overline{\mathcal{M}}_{0,n+1}(V, \beta) \to \overline{\mathcal{M}}_{0,n}(V, \beta)$ be defined by forgetting the ith marked point. First we explain the push forward. Suppose that Γ' represents a stratum in $\overline{\mathcal{M}}_{0,n+1}(V, \beta)$. If removing the tail labeled by i from Γ' does not destabilize it, then $\pi_*([\overline{\mathcal{M}}'_\Gamma]) = 0$. On the other hand, if removing the tail labeled i does destabilize Γ', then

$$\pi_*([\overline{\mathcal{M}}_{\Gamma'}]) = \frac{|\operatorname{Aut}\Gamma|}{|\operatorname{Aut}\Gamma'|}[\overline{\mathcal{M}}_\Gamma],$$

where Γ is obtained by stabilization [44].

Next we explain how to pull back the cohomology classes represented by graphs. Let $\overline{\mathcal{M}}_\Gamma$ be a stratum in $\overline{\mathcal{M}}_{0,n}(V, \beta)$. The pull back of its fundamental class is a subvariety of $\overline{\mathcal{M}}_{0,n+1}(V, \beta)$ corresponding to the sum of $|V(\Gamma)|$ graphs with rational coefficients. Each of these graphs is obtained by attaching a tail numbered i to a vertex of Γ [44], call this graph Γ', and the corresponding coefficient is $|\operatorname{Aut}\Gamma'|/|\operatorname{Aut}\Gamma|$.

These rules for pushing forward and pulling back cohomology classes on the moduli space of stable maps are analogous to the rules for the moduli space of curves.

In the sequel we will need trees whose vertices are decorated with more than its degree. In addition, we will decorate the vertices with cohomology classes. Let Γ be a stable tree that determines a subvariety of $\overline{\mathcal{M}}_{0,n}(V,\beta)$, and let $\gamma_v \in H^\bullet(\overline{\mathcal{M}}(v))$, $v \in V(\Gamma)$. We denote the cohomology class

$$\frac{1}{|\operatorname{Aut}\Gamma|}\rho_{\Gamma *} j_\Gamma^*(\otimes_v \gamma_v)$$

by the picture of Γ where each vertex v is in addition labeled by the cohomology class γ_v. We omit this label if $\gamma_v = 1$, i.e., γ_v is the orbifold fundamental class of $\overline{\mathcal{M}}(v)$. In particular, $[\overline{\mathcal{M}}_\Gamma]$ is represented by Γ with no additional labels attached.

Lemma 4.1. *Let $\pi : \overline{\mathcal{M}}_{0,n+1}(V,\beta) \to \overline{\mathcal{M}}_{0,n}(V,\beta)$ be the universal curve determined by forgetting the tail labeled i. Let Γ' be a tree that determines a subvariety in $\overline{\mathcal{M}}_{0,n+1}(V,\beta)$ and Γ be the tree obtained from Γ' by removing the tail labeled i. We assume that removing the tail labeled i does not destabilize Γ', and for each vertex v' of Γ' we denote by v the corresponding vertex of Γ. Let w' be the vertex from which the tail numbered i emanates, and let $\pi(w) : \overline{\mathcal{M}}(w') \to \overline{\mathcal{M}}(w)$ be the universal curve over $\overline{\mathcal{M}}(w)$ which forgets i.*

If $\gamma' \in H^\bullet(\overline{\mathcal{M}}_{0,n+1}(V,\beta))$ is represented by Γ' decorated with the cohomology classes $\gamma'_{v'}$, $v' \in V(\Gamma')$, then its push forward under π is the cohomology class γ represented by the multiple $|\operatorname{Aut}\Gamma|/|\operatorname{Aut}\Gamma'|$ of Γ decorated with the cohomology classes γ_v, where $\gamma_v = \gamma'_{v'}$ when $v \neq w$, and $\gamma_w = \pi(w)_*(\gamma_{w'})$.

Proof. For each vertex v' of Γ' there exists the corresponding vertex v of Γ, and $\overline{\mathcal{M}}(v')$ is isomorphic to $\overline{\mathcal{M}}(v)$. These isomorphisms can be chosen so that they respect the fibered products. We denote by X the product $\prod_{v\neq w}\overline{\mathcal{M}}(v)$ and by X_V the fibered product $\widehat{\prod}_{v\neq w}\overline{\mathcal{M}}(v)$. Consider the following commutative diagram where the horizontal arrows are natural morphisms, and two left vertical morphisms are induced by $\pi(w)$:

$$\begin{array}{ccccc}
X \times \overline{\mathcal{M}}(w') & \xleftarrow{\ j_{\Gamma'}\ } & X_V \times_V \overline{\mathcal{M}}(w') & \xrightarrow{\ \rho_{\Gamma'}\ } & \overline{\mathcal{M}}_{0,n+1}(V,\beta) \\
\downarrow{\scriptstyle \widetilde{\pi}} & & \downarrow{\scriptstyle \widetilde{\pi}} & & \downarrow{\scriptstyle \pi} \\
X \times \overline{\mathcal{M}}(w) & \xleftarrow{\ j_{\Gamma}\ } & X_V \times_V \overline{\mathcal{M}}(w) & \xrightarrow{\ \rho_{\Gamma}\ } & \overline{\mathcal{M}}_{0,n}(V,\beta).
\end{array}$$

In order to prove the statement of the lemma one needs to show that

$$\pi_*\,\rho_{\Gamma' *}\,j_{\Gamma'}^* = \rho_{\Gamma *}\,j_\Gamma^*\,\widetilde{\pi}_*.$$

This reduces to showing that $\widetilde{\pi}_* \, j^*_{\Gamma'} = j^*_{\Gamma}\widetilde{\pi}_*$. The latter follows from [15, 6.2], [16]. □

Remark . If in the lemma above one assumes that Γ' destabilizes after forgetting the tail labeled by i, then $\overline{\mathcal{M}}(w')$ is isomorphic to a point. If the degree of $\gamma_{w'}$ is greater than 0 attached to w', then the decorated graph represents 0 in cohomology. If $\gamma_{w'} = 1$ then one pushes the corresponding cohomology class down as in the case without decorations — by stabilizing and dividing by the orders of the appropriate automorphism groups.

For example, consider the cohomology class γ in $\overline{\mathcal{M}}_{0,5}(V, \beta)$ (where β is assumed to be nonzero) represented by the decorated stable tree

$\gamma :=$ 1, 2, 3 — γ'' β_1 — γ' β_2 — 4, 5

where it is understood that $\beta_1 + \beta_2 = \beta$ and γ' is a cohomology class on the space $\overline{\mathcal{M}}_{0,3}(V, \beta_2)$ associated to the right vertex and γ'' is a cohomology class on the space $\overline{\mathcal{M}}_{0,4}(V, \beta_1)$ associated to the left vertex. Consider the morphism $\pi : \overline{\mathcal{M}}_{0,5}(V, \beta) \rightarrow \overline{\mathcal{M}}_{0,4}(V, \beta)$ which forgets the 5th marked point. If β_2 is nonzero then

$\pi_*(\gamma) :=$ 1, 2, 3 — γ'' β_1 — $\widetilde{\pi}_*(\gamma')$ β_2 — 4

where $\widetilde{\pi} : \overline{\mathcal{M}}_{0,3}(V, \beta_2) \rightarrow \overline{\mathcal{M}}_{0,2}(V, \beta_2)$ is the projection forgetting the point labeled by 5 associated to the right vertex. On the other hand, if $\beta_2 = 0$ then the right hand side of the previous equation is now unstable. If γ' has nonzero dimension then $\pi_*(\gamma)$ vanishes. If γ' is the unit element then

$\pi_*(\gamma) :=$ 1, 2, 3 — γ'' β — 4

In all of these cases, the automorphism group is trivial so the prefactor does not arise.

5. Tautological Classes

In this section we introduce the tautological ψ and κ classes on the moduli spaces of stable maps which generalize the corresponding tautological classes on the moduli spaces of stable curves. We will also discuss the properties of the pull back of the ψ classes under the stabilization morphism, and derive the restriction properties of the tautological classes to the boundary strata.

Let $\pi : \overline{\mathcal{M}}_{0,n+1}(V,\beta) \to \overline{\mathcal{M}}_{0,n}(V,\beta)$ be the universal curve. We assume that π 'forgets' the $(n+1)^{\text{st}}$ marked point. The morphism π has n canonical sections $\sigma_1, \ldots, \sigma_n$. We denote their images by $D_1, \ldots, D_n$. If $g = 0$ and V is convex each D_i is a divisor. We denote by ω the relative dualizing sheaf of π. It is an invertible sheaf, that is, it determines a line bundle on $\overline{\mathcal{M}}_{0,n+1}(V,\beta)$ [5].

Definition 5.1. For each $i = 1, \ldots, n$ the *tautological line bundle* $\mathcal{L}_i$ on $\overline{\mathcal{M}}_{0,n}(V,\beta)$ is $\sigma_i^*\omega$. The *tautological class* $\psi_i \in H^2(\overline{\mathcal{M}}_{0,n}(V,\beta))$ is equal to the first Chern class $c_1(\mathcal{L}_i)$.

One can also pull back cohomology classes from V to $\overline{\mathcal{M}}_{0,n}(V,\beta)$ using the evaluation maps. The definition of the κ classes involves both, powers of the ψ classes and these pull backs.

Definition 5.2. (cf. [43, VI.3.10]) The *tautological class* κ_a in $H^\bullet(\overline{\mathcal{M}}_{0,n}(V,\beta)) \otimes H^\bullet(V)^*$ for $a \geq -1$ is defined as follows. For each $\gamma \in H^\bullet(V)$, the cohomology class $\kappa_a(\gamma)$ is the push forward of $\psi_{n+1}^{a+1} ev_{n+1}^*(\gamma)$ with respect to the projection $\pi : \overline{\mathcal{M}}_{0,n+1}(V,\beta) \to \overline{\mathcal{M}}_{0,n}(V,\beta)$ which forgets the $(n+1)^{\text{st}}$ marked point. In particular, if γ has definite degree $|\gamma|$ then $\kappa_a(\gamma)$ has degree $2a + |\gamma|$. If $\{ e_\alpha \}_{\alpha \in A}$, is a homogeneous basis for $H^\bullet(V)$ such that e_0 is the identity element then $\kappa_{a,\alpha}$ denotes the cohomology class $\kappa_a(e_\alpha)$.

Remark . The class $\kappa_{-1,0}$ vanishes due to dimensional reasons. In addition, all classes $\kappa_{-1}(\gamma)$ vanish on $\overline{\mathcal{M}}_{0,n}(V,0)$.

Our definition corresponds to the 'modified' κ classes defined by Arbarello and Cornalba [1] rather than the 'classical' κ classes defined by Mumford [44]. It is easy to see that $\psi_{n+1} = c_1(\omega(D_1 + \ldots + D_n))$, and therefore one can also define the κ classes using the relative dualizing sheaf.

Next we will show that the ψ classes restrict to the strata in the expected manner. For a more general treatment, we refer to [19, 43].

Lemma 5.3. *Let Γ be a stable tree which determines a stratum in $\overline{\mathcal{M}}_{0,n}(V,\beta)$. Suppose that the tail labeled i is attached to $w \in V(\Gamma)$. Denote the class ψ_i on $\overline{\mathcal{M}}_{0,n}(V,\beta)$ by ψ_i, and on $\overline{\mathcal{M}}(w)$ by ψ_i'. Then $\rho_\Gamma^*(\psi_i) = j_\Gamma^*\psi_i'$.*

Proof. Let $\pi : \overline{\mathcal{M}}_{0,n+1}(V,\beta) \to \overline{\mathcal{M}}_{0,n}(V,\beta)$ and $\pi(w) : \overline{\mathcal{M}}(w') \to \overline{\mathcal{M}}(w)$ be the universal curves forgetting the marked point labeled $n+1$, and σ_i be the corresponding sections which we denote by the same symbol. We use the same notation as in the proof of Lemma 4.1. Consider the commutative diagram

$$\begin{array}{ccccc} X \times \overline{\mathcal{M}}(w') & \xleftarrow{j_{\Gamma'}} & X_V \times_V \overline{\mathcal{M}}(w') & \xrightarrow{\rho_{\Gamma'}} & \overline{\mathcal{M}}_{0,n+1}(V,\beta) \\ \downarrow \tilde{\pi} & & \downarrow \tilde{\pi} & & \downarrow \pi \\ X \times \overline{\mathcal{M}}(w) & \xleftarrow{j_{\Gamma}} & X_V \times_V \overline{\mathcal{M}}(w) & \xrightarrow{\rho_{\Gamma}} & \overline{\mathcal{M}}_{0,n}(V,\beta). \end{array}$$

Let us denote by ω with a subscript the relative dualizing sheaf of the morphism corresponding to the subscript. It is clear that $\rho_1^* \omega_\pi = \omega_{1 \times_V \tilde{\pi}} = j_1^* \omega_{1 \times \tilde{\pi}}$ when restricted to the image of σ_i. It follows that

$$\rho_\Gamma^* \psi_i = \sigma_i^* \rho_{\Gamma'}^* (\omega_\pi) = \sigma_i^* j_{\Gamma'}^* (\omega_{\tilde{\pi}}) = j_\Gamma^* \psi_i'.$$

□

It is easy to verify that the pull backs of the cohomology classes on V under the evaluation maps restrict in the same manner.

Lemma 5.4. *Let Γ be a stable tree which determines a stratum in $\overline{\mathcal{M}}_{0,n}(V,\beta)$. Denote the class $\kappa_{a,\alpha}$ on $\overline{\mathcal{M}}_{0,n}(V,\beta)$ (resp., $\overline{\mathcal{M}}(v)$, where $v \in V(\Gamma)$) by κ (resp., $\kappa(v)$). Then*

$$\rho_\Gamma^*(\kappa) = j_\Gamma^* \sum_{v \in V(\Gamma)} \kappa(v).$$

Proof. Consider the set $\mathcal{G}$ of cardinality $|V(\Gamma)|$ whose elements are graphs Γ' each of which is obtained from Γ by attaching a tail labeled $n+1$ to one of the vertices of Γ. Each $\Gamma' \in \mathcal{G}$ determines a stratum of $\overline{\mathcal{M}}_{0,n+1}(V,\beta)$, and there are natural morphisms

$$\prod_{v \in V(\Gamma')} \overline{\mathcal{M}}(v) \to \prod_{v \in V(\Gamma)} \overline{\mathcal{M}}(v),$$

and the induced one on the fibered products. We denote both of them by $\tilde{\pi}$.

Consider the following commutative diagram

$$\begin{array}{ccccc} \coprod_{\Gamma' \in \mathcal{G}} \prod_{v \in \Gamma'} \overline{\mathcal{M}}(v) & \xleftarrow{j_{\Gamma'}} & \coprod_{\Gamma' \in \mathcal{G}} \widehat{\prod}_{v \in \Gamma'} \overline{\mathcal{M}}(v) & \xrightarrow{\rho_{\Gamma'}} & \overline{\mathcal{M}}_{0,n+1}(V,\beta) \\ \downarrow \tilde{\pi} & & \downarrow \tilde{\pi} & & \downarrow \pi \\ \prod_{v \in \Gamma} \overline{\mathcal{M}}(v) & \xleftarrow{j_{\Gamma}} & \widehat{\prod}_{v \in \Gamma} \overline{\mathcal{M}}(v) & \xrightarrow{\rho_{\Gamma}} & \overline{\mathcal{M}}_{0,n}(V,\beta). \end{array}$$

Note that the left square is a fibered square, and the right square is a normalization of the corresponding fibered square, and π, $\tilde{\pi}$ are flat. Therefore

[15, 16]:

$$\begin{aligned}\rho_\Gamma^*(\kappa) &= \rho_\Gamma^*\pi_*(\psi_{n+1}^{a+1}ev_{n+1}(e_\alpha)) = \widetilde{\pi}_*\rho_{\Gamma'}^*(\psi_{n+1}^{a+1}ev_{n+1}(e_\alpha)) \\ &= \widetilde{\pi}_*j_{\Gamma'}^*(\psi_{n+1}^{a+1}ev_{n+1}(e_\alpha)) = j_\Gamma^*\widetilde{\pi}_*(\psi_{n+1}^{a+1}ev_{n+1}(e_\alpha)) \\ &= j_\Gamma^*\sum_v \kappa(v).\end{aligned}$$

□

Our next goal is to express the ψ classes via the boundary strata. Since the action of the symmetric group on the moduli spaces of stable maps interchanges the ψ classes it is enough to give a presentation of ψ_1 via the boundary strata. We will denote ψ_1 on $\overline{\mathcal{M}}_{0,n}(V,\beta)$ by $\psi_{(n,\beta)}$. We also assume that if a graph is not stable then it represents the zero cohomology class.

Lemma 5.5. *If $n \geq 3$, $a \geq 1$, then the following holds in $H^\bullet(\overline{\mathcal{M}}_{0,n}(V,\beta))$*

$$\psi_{(n,\beta)}^a = \sum_{\substack{\beta_1+\beta_2=\beta \\ I\sqcup J=[n-2] \\ 1\in J}}$$

(graph: vertex β_1 with legs $n-1$, n and I; edge to vertex β_2 with legs 1 (marked ψ^{a-1}) and J)

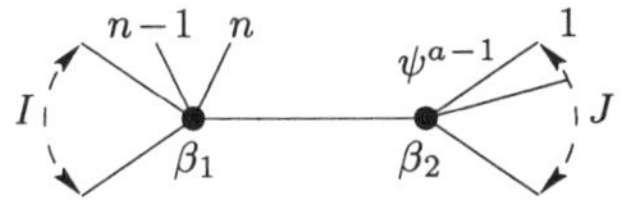

Proof. Getzler [19] proved the lemma for all projective manifolds in case $a = 1$ using techniques involving algebraic stacks. Pandharipande [47] informed us that he also has a proof of this lemma (unpublished). We present a proof in case of convex V that does not use stacks.

It is enough to prove the lemma for $a = 1$. This and Lemma 5.3 will imply the lemma for all $a \geq 1$.

First we prove that

$$\psi_{(3,\beta)} = \sum_{\beta_1+\beta_2=\beta} [D_{\beta_1,\beta_2}],$$

where

$$[D_{\beta_1,\beta_2}] = \sum_{\substack{\beta_1+\beta_2=\beta \\ I\sqcup J=[n-2] \\ 1\in J}}$$

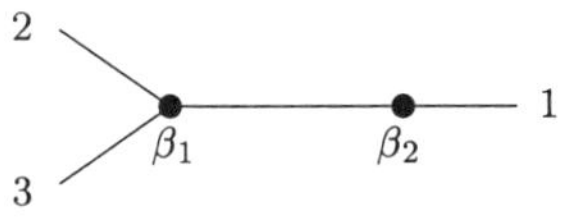

(graph: vertex β_1 with legs 2, 3; edge to vertex β_2 with leg 1)

and D_{β_1,β_2} is the corresponding divisor. The stability condition implies that $\beta_2 \neq 0$.

First note that the restriction of $\psi_{(3,\beta)}$ to $U := \overline{\mathcal{M}}_{0,3}(V,\beta) - \bigcup D_{\beta_1,\beta_2}$ is zero. Indeed, consider the universal map $\overline{\mathcal{M}}_{0,4}(V,\beta) \to \overline{\mathcal{M}}_{0,3}(V,\beta)$ restricted to U. Let $x \in U$, and C_x be the inverse image of x under the universal map. The image of the section of the universal map determined by the first marked point lies on the irreducible component of C_x which remains stable after forgetting the map $f_x : C_x \to V$. This shows that $\psi_{(3,\beta)}$ restricted to U is the pull-back from $\overline{\mathcal{M}}_{0,3}$, and therefore trivial.

It follows from the previous paragraph that

$$\psi_{(3,\beta)} = \sum_{\beta_1+\beta_2=\beta,\, i} k_{\beta_1,\beta_2,i}\,[D_{\beta_1,\beta_2,i}], \tag{2}$$

where $D_{\beta_1,\beta_2,i}$ are (disjoint) irreducible components of D_{β_1,β_2}. In order to determine the coefficients it suffices to intersect (2) with an arbitrary complete non-singular curve S in $\overline{\mathcal{M}}_{0,3}(V,\beta)$. Let $T \to S$ be a family of 3-pointed stable maps over S, and ν be the corresponding morphism $S \to \overline{\mathcal{M}}_{0,3}(V,\beta)$. We can assume that the image of ν intersects each codimension one boundary stratum transversally, and it is disjoint from the intersection of the boundary strata. Then $[S]\cdot\sum[D_{\beta_1,\beta_2}] = \sum_{\beta_1,\beta_2,i} l_{\beta_1,\beta_2,i}$, where $l_{\beta_1,\beta_2,i}$ is the number of points in the intersection of S with $D_{\beta_1,\beta_2,i}$. The stabilization morphism, i.e., the morphism determined by forgetting the map to V, determines a morphism $T \to S\times\mathbb{CP}^1$. It is the blow up at the points corresponding to contracted rational components. The image of the first section contains exactly those points which lie in the fibers over $D_{\beta_1,\beta_2,i}\cap S$. It follows that $\psi_{(3,\beta)}\cdot[S] = \sum[D_{\beta_1,\beta_2}]\cdot[S]$, and therefore $\psi_{(3,\beta)} = \sum[D_{\beta_1,\beta_2}]$.

In order to obtain the statement of the lemma for $n \geq 4$ one can first show that if $\pi : \overline{\mathcal{M}}_{0,n+1}(V,\beta) \to \overline{\mathcal{M}}_{0,n}(V,\beta)$ is the universal map, then $\psi_{(n+1,\beta)} = \pi^*\psi_{(n,\beta)} + [D_{1,n+1}]$, where $D_{1,n+1}$ is the image of the first section of the universal map. This can be done by using an argument similar to the above. (See also [33].) Then one needs to pull back cohomology classes represented by the boundary strata using the description from Sec. 4. $\square$

As a corollary using Lemma 4.1 we express the κ classes via the boundary strata and the κ_{-1} classes. We denote the class $\kappa_{a,\alpha}$ on $\overline{\mathcal{M}}_{0,n}(V,\beta)$ by $\kappa_{a,\alpha,(n,\beta)}$.

Corollary 5.6. *If $n \geq 2$, $a \geq 1$, then the following holds in $H^\bullet(\overline{\mathcal{M}}_{0,n}(V,\beta))$*

$$\kappa_{a,\alpha,(n,\beta)} = \sum_{\substack{\beta_1+\beta_2=\beta \\ I\sqcup J=[n-2]}}$$

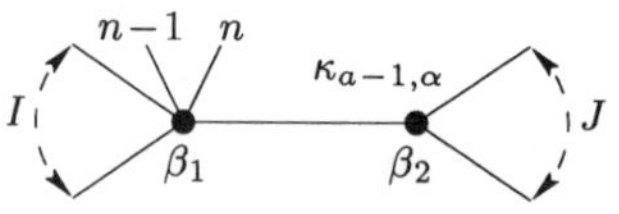

If $n \geq 2$ then the following holds in $H^\bullet(\overline{\mathcal{M}}_{0,n}(V,\beta))$

$$\kappa_{0,\alpha,(n,\beta)} = \sum_{\substack{\beta_1+\beta_2=\beta \\ I\sqcup J=[n-2]}} \quad I \quad n-1 \quad n \quad \beta_1 \quad \kappa_{-1,\alpha} \quad \beta_2 \quad J$$

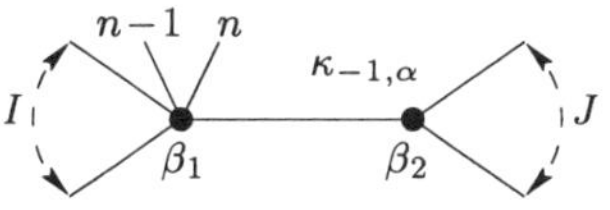

$$+ \sum_{i\in[n-2]} \quad [n-2]-i \quad n-1 \quad n \quad \mathrm{ev}_i^* e_\alpha \quad \beta \quad i$$

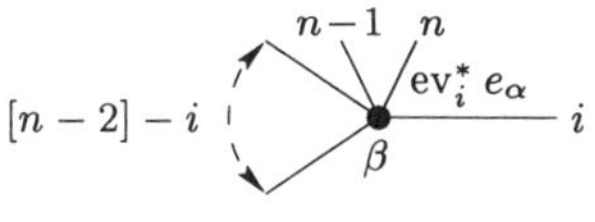

It follows from the above presentation of the κ classes that $\kappa_{0,0} = n-2 \in H^0(\overline{\mathcal{M}}_{0,n}(V,\beta))$. This can be also seen directly by integrating over the fibers of the universal curve. One can also check that if $\gamma \in H^2(V)$, then $\kappa_{-1}(\gamma) = \int_\beta \gamma$ on $\overline{\mathcal{M}}_{0,n}(V,\beta)$. We will need this in Sec. 7.

6. Families of Cohomological Field Theories

In this section we define the notion of a cohomological field theory and prove that the Gromov–Witten invariants twisted by the κ classes endows $H^\bullet(V)$ with the a family of genus zero CohFT structures. This is equivalent to endowing $H^\bullet(V)$ with a family of formal Frobenius manifold structures (with nonflat identity element, in general) arising from the Poincaré pairing and deformations of the cup product on V. Our deformations contain quantum cohomology as a special case.

Notation . Let V be a topological space and let $H^\bullet(V,\mathbb{C})$ be given a homogeneous basis $\mathbf{e} := \{ e_\alpha \}_{\alpha\in A}$ and let e_0 denote the identity element. Let $\mathbf{s} := \{ s_r^\alpha \,|\, r \geq -1, \alpha \in A \}$ be a collection of formal variables with grading $|s_r^\alpha| = 2r + |e_\alpha|$. All formal power series and polynomials in a collection of variables (*e.g.*, $\mathbf{s}$) are in the $\mathbb{Z}_2$-graded sense.

Definition 6.1. Let Λ consist of formal symbols q^β for all $\beta \in B(V)$ together with the multiplication $(q^\beta\, q^{\beta'}) \mapsto q^{\beta+\beta'}$. Let $\mathbb{C}[[\Lambda]]$ consist of formal sums $\sum_{\beta\in B(V)} a_\beta\, q^\beta$ where a_β are elements in $\mathbb{C}$. Assign to each q^β, the degree $-2c_1(V) \cap \beta$. The product is well defined according to [34, Prop. II.4.8]. This endows Λ with the structure of a semigroup with unit. Furthermore, let $\mathbb{C}[[\Lambda,\mathbf{s}]] := \mathbb{C}[[\Lambda]][[\mathbf{s}]]$, formal power series in the variables $\mathbf{s}$ with coefficients in $\mathbb{C}[[\Lambda]]$.

Suppose that $H_2(V,\mathbb{Z})$ has no torsion then the semigroup Λ can be realized concretely by choosing a basis $\{ e^\epsilon \}$ for $H_2(V,\mathbb{Z})$ (say the one which is a subset of the basis dual via the Poincaré pairing to $\{ e_\alpha \}$), introducing a formal variable q_ϵ for each such basis element, writing $\beta = \sum_\epsilon \beta_\epsilon e^\epsilon$, and defining $q^\beta = \prod_\epsilon q_\epsilon^{\beta_\epsilon}$. The variable q_ϵ is then assigned the degree $-2c_1(V) \cap e^\epsilon$. If $H_2(V,\mathbb{Z})$ contains a torsion subgroup decomposed into a product of cyclic groups then one can do the same by introducing a 'root of unity' for each generator of a cyclic subgroup in the obvious manner.

Theorem 6.2. *Let V be a smooth, projective, convex variety. For each $n \geq 3$, let $\Omega_{0,n}$ be elements in $(H^\bullet(\overline{\mathcal{M}}_{0,n}) \otimes T^n H^\bullet(V)^*)[[\Lambda, \mathbf{s}]]$ defined by*

$$\Omega_{0,n}(\gamma_1, \dots, \gamma_n) := \sum_\beta \mathrm{st}_*(ev_1^*\gamma_1 \cdots ev_n^*\gamma_n \exp(\mathbf{s}\boldsymbol{\kappa}))\, q^\beta \tag{3}$$

where $\mathrm{st} : \overline{\mathcal{M}}_{0,n}(V, \beta) \to \overline{\mathcal{M}}_{0,n}$ is the forgetful map, $\gamma_1, \gamma_2, \dots, \gamma_n$ are elements in $H^\bullet(V, \mathbb{C})$. The $\Omega := \{\Omega_{0,n}\}$ endows $(H^\bullet(V, \mathbb{C}[[\Lambda, \mathbf{s}]]), \eta)$ with the structure of a CohFT where η is the Poincaré pairing extended $\mathbb{C}[[\Lambda, \mathbf{s}]]$-linearly. In particular, if $H^\bullet(V, \mathbb{C})$ consists of only even dimensional cohomology classes then for all values of $\{ s_r^\alpha \}$, $\Omega := \{\Omega_{0,n}\}$ endows $(H^\bullet(V, \mathbb{C}[[\Lambda]]), \eta)$ with the structure of a CohFT where η is the Poincaré pairing.

Proof. Let $n \geq 3$, and let Γ be a stable graph for $\overline{\mathcal{M}}_{0,n}$. Let $\mathcal{G}$ be the set of graphs Γ' obtained by decorating the vertices of Γ with elements $\beta(v) \in H_2(V)$ such that $\sum_{v \in V(\Gamma)} \beta(v) = \beta$. Clearly, each $\Gamma' \in \mathcal{G}$ determines a stratum in $\overline{\mathcal{M}}_{0,n}(V, \beta)$. It follows from the deformation theory considerations in [17] that the union of $\overline{\mathcal{M}}_{\Gamma'}$, $\Gamma' \in \mathcal{G}$, is the scheme theoretic preimage of $\overline{\mathcal{M}}_\Gamma$ under the morphism st.

Consider the following commuting diagram:

$$\begin{array}{ccccc}
\coprod_{\Gamma' \in \mathcal{G}} \prod_{v \in V(\Gamma')} \overline{\mathcal{M}}(v) & \xleftarrow{j_{\Gamma'}} & \coprod_{\Gamma' \in \mathcal{G}} \widehat{\prod}_{v \in V(\Gamma')} \overline{\mathcal{M}}(v) & \xrightarrow{\rho_{\Gamma'}} & \overline{\mathcal{M}}_{0,n}(V, \beta) \\
\downarrow{\scriptstyle \widetilde{\mathrm{st}}} & & \downarrow{\scriptstyle \widetilde{\mathrm{st}}} & & \downarrow{\scriptstyle \mathrm{st}} \\
\prod_{v \in V(\Gamma)} \overline{\mathcal{M}}(v) & \xleftarrow{\mathrm{id}} & \prod_{v \in V(\Gamma)} \overline{\mathcal{M}}_{0,n(v)} & \xrightarrow{\rho_\Gamma} & \overline{\mathcal{M}}_{0,n}.
\end{array}$$

One has to check that the element $\Omega_{0,n}(\gamma_1, \dots, \gamma_n)$ defined by (3) possesses the restriction property (1). It suffices to check it for each β. Since the right square of the commutative diagram is bi-rational to the corresponding fibered square one has that $\rho_\Gamma^* \,\mathrm{st}_* = \sum_{\Gamma' \in \mathcal{G}} \widetilde{\mathrm{st}}_* \, j_{\Gamma' *} \, \rho_{\Gamma'}^*$ [15, 16]. The required restriction property follows from the following facts. Firstly, for each class $\kappa = \kappa_{a,\alpha}$ one has $\rho_{\Gamma'}^* \,\kappa = \sum_{v \in V(\Gamma')} j_{\Gamma'}^* \,\kappa(v)$, where $\kappa(v)$ is the corresponding class on $\overline{\mathcal{M}}(v)$, as shown in Lemma 5.4. Secondly, $\rho_{\Gamma'}^* ev_i^* \gamma_i = j_{\Gamma'}^* \, ev_{(v),i}^*(\gamma_i)$, where v is the vertex of Γ' to which the tail labeled i attached, and $ev_{(v),i}$ is the corresponding evaluation morphism. Thirdly, the composition $j_{\Gamma' *} \, j_{\Gamma'}^*$ is the multiplication of by the Thom class of V in $V \times V$ [15, 16].

Finally, suppose that $H^\bullet(V)$ consists of only even dimensional classes. By dimensional considerations and the fact that $\kappa_{0,0} = n-2$ and $\kappa_{-1}(\gamma) = \int_\beta \gamma$

when γ has degree 2, the coefficient of q^β in $\Omega_{0,n}$ converges when one plugs in numbers for s_n^α for all $n = -1, 0, \ldots$ and $\alpha = 0, 1, \ldots, r$. □

As before, one can extend the ground ring $\mathbb{C}$ above to $\mathbb{C}[[\Lambda, \mathbf{s}]]$ in the definition of the potential of a genus zero CohFT and the above theorem extends, as well. In our setting, the potential is a formal function on $\mathcal{H} := H^\bullet(V, \mathbb{C}[[\Lambda, \mathbf{s}]])$ of the following kind. Φ belongs to $\mathbb{C}[[\Lambda, \mathbf{s}]][[x^0, \ldots, x^r]]$ where we have used

$$I_{0,n}(e_{\alpha_1}, \ldots, e_{\alpha_n}) = \int_{\overline{\mathcal{M}}_{0,n}} \langle \Omega_{0,n}, e_{\alpha_1} \otimes e_{\alpha_2} \otimes \cdots \otimes e_{\alpha_n} \rangle .$$

Again, if $H^\bullet(V)$ consists entirely of even dimensional classes then plugging in numbers for all s_n^α where $n = -1, 0, 1, \ldots$ and $\alpha = 0, 1, \ldots, r$, one obtains families of CohFT structures on $H^\bullet(V, \mathbb{C}[[\Lambda]])$.

7. Topological Recursion Relations

In this section we will derive differential equations for the generating function which incorporates the intersection numbers of the ψ and κ classes.

Notation . Let $\mathcal{S}_k^r$ be the set of infinite sequences of non-negative integers

$$\mathbf{m} = (m_k^0, m_k^1, \ldots, m_k^r, m_{k+1}^0, \ldots, m_{k+1}^r, m_{k+2}^0, \ldots)$$

such that $m_a^\alpha = 0$ for all a sufficiently large. We denote by $\boldsymbol{\delta}_a^\alpha$ the infinite sequence whose only non-zero entry is $m_a^\alpha = 1$. We use the Latin characters for the lower index and the Greek characters for the upper index. We will use the notation of the type

$$\mathbf{m}! := \prod_{\alpha=0}^{r} \prod_{a \geq k} m_a^\alpha! \quad \text{and} \quad \binom{\mathbf{m}}{\mathbf{m}'} := \frac{\mathbf{m}!}{\mathbf{m}'!\,(\mathbf{m} - \mathbf{m}')!} .$$

We also consider $\mathbf{t} = \{t_d^\mu\}$, $d \geq 0$, $\mu = 0, \ldots, r$, $\mathbf{s} = \{s_a^\alpha\}$, $a \geq -1$, $\alpha = 0, \ldots, r$, be a collection of independent formal variables with grading $|t_a^\alpha| = 2a - 2 + |e_\alpha|$ and $|s_a^\alpha| = 2a + |e_\alpha|$.

Differential Equations

We start with introducing the notation for the intersection numbers. This notation generalizes the one from [54, 26].

Definition 7.1. Let $\beta \in H_2(V,\mathbb{Z})$, and $\{e_\alpha\}$, $\alpha = 0,\dots,r$ is a basis of $H^\bullet(V)$. If $\beta \neq 0$, or $\beta = 0$ and $n \geq 3$, then

$$
\begin{aligned}
&\langle \tau_{d_1}^{\mu_1}\tau_{d_2}^{\mu_2}\dots\tau_{d_n}^{\mu_n}\kappa_{a_1,\alpha_1},\kappa_{a_2,\alpha_2}\dots\kappa_{a_l,\alpha_l}\rangle_\beta \\
&\quad := \int_{\overline{\mathcal{M}}_{0,n}(V,\beta)} t_{d_1}^{\mu_1}\psi_1^{d_1}ev_1^*(e_{\mu_1})\,t_{d_2}^{\mu_2}\psi_2^{d_2}ev_2^*(e_{\mu_2})\dots t_{d_n}^{\mu_n}\psi_n^{d_n}ev_n^*(e_{\mu_n}) \\
&\qquad\qquad \times s_{a_1}^{\alpha_1}\kappa_{a_1,\alpha_1}\,s_{a_2}^{\alpha_2}\kappa_{a_2,\alpha_2}\dots s_{a_l}^{\alpha_l}\kappa_{a_l,\alpha_l} = (-1)^\epsilon t_{d_1}^{\mu_1}\dots t_{d_n}^{\mu_n}\,s_{a_1}^{\alpha_1}\dots s_{a_l}^{\alpha_l} \\
&\quad\times \int_{\overline{\mathcal{M}}_{0,n}(V,\beta)} \psi_1^{d_1}ev_1^*(e_{\mu_1})\dots\psi_n^{d_n}ev_n^*(e_{\mu_n})\,\kappa_{a_1,\alpha_1}\dots\kappa_{a_l,\alpha_l}.
\end{aligned}
$$

If $\beta = 0$ and $n < 3$ we define all intersection numbers to be equal to 0.

Let the sequence $(d_1,\mu_1),(d_2,\mu_2),\dots,(d_n,\mu_n)$ contain m_d^μ pairs (d,μ), where $d \geq 0$, $0 \leq \mu \leq r$, and the sequence $(a_1,\alpha_1),(a_2,\alpha_2),\dots,(a_l,\alpha_l)$ contain p_a^α pairs (a,α) where $a \geq -1$, $0 \leq \alpha \leq r$. Then we also denote the intersection number above by $\langle \boldsymbol{\tau}^{\mathbf{m}}\boldsymbol{\kappa}^{\mathbf{p}}\rangle_\beta$.

We also set

$$
\langle \boldsymbol{\tau}^{\mathbf{m}}\boldsymbol{\kappa}^{\mathbf{p}}\rangle := \sum_\beta \langle \boldsymbol{\tau}^{\mathbf{m}}\boldsymbol{\kappa}^{\mathbf{p}}\rangle_\beta\, q^\beta,
$$

where q is the formal variable introduced in the previous section.

Remark . We put the formal variables in the definition of the intersection numbers in order to take care of the signs provided V has odd cohomology classes. If V has only even cohomology classes, then ϵ in the formula above is 0. Then setting all formal t variables equal to 1 and all formal s variables equal to 0 we obtain the definition due to Witten [54]. In this paper we assume that V may have odd cohomology classes even though we do not know examples of convex manifolds with odd cohomology classes.

Definition 7.2. We define $H(\mathbf{t};\mathbf{s})$ in $\mathbb{C}[[\Lambda,\mathbf{t},\mathbf{s}]]$ by

$$
H(\mathbf{t};\mathbf{s}) := \sum_{\mathbf{m}\in S_0^r\ \mathbf{p}\in S_{-1}^r} \frac{1}{\mathbf{m}!}\frac{1}{\mathbf{p}!}\langle \boldsymbol{\tau}^{\mathbf{m}}\boldsymbol{\kappa}^{\mathbf{p}}\rangle
$$

Notice that H has grading $2(\dim_{\mathbb{C}} V - 3)$.

We are going to exploit the presentations of the ψ classes and the κ classes in Lemma 5.5 and Corollary 5.6 in order to obtain a system of differential equations for H. The explicit presentation of the tautological classes allows one to obtain recursion relations for the intersection numbers, and these recursion relations in turn provide the corresponding differential equations. The technical details are standard (cf. [54, 30, 26]), and we omit them.

We recall that we denote by η the Poincaré pairing on $H^\bullet(V)$. We denote $\eta(e_{\alpha_1}, e_{\alpha_2})$ by η_{α_1,α_2}, and by η^{α_1,α_2} the coefficients of the inverse metric. We denote by $c^{\alpha_1}_{\alpha_2,\alpha_3}$ the coefficients of the multiplication tensor, that is

$$e_{\alpha_2} e_{\alpha_3} = \sum_{\alpha_1=0}^{r} c^{\alpha_1}_{\alpha_2,\alpha_3} e_{\alpha_1}.$$

Proposition 7.3. *The intersection numbers defined above satisfy the recursion relations below called the topological recursion relations. The first two relations are valid when $a_1 \geq 1$.*

$$(t^{\alpha_1}_{a_1})^{-1}\langle \boldsymbol{\tau}^{\mathbf{m}+\delta^{\alpha_1}_{a_1}+\delta^{\alpha_2}_{a_2}+\delta^{\alpha_3}_{a_3}} \boldsymbol{\kappa}^{\mathbf{p}}\rangle_\beta = (t^{\alpha_1}_{a_1-1})^{-1}(t^{\sigma_1}_0)^{-1}(t^{\sigma_2}_0)^{-1}$$
$$\times \sum \binom{\mathbf{m}}{\mathbf{m}'}\binom{\mathbf{p}}{\mathbf{p}'} \langle \boldsymbol{\tau}^{\mathbf{m}'+\delta^{\alpha_1}_{a_1-1}+\delta^{\sigma_1}_0} \boldsymbol{\kappa}^{\mathbf{p}'}\rangle_{\beta_1} \eta^{\sigma_1,\sigma_2} \langle \boldsymbol{\tau}^{\mathbf{m}''+\delta^{\sigma_2}_0+\delta^{\alpha_2}_{a_2}+\delta^{\alpha_3}_{a_3}} \boldsymbol{\kappa}^{\mathbf{p}''}\rangle_{\beta_2}$$

$$(s^{\alpha_1}_{a_1})^{-1}\langle \boldsymbol{\tau}^{\mathbf{m}+\delta^{\alpha_2}_{a_2}+\delta^{\alpha_3}_{a_3}} \boldsymbol{\kappa}^{\mathbf{p}+\delta^{\alpha_1}_{a_1}}\rangle_\beta = (s^{\alpha_1}_{a_1-1})^{-1}(t^{\sigma_1}_0)^{-1}(t^{\sigma_2}_0)^{-1}$$
$$\times \sum \binom{\mathbf{m}}{\mathbf{m}'}\binom{\mathbf{p}}{\mathbf{p}'} \langle \boldsymbol{\tau}^{\mathbf{m}'+\delta^{\sigma_1}_0} \boldsymbol{\kappa}^{\mathbf{p}'+\delta^{\alpha_1}_{a_1-1}}\rangle_{\beta_1} \eta^{\sigma_1,\sigma_2} \langle \boldsymbol{\tau}^{\mathbf{m}''+\delta^{\sigma_2}_0+\delta^{\alpha_2}_{a_2}+\delta^{\alpha_3}_{a_3}} \boldsymbol{\kappa}^{\mathbf{p}''}\rangle_{\beta_2}$$

$$(s^{\alpha_1}_{0})^{-1}\langle \boldsymbol{\tau}^{\mathbf{m}+\delta^{\alpha_2}_{a_2}+\delta^{\alpha_3}_{a_3}} \boldsymbol{\kappa}^{\mathbf{p}+\delta^{\alpha_1}_{0}}\rangle_\beta = (s^{\alpha_1}_{-1})^{-1}(t^{\sigma_1}_0)^{-1}(t^{\sigma_2}_0)^{-1}$$
$$\times \sum \binom{\mathbf{m}}{\mathbf{m}'}\binom{\mathbf{p}}{\mathbf{p}'} \langle \boldsymbol{\tau}^{\mathbf{m}'+\delta^{\sigma_1}_0} \boldsymbol{\kappa}^{\mathbf{p}'+\delta^{\alpha_1}_{-1}}\rangle_{\beta_1} \eta^{\sigma_1,\sigma_2} \langle \boldsymbol{\tau}^{\mathbf{m}''+\delta^{\sigma_2}_0+\delta^{\alpha_2}_{a_2}+\delta^{\alpha_3}_{a_3}} \boldsymbol{\kappa}^{\mathbf{p}''}\rangle_{\beta_2}$$
$$+ \sum_{a=0}^{\infty}\sum_{\alpha,\nu=0}^{r} m^\alpha_a c^\nu_{\alpha,\alpha_1} t^\alpha_a (t^\nu_a)^{-1} \langle \boldsymbol{\tau}^{\mathbf{m}-\delta^\alpha_a+\delta^\nu_a+\delta^{\alpha_2}_{a_2}+\delta^{\alpha_3}_{a_3}} \boldsymbol{\kappa}^{\mathbf{p}}\rangle_\beta$$

where the summation is over $\beta_1 + \beta_2 = \beta, \mathbf{m}' + \mathbf{m}'' = \mathbf{m}, \mathbf{p}' + \mathbf{p}'' = \mathbf{p}$, and σ_1, σ_2 varying from 0 to r.

Equivalently, the function $H(\mathbf{t};\mathbf{s})$ satisfies the system of differential equations below. The first two equations are valid when $a_1 \geq 1$.

$$\frac{\partial^3 H}{\partial t^{\alpha_1}_{a_1}\partial t^{\alpha_2}_{a_2}\partial t^{\alpha_3}_{a_3}} = \sum_{\sigma_1,\sigma_2} \frac{\partial^2 H}{\partial t^{\alpha_1}_{a_1-1}\partial t^{\sigma_1}_0}\eta^{\sigma_1,\sigma_2}\frac{\partial^3 H}{\partial t^{\sigma_2}_0\partial t^{\alpha_2}_{a_2}\partial t^{\alpha_3}_{a_3}},$$

$$\frac{\partial^3 H}{\partial s^{\alpha_1}_{a_1}\partial t^{\alpha_2}_{a_2}\partial t^{\alpha_3}_{a_3}} = \sum_{\sigma_1,\sigma_2} \frac{\partial^2 H}{\partial s^{\alpha_1}_{a_1-1}\partial t^{\sigma_1}_0}\eta^{\sigma_1,\sigma_2}\frac{\partial^3 H}{\partial t^{\sigma_2}_0\partial t^{\alpha_2}_{a_2}\partial t^{\alpha_3}_{a_3}},$$

$$\frac{\partial^3 H}{\partial s^{\alpha_1}_{0}\partial t^{\alpha_2}_{a_2}\partial t^{\alpha_3}_{a_3}} = \sum_{\sigma_1,\sigma_2} \frac{\partial^2 H}{\partial s^{\alpha_1}_{-1}\partial t^{\sigma_1}_0}\eta^{\sigma_1,\sigma_2}\frac{\partial^3 H}{\partial t^{\sigma_2}_0\partial t^{\alpha_2}_{a_2}\partial t^{\alpha_3}_{a_3}}$$
$$+ \sum_{a=0}^{\infty}\sum_{\alpha,\nu=0}^{r} c^\nu_{\alpha,\alpha_1} t^\alpha_a \frac{\partial^3 H}{\partial t^\nu_a \partial t^{\alpha_2}_{a_2}\partial t^{\alpha_3}_{a_3}},$$

where the indices σ_1, σ_2 vary from 0 to r. □

Remark. If one considers the function $F(\mathbf{t}) := H(\mathbf{t};\mathbf{0})$, then the first equations reduce to the equations first written by Witten in [54]. Witten showed [54] that the WDVV equations in Th. 2.10 follow from the differential equation for $F(\mathbf{t})$.

Corollary 7.4. *Let H be as above. Let $x^\alpha := t_0^\alpha$, $\mathbf{x} := (x^0, \ldots, x^r)$, and let $\mathbf{t} := (\mathbf{x}, \widetilde{\mathbf{t}})$ where $\widetilde{\mathbf{t}}$ consists of all $t_{\underline{n}}^\alpha$ such that $n \geq 1$. Let $\Phi(\mathbf{x}) := H(\mathbf{x}, \widetilde{\mathbf{t}}; \mathbf{s})$ be regarded as an element of $\mathbb{C}[[\Lambda, \widetilde{\mathbf{t}}, \mathbf{s}]][[\mathbf{x}]]$ then Φ endows $H^\bullet(V, \mathbb{C}[[\Lambda, \widetilde{\mathbf{t}}, \mathbf{s}]])$ with the structure of a CohFT (over the ground ring $\mathbb{C}[[\Lambda, \widetilde{\mathbf{t}}, \mathbf{s}]]$).*

Proof. Our proof follows that in [54] but keeping track of signs. Differentiate the first set of equations in the last theorem with respect to the variable $t_{a_4}^{\alpha_4}$, setting $a_1 = 1$ and $a_2 = a_3 = a_4 = 0$ and noting that the left hand side is essentially (graded) symmetric under permutation of the subscripts $2, 3, 4$. Therefore, the antisymmetric part of the right hand side must vanish. This is nothing more than the WDVV equations (with the correct signs). This implies (by Theorem 2.10) that Φ is a potential for a genus zero CohFT. □

Puncture and Dilaton Equations

Here we introduce analogs of the puncture and the dilaton equations. Using these equations and the topological recursion relations we will show that the intersection numbers defined above can be expressed in terms of Gromov–Witten invariants.

For the rest of this subsection let us fix the following notation. Let $\pi : \overline{\mathcal{M}}_{0,n+1}(V, \beta) \to \overline{\mathcal{M}}_{0,n}(V, \beta)$ denote the universal stable map which 'forgets' the $n + 1^{\text{st}}$ marked point. To distinguish the ψ and the κ classes upstairs and downstairs we put 'hats' over the former. Since the pull back of $ev_i(e_\alpha)$ is still $ev_i(e_\alpha)$ we denote these classes by the same symbol upstairs and downstairs.

In addition, if $\mathbf{m} = \{m_a^\alpha\} \in S_k^r$, $k = -1, 0$, then

$$|\mathbf{m}| := \sum_{\alpha=0}^{r} \sum_{a \geq 0} a m_a^\alpha, \quad \|\mathbf{m}\| := \sum_{\alpha=0}^{r} \sum_{a \geq 0} m_a^\alpha,$$

and if $\mathbf{m} = \{m_a^\alpha\} \in S_0^r$, then

$$\begin{aligned} e(\mathbf{m}) &:= e_0^{m_0^0} \ldots e_r^{m_0^r} e_0^{m_1^0} \ldots e_r^{m_1^r} e_0^{m_2^0} \ldots, \\ \mathbf{s}^{\mathbf{m}} &:= \ldots (s_2^0)^{m_2^0} (s_1^r)^{m_1^r} \ldots (s_1^0)^{m_1^0} (s_0^r)^{m_0^r} \ldots (s_0^0)^{m_0^0}. \end{aligned}$$

Note that in both cases we have summation over $a \geq 0$, not over $a \geq k$, and that the order of the terms in the products is the opposite. We denote by $c_{e(\mathbf{m}),\mu}^\nu$ the coefficients of e_ν in the product $e(\mathbf{m})e_\mu$. We also say that $\mathbf{m} \in S_{0,0}^r$ if $\mathbf{m} \in S_0^r$, and $m_a^\alpha = 0$ provided $a \geq 1$.

It is easy to derive the following formulas for the pull back of the ψ and κ classes with respect to π (cf. [45]), namely,

$$\widehat{\psi}_i^a = \pi^* \psi_i^a + \pi^* \psi_i^{a-1} D_i \quad \text{and} \quad \widehat{\kappa}_{a,\alpha} = \pi^* \kappa_{a,\alpha} + \widehat{\psi}_{n+1}^a ev_{n+1}^*(e_\alpha).$$

The pullback formulas above induce the recursion relations (4) below. These are analogs of the puncture and dilaton equations. The proof is standard, and we omit the details (cf. [26, Sec. 3.1]). The variables take care of the signs as in the case of the topological recursion relations. When $a \geq 1$, or when $a = 0$ and $\mathbf{m} \in \mathcal{S}_{0,0}^r$, one has

$$\begin{aligned}
&(t_a^\alpha)^{-1} \langle \boldsymbol{\tau}^{\mathbf{m}+\delta_a^\alpha} \boldsymbol{\kappa}^{\mathbf{p}} \rangle_\beta \\
&= \sum_{\substack{\mathbf{p}'+\mathbf{p}''=\mathbf{p} \\ \mathbf{p}'' \in \mathcal{S}_0^r}} \sum_{\nu=o}^{r} \binom{\mathbf{p}}{\mathbf{p}'} c_{e(\mathbf{p}''),\alpha}^\nu \mathbf{s}^{\mathbf{p}''} (s_{|\mathbf{p}''|+a-1}^\nu)^{-1} \langle \boldsymbol{\tau}^{\mathbf{m}} \boldsymbol{\kappa}^{\mathbf{p}'+\delta_{|\mathbf{p}''|+a-1}^\nu} \rangle_\beta. \quad (4)
\end{aligned}$$

Here we take $\mathbf{p}'' \in \mathcal{S}_{0,r}$ since $\pi^*(\kappa_{-1,\alpha}) = \widehat{\kappa}_{-1,\alpha}$. Of course, one can not apply (4) when $\beta = 0$ and $\|\mathbf{m} + \delta_a^\alpha\| = 3$ since the space $\overline{\mathcal{M}}_{0,2}(V, 0)$ does not exist. One can also derive a recursion relation in case when $a = 0$, and $\mathbf{m}$ is arbitrary, but it is rather messy, and we will not need it in the sequel. The recursion relations (4) are equivalent to the differential equations

$$\frac{\partial(H - H_{in})}{\partial t_a^\alpha} = \sum_{\mathbf{p} \in \mathcal{S}_0^r} \sum_{\nu=0}^{r} c_{e(\mathbf{p}),\alpha}^\nu \frac{\mathbf{s}^{\mathbf{p}}}{\mathbf{p}!} \frac{\partial H}{\partial s_{|\mathbf{p}|+a-1}^\nu}$$

valid for all $a \geq 1$, and for $a = 0$ if instead of H we consider the function H_1 which is obtained from H by setting all variables t_i^μ with $i \geq 1$ to zero. The function H_{in} is the part of H which corresponds intersection numbers on $\overline{\mathcal{M}}_{0,3}(V, 0)$. It is a function in variables t_0^α, s_0^α, $\alpha = 0, \ldots, r$, determined by the cup-product on V since $\kappa_{0,\alpha}$ on $\overline{\mathcal{M}}_{0,3}(V, 0)$ is equal to $ev^*(d_\alpha)$. (All evaluation maps are equal when $\beta = 0$.)

We want to show that (4) and Prop. 7.3 allow us to compute inductively all intersection numbers of the ψ and the κ classes provided we know the Gromov–Witten invariants of V. Notice that (4) allows one to eliminate τ_a^α with $a \geq 1$ from the intersection numbers. Therefore we reduced the calculation to the case of the intersection numbers of the κ classes and the pull backs of the cohomology classes from V if no ψ classes are present.

First we want to reduce everything to the case of the intersection numbers of $\kappa_{-1,\alpha}$ and $ev_i^*(e_\mu)$. Suppose that we have an intersection number of the type $\langle \boldsymbol{\tau}^{\mathbf{m}} \boldsymbol{\kappa}^{\mathbf{p}} \rangle_\beta$, where $\mathbf{m} \in \mathcal{S}_{0,0}^r$, and there exists $p_a^\alpha > 0$ for some $a \geq 0$. If $\|\mathbf{m}\| \geq 2$, then the second or third recursion relation in Prop. 7.3 allows us to express it in terms of the intersection numbers of the type $\langle \boldsymbol{\tau}^{\mathbf{m}'} \boldsymbol{\kappa}^{\mathbf{p}'} \rangle_{\beta_1}$, where $\mathbf{m}' \in \mathcal{S}_{0,0}^r$, $\|\mathbf{m}'\| \geq 1$, and either $\|\mathbf{p}'\| < \|\mathbf{p}\|$, or $\|\mathbf{p}'\| = \|\mathbf{p}\|$, and $|\mathbf{p}'| < |\mathbf{p}|$.

However, if $\|\mathbf{m}\| = 1$, we can not apply the recursion relations from Prop. 7.3. Here we have to use (4). Namely, pick α such that $|e_\alpha| = 2$ and $\int_\beta \alpha \neq 0$. (Clearly, such α exists if V is not a point.) When we apply (4) to $\langle \boldsymbol{\tau}^{\mathbf{m}+\delta_0^\alpha} \boldsymbol{\kappa}^{\mathbf{p}} \rangle_\beta$ we can split the sum in the right hand side into two parts. One term corresponds to $\mathbf{p}' = \mathbf{p}$, and $\mathbf{p}'' = \mathbf{0}$. This term is

$$\langle \boldsymbol{\tau}^{\mathbf{m}} \boldsymbol{\kappa}^{\mathbf{p}} \kappa_{-1,\alpha} \rangle_\beta = \langle \boldsymbol{\tau}^{\mathbf{m}} \boldsymbol{\kappa}^{\mathbf{p}} \rangle_\beta \int_\beta \alpha$$

Here and in the rest of the argument we disregard the formal variables since they just take care of the signs which do not play an important role in the our argument. The second term is

$$\sum_{\substack{\mathbf{p}'+\mathbf{p}''=\mathbf{p} \\ 0 \neq \mathbf{p}'' \in \mathcal{S}_0^r}} \sum_{\nu=0}^{r} \binom{\mathbf{p}}{\mathbf{p}'} c^{\nu}_{e(\mathbf{p}''),\alpha} \langle \boldsymbol{\tau}^{\mathbf{m}} \boldsymbol{\kappa}^{\mathbf{p}'+\delta^{\nu}_{|\mathbf{p}''|-1}} \rangle_\beta .$$

For all these terms

$$\|\mathbf{p}' + \delta^{\nu}_{|\mathbf{p}''|-1}\| = \|\mathbf{p}'\| + 1 \leq \|\mathbf{p}\|.$$

The equality is attained only when $\mathbf{p}'' = \delta^{\mu}_{j}$ for $j \geq 1$. But for such terms $|\mathbf{p}' + \delta^{\nu}_{|\mathbf{p}''|-1}| < |\mathbf{p}|$. It follows that we can express each intersection number $\langle \boldsymbol{\tau}^{\mathbf{m}} \boldsymbol{\kappa}^{\mathbf{p}} \rangle_\beta$ with $\mathbf{m} \in \mathcal{S}^r_{0,0}$, $\|\mathbf{m}\| \geq 1$ in terms of the intersection numbers $\langle \boldsymbol{\tau}^{\mathbf{m}'} \boldsymbol{\kappa}^{\mathbf{p}'} \rangle_{\beta_1}$, where either $\|\mathbf{p}'\| < \|\mathbf{p}\|$, or $\|\mathbf{p}'\| = \|\mathbf{p}\|$ and $|\mathbf{p}'| < |\mathbf{p}|$. Note, that it is easy to extend our argument to the case when $\|\mathbf{m}\| = 0$.

Inductively this reduces the computation of the intersection numbers to the integrals of the type

$$\int_{\overline{\mathcal{M}}_{0,n}(V,\beta)} ev_1^*(\gamma_1) \dots ev_n^*(\gamma_n) \kappa_{-1}(\gamma_{n+1}) \dots \kappa_{-1}(\gamma_{n+l}), \tag{5}$$

where γ_i are cohomology classes of V. Recall that $\pi^*(ev_i^*(\gamma)) = ev_i^*(\gamma)$ and $\pi^*(\kappa_{-1}(\gamma)) = \kappa_{-1}(\gamma)$. Iterating this equality one obtains that the intersection number (5) is equal to

$$\int_{\overline{\mathcal{M}}_{0,n+l}(V,\beta)} ev_1^*(\gamma_1) \dots ev_n^*(\gamma_n) ev_{n+1}^*(\gamma_{n+1}) \dots ev_{n+l}^*(\gamma_{n+l}).$$

The integral above is the Gromov–Witten invariant evaluated on the classes $\gamma_1, \dots, \gamma_{n+l}$.

Remark. In [37] the authors show how to compute the intersection numbers of the ψ classes and the pull backs of the cohomology classes from V provided that the Gromov–Witten invariants are known. One can show that (4) allows one to eliminate the κ classes from the intersection numbers. This provides an alternative way of calculating these intersection numbers via the Gromov–Witten invariants.

An Example

The simplest application of the above equations is to the case where V is a point. This situation is discussed in some detail in [26] in genus zero and one.

The next simplest case is when $V = \mathbb{CP}^1$. Let $\{ e_0,\, e_1 \}$ be a basis for $H^\bullet(\mathbb{CP}^1)$ where e_0 is the identity element and e_1 is the element in $H^2(\mathbb{CP}^1)$ such that the Poincaré metric $\eta_{0,1} := 1$. By abuse of notation, an element β in $H_2(\mathbb{CP}^1, \mathbb{Z})$ is sometimes regarded as an integer (also denoted by β) times the fundamental class of $\mathbb{CP}^1$. The moduli space of stable maps $\overline{\mathcal{M}}_{0,n}(V,\beta)$, has complex dimension $-2+n+2\beta$. On $\overline{\mathcal{M}}_{0,n}(V,\beta)$, there are the identities $\kappa_{0,0} = (n-2)e_0$ and $\kappa_{-1,0} = 0$, and $\kappa_{-1,1} = \beta$.

Let $x^\alpha := t_0^\alpha$ for $\alpha = 0, 1$. The generating function $H(\mathbf{s}, \mathbf{t})$ has degree -4. We will now present an explicit formula for $H(x^0, x^1, s_{-1}^1, s_0^0, s_0^1; q)$ where the q dependence has been made explicit and it is understood that all other variables have been set to zero. In particular, H contains no ψ classes at all.

The generating function $\widetilde{H}$ consists of all terms in H except those arising from intersection numbers on $\overline{\mathcal{M}}_{0,3}(V,0)$. One can directly verify that

$$\begin{aligned} H(x^0, x^1, s_{-1}^1, s_0^0, s_0^1; q) &= \widetilde{H}(x^0, x^1, s_{-1}^1, s_0^0, s_0^1; q) \\ &\quad + e^{s_0^0}\left(\tfrac{1}{2}\,(x^0)^2\, x^1 \,+\, \tfrac{1}{6}\,(x^0)^3\, s_0^1 \right) \end{aligned} \tag{6}$$

where the usual cup product on $H^\bullet(\mathbb{CP}^1)$ corresponds to the special case where all of the s_a^α variables have been set to zero.

The puncture/dilaton equations for $\mathbb{CP}^1$ acting upon the generating function $\widetilde{\mathrm{H}} := \widetilde{\mathrm{H}}(x^0, x^1, s_{-1}^1, s_0^0, s_0^1; q)$ are

$$\frac{\partial\, \widetilde{\mathrm{H}}}{\partial x^0}(x^0, x^1, s_{-1}^1, s_0^0, s_0^1; q) \;=\; e^{s_0^0}\, s_0^1\, \frac{\partial\, \widetilde{\mathrm{H}}}{\partial s_{-1}^1}$$

and

$$\frac{\partial\, \widetilde{\mathrm{H}}}{\partial x^1}(x^0, x^1, s_{-1}^1, s_0^0, s_0^1; q) \;=\; e^{s_0^0}\, \frac{\partial\, \widetilde{\mathrm{H}}}{\partial s_{-1}^1}.$$

The relevant topological recursion relations for the generating function $H := H(x^0, x^1, s_{-1}^1, s_0^0, s_0^1; q)$ becomes

$$\frac{\partial H_{\alpha_1\alpha_2}}{\partial s_0^0} \;=\; x^0 H_{0\alpha_1\alpha_2} \,+\, x^1 H_{1\alpha_1\alpha_2}$$

and

$$\frac{\partial H_{\alpha_1\alpha_2}}{\partial s_0^1} = \frac{\partial H_0}{\partial s_{-1}^1} H_{1\alpha_1\alpha_2} + \frac{\partial H_1}{\partial s_{-1}^1} H_{0\alpha_1\alpha_2} + x^0 H_{1\alpha_1\alpha_2}$$

for all $\alpha_1, \alpha_2 = 0, 1$. Furthermore, H_i denotes the derivative of H with respect to x^i, H_{ij} is the derivative of H with respect to the variables x^i and x^j, and similarly for H_{ijk}.

The puncture/dilaton equations yields a solution of the form

$$\widetilde{\mathrm{H}}(x^0, x^1, s_{-1}^1, s_0^0, s_0^1; q) = \sum_{n=1}^{\infty} e^{-2s_0^0}\, \widetilde{q}^n \, \frac{(s_0^1)^{2n-2}}{(2n-2)!}\, h_n \tag{7}$$

where $\widetilde{q} := q\exp(s_{-1}^1 + e^{s_0^0}(x^1 + s_0^1 x^0))$ and h_n are numbers. The form of $\widetilde{\mathrm{H}}(x^0, x^1, s_{-1}^1, s_0^0, s_0^1; q)$ above implies that it is enough to know the function $\widetilde{\mathrm{H}}(x^0, x^1, s_0^1; q)$ (where s_{-1}^1 and s_0^0 have both been set to zero) since one can recover $\widetilde{\mathrm{H}}(x^0, x^1, s_{-1}^1, s_0^0, s_0^1; q)$ from it. *Henceforth, unless otherwise stated, assume that H and $\widetilde{\mathrm{H}}$ depend only upon the variables x^0, x^1, s_0^1 and q.*

The topological recursion relations can then all be reduced to the single equation

$$\frac{\partial\, \widetilde{\mathrm{H}}''}{\partial s_0^1} = 2\, s_0^1\, \widetilde{\mathrm{H}}''\, \widetilde{\mathrm{H}}''' + x^0\, \widetilde{\mathrm{H}}'''$$

where $'$ represents the derivative $q(\partial/\partial q)$.

This equation gives rise to the recursion relations

$$h_{n+1} = \sum_{l=1}^{n} \frac{(2n-1)!\, l^2\, (n+1-l)^2}{(2l-2)!\,(2(n-l))!\,(n+1)}\, h_l\, h_{n+1-l}$$

and $h_1 = 1$.

The first few terms of the potential are readily seen to be

$$\begin{aligned} &H(x^0, x^1, s_{-1}^1, s_0^0, s_0^1; q) \\ &= \exp(s_0^0)\left(\tfrac{1}{2}(x^0)^2 x^1 + \tfrac{1}{6}(x^0)^3 s_0^1\right) \\ &\quad + \exp(-2s_0^0)\left(\widetilde{q} + \widetilde{q}^2\,\frac{(s_0^1)^2}{4} + \widetilde{q}^3\,\frac{(s_0^1)^4}{6} + \widetilde{q}^4\,\frac{(s_0^1)^6}{6} + \mathrm{o}(q^5 (s_0^1)^8)\right). \end{aligned} \tag{8}$$

As expected, after setting all s_a^α variables to zero one recovers

$$H(x^0, x^1; q) = \frac{1}{2}(x^0)^2 x^1 + q\exp(x^1),$$

the usual Gromov–Witten potential of $\mathbb{CP}^1$. More generally, the potential H parametrizes Frobenius algebras structures on $H^\bullet(\mathbb{CP}^1)$ which are further deformations of the quantum cup product on $H^\bullet(\mathbb{CP}^1)$.

It is amusing to note that integrating equation 7 twice, one can obtain an expression for $\widetilde{\mathrm{H}}'$ in the form of a transcendental equation, namely,

$$(1-u)\,\exp(u) \;=\; q(s_0^1)^2\,\exp(1+s_0^1x^0+s^1)$$

where

$$u \;:=\; \sqrt{1-2(s_0^1)^2\,\widetilde{\mathrm{H}}'(x^0,x^1,s_0^1;q)}.$$

Acknowledgments

We are grateful to the Max Planck Institut für Mathematik, where this work was initiated, for their financial support and for providing a wonderfully stimulating atmosphere. We would like to thank D. Abramovich, P. Belorousski, E. Getzler, Yu. Manin, R. Pandharipande, and D. Zagier for useful conversations. We also thank M. Polito for pointing out at an erroneous statement in an earlier version of this paper. We would like to continue to thank K. Belabas for his TEXnical assistance and for providing the music.

References

[1] Arbarello, E. and Cornalba, M.: Combinatorial and algebro-geometric cohomology classes on the moduli space of curves, *J. Algebraic Geom.* 5 (1996) 705–749.

[2] Batyrev, V. V.: Stringy Hodge numbers and Virasoro algebra, `alg-geom/9711019`.

[3] Behrend, K.: Gromov–Witten invariants in algebraic geometry, *Invent. Math.* 127 (1997) 601–617.

[4] Behrend, K. and Fantechi, B.:The intrinsic normal cone, *Invent. Math.* 128 (1997) 45–88.

[5] Behrend, K. and Manin, Yu. I.: Stacks of stable maps and Gromov–Witten invariants. *Duke Math. J.* 85 (1996) 1–60.

[6] Cohen, F. R.:The homology of $\mathcal{C}_{n+1}$ spaces, $n \geq 0$, *The homology of iterated loops spaces.* LNM 533. Berlin, Heidelberg, New York: Springer, 1976, 207–351.

[7] ______ :Artin's braid groups, classical homotopy theory and sundry other curiosities. *Contemp. Math.* 78 (1988) 167–206.

[8] Cornalba, M.: On the projectivity of the moduli spaces of curves, *J. Reine Angew. Math.* 443 (1993) 11–20.

[9] Deligne, P. and Mumford, D.: The irreducibility of the space of curves of given genus, *Publ. Math. Inst. Hautes Études Sci.* 36 (1969) 75–110.

[10] Di Francesco, P. and Itzykson, C.: Quantum intersection rings, in *T*he moduli space of curves (Texel Island, 1994) (R. Dijkgraaf, C. Faber and G. van der Geer, Eds.), Progr. Math., 129, Birkhaüser, Boston 1995, pp. 81–148,

[11] Dijkgraaf, R.: Intersection theory, integrable hierarchies and topological field theory, in *New Symmetry Principles in Quantum Field Theory* (G. Mack, Ed.), Plenum, 1993, pp. 95-158.

[12] Dubrovin, B.: Geometry of 2D topological field theories, in *Integrable systems and Quantum Groups,* Lecture Notes in Math. 1620, Springer, Berlin, 1996, pp. 120–348.

[13] Eguchi, T., Hori, K. and Xiong, C. S.: Quantum cohomology and the Virasoro algebra, *Phys. Letters B* 402 (1997) 71–80.

[14] Eguchi, T., Jinzenji, M. and Xiong, C. S.: Quantum cohomology and free field representation, *Nuclear Phys. B* 510 (1998) 608–622.

[15] Fulton, W.: *Intersection Theory,* Springer-Verlag, Berlin, Heidelberg, 1984.

[16] Fulton, W. and MacPherson, R.: Categorical framework for the study of singular spaces, *Mem. Amer. Math. Soc.* 243 (1981).

[17] Fulton, W. and Pandharipande, R.: Notes on stable maps and quantum cohomology, in *Algebraic Geometry (Santa Cruz, 1995)* (J. Kollár, R. Lazarsfeld, and D. Morrison, Eds.), Proc. Symp. Pure Math. 62, II, Amer. Math. Soc. 1997, pp. 45–96.

[18] Getzler, E.: Batalin-Vilkovisky algebras and two-dimensional topological field theories. *Commun. Math. Phys.* 159 (1994) 265–285.

[19] ______: Topological recursion relations in genus 2, in *Integrable systems and algebraic geometry (Kobe/Kyoto, 1997),* World Sci. Publishing, River Edge, NJ, 1998, pp. 73–106.

[20] ______: The Virasoro conjecture for Gromov–Witten invariants, `math.AG/9812026`.

[21] Getzler, E. and Kapranov, M.:Modular operads, *Compositio Math.* 110 (1998) 65–126.

[22] ______: Cyclic operads and cyclic homology, in *Geometry, topology, & physics,* Conf. Proc. Lecture Notes Geom. Topology, VI, Internat. Press, Cambridge, MA, 1995, pp. 167–201.

[23] Getzler, E. and Pandharipande, R.: Virasoro constraints and the Chern classes of the Hodge bundle, *Nuclear Phys. B* 530 (1998) 701–714.

[24] Ginzburg, V. and Kapranov, M. M.: Koszul duality for operads, *Duke Math. J.* 76 (1994), 203–272.

[25] Hitchin, N.: Frobenius manifolds, in *Gauge theory and Symplectic Geometry* (J. Hurtubise and F. Lalonde, Eds.), NATO-ASO Series C 488, Kluwer, Boston, 1997, pp. 69–112.

[26] Kabanov, A. and Kimura, T.: Intersection numbers and rank one cohomological field theories in genus one, *Comm. Math. Phys.* 194 (1998) 651–674.

[27] ______: A Change of Coordinates on the Large Phase Space of Quantum Cohomology, `math.AG/9907096`.

[28] Kac, V. G. and Schwartz, A.: Geometric interpretation of the partition function of 2D gravity, *Phys. Lett.* 257 (1991) 329–334.

[29] Katz, S.: Virasoro constraints on Gromov–Witten invariants, unpublished manuscript.

[30] Kaufmann, R., Manin, Yu. I. and Zagier, D.: Higher Weil-Petersson volumes of moduli spaces of stable n-pointed curves, *Comm. Math. Phys.* 181 (1996) 763–787.

[31] Kimura, T., Stasheff, J. and Voronov, A. A.: On operad strucures on moduli spaces and string theory, *Comm. Math. Phys.*171 (1995), 1–25.

[32] Keel, S.: Intersection theory of moduli spaces of stable n-pointed curves of genus zero, *Trans. Amer. Math. Soc.* 330 (1992) 545–574.

[33] Knudsen, F. F.: The projectivity of the moduli space of stable curves, II, III, *Math. Scandinavica* 52 (1983) 161–199, 200–212.

[34] Kollár, J.: *Rational curves on algebraic varieties,* Springer-Verlag, Berlin, 1995.

[35] Kontsevich, M.: Intersection theory on the moduli space of curves and the matrix Airy function, *Comm. Math. Phys.* 147 (1992) 1–23.

[36] Kontsevich, M. and Manin, Yu. I.: Gromov–Witten classes, quantum cohomology, and enumerative geometry, *Comm. Math. Phys.* 164 (1994) 525–562.

[37] ______: Relations between the correlators of the topological sigma–model coupled to gravity, *Comm. Math. Phys.* 196 (1998) 385–398.

[38] Kontsevich, M. and Manin Yu. I. (with Appendix by R. Kaufmann): Quantum cohomology of a product, *Invent. Math.* 124 (1996) 313–340.

[39] Li, J. and Tian, G.: Virtual moduli cycles and Gromov–Witten invariants of algebraic varieties, *J. Amer. Math Soc.* 11 (1998) 119–174.

[40] Lian, B. and Zuckerman, G. J.: New perspectives on the BRST-algebraic structure of string theory, *Com. Math. Phys* 154 (1993), 613 – 646.

[41] Liu, X. and Tian, G.: Virasoro constraints for quantum cohomology. *J. Differential Geom.* 50 (1998) 537–590.

[42] Looijenga, E.: Intersection theory on Deligne-Mumford compactifications [after Witten and Kontsevich], *Astérisque* 216 (1993) 187–212.

[43] Manin, Yu. I.: *Frobenius manifolds, quantum cohomology, and moduli spaces,* Amer. Math. Soc. Colloquium Publications 47, Amer. Math. Soc., Providence, RI, 1999.

[44] Mumford, D.: Towards an enumerative geometry of the moduli space of curves, in *Arithmetic and Geometry* (M. Artin and J. Tate, Eds.), Part II, Progress in Math. 36, Birkhäuser, Basel, 1983, 271–328.

[45] Pandharipande, R.: A reconstruction theorem for gravitational descendents, *Mittag-Leffler preprint.*

[46] ———: A geometric construction of Getzler's elliptic relation, *Math. Ann.* 313 (1999) 715–729.

[47] ———: Private communication.

[48] Ruan, Y. and Tian, G.: A mathematical theory of quantum cohomology, *J. Diff. Geom.* 42 (1995) 259–367.

[49] Segal, G.: The definition of a conformal field theory. Preprint. Oxford.

[50] ———: Two dimensional conformal field theories and modular functors, IXth Int. Congr. on Math. Phys. (Bristol; Philadephia) (B. Simon, A. Gruman, and I. M. Davies, eds.), IOP Publishing Ltd, 1989, 22–37.

[51] ———: Topology from the point of view of Q.F.T., Lectures at Yale University, March 1993.

[52] Stasheff, J. D.: On the homotopy associativity of H-spaces, I, *Trans. Amer. Math. Soc.* 108 (1963), 272–292.

[53] ———: On the homotopy associativity of H-spaces, II, *Trans. Amer. Math. Soc.* 108 (1963), 293 – 312.

[54] Witten, E.: Two-dimensional gravity and intersection theory on moduli space, *Surveys in Diff. Geom.* 1 (1991) 243–310.

D-BRANE ACTIONS ON KÄHLER MANIFOLDS

Akishi KATO
Graduate School of Mathematical Sciences, University of Tokyo
3-8-1 Komaba, Meguro-ku, Tokyo 153-8914, Japan
akishi@ms.u-tokyo.ac.jp

Abstract We discuss how to formulate the matrix model actions describing the multiple D-brane dynamics on curved (Kähler) manifolds along the lines of [1], [2]. We briefly review the D-geometry axioms, the requirement that the action reproduce the masses of stretched open strings. In the case of three complex dimensions we show that the axioms cannot be satisfied unless the metric is Ricci flat, and argue that such actions do exist when the metric is Ricci flat. The results provide a noncommutative extension of Kähler geometry.

Mathematics Subject Classification (2000): 81T20, 81T40, 81T75

Keywords: matrix model, D-brane, open string, nonlinear sigma model, noncommutative geometry

1. Introduction

The unification of gravity and quantum field theory is a longstanding problem in particle physics, and much effort has been made in order to grasp this problem by using the string theory approach.

During the past 10 years it became apparent that the so called string duality plays an important rôle in understanding the relations between the various formulations of string theory. This duality requires non-perturbative collective excitations in addition to ordinary perturbative string states. At weak string coupling, such solitonic excitations, now known as D-branes, can be exactly

Y. Maeda, et al. (eds.), Noncommutative Differential Geometry and Its Applications to Physics, 99–121.

described by boundary CFT [3]. In particular, flat background corresponds to free fields theory and gives us a lot of insights into the rôle of string duality.

D-brane dynamics is very interesting for several reasons. D-branes can probe shorter scales than string scale. Considering the low energy world volume theory of branes, generally super-Yang–Mills theory [4] with various gauge group and matter are included, and a property which could be used as a starting point to describe particle physics phenomenology.

When branes get close to each other, open strings stretched between branes become massless and lead to non-Abelian gauge symmetry enhancement.

In type IIA string theory, D 0-branes become the lightest states in the strong coupling limit, where the theory tends to M-theory with eleven dimensional Lorentz invariance. Based on this fact, the authors of [5] proposed the so called Matrix conjecture:

- M-theory formulated in the infinite momentum frame is exactly equivalent to $N \to \infty$ limit of a supersymmetric matrix quantum mechanics of matrices of size N.

We do not give a precise formulation of this conjecture here, but it seems to be the first concrete, quantitative proposal to what 'quantum geometry' means in string theory. The conjecture has passed several nontrivial tests. For example, the effective potential for graviton scattering is correctly reproduced [6],[7], [8].

In this new development, a key ingredient is that the space-time points are not just a collection of numbers, but eigenvalues of $N \times N$ matrices. This is somewhat similar to the idea of algebraic geometry where 'points' are replaced by spectra of an commutative algebra. In brane physics points are spectra of large matrices.

In the physics literature matrix models can be classified into two types. The first type is the 'old' matrix model, which was extensively used as a regularization of two dimensional quantum gravity around the early nineties [9]. This is defined by the partition function

$$Z_N = \int_{\mathcal{H}_N} dX \exp\{\mathrm{Tr}\, V(X)\}\,.$$

The integral is taken over the space $\mathcal{H}_N$ of all hermitian matrices of size N; dX is the volume form on $\mathcal{H}_N$ induced from metric $\mathrm{Tr}\,(dX)^2$. $V(X)$ is a polynomial of variable X. Both the classical action $\mathrm{Tr}\, V(X)$ and the integral measure dX is invariant under $U(N)$ gauge action

$$X \mapsto X^U := U^{-1}XU, \tag{1}$$

As is well known, the integral can be reduced to an N dimensional integral by diagonalizing X by the gauge action (1); the Jacobian is given by the van der

Monde determinant:

$$\begin{aligned} Z_N &= \text{const} \int_{\mathbb{R}^N} \prod_{i=1}^{N} d\lambda_i \prod_{i<j} (\lambda_i - \lambda_j)^2 \exp\left\{ \sum_{i=1}^{N} V(\lambda_i) \right\} \\ &= \text{const} \int_{\mathbb{R}^N} \prod_{i=1}^{N} d\lambda_i \exp\left\{ \sum_{i=1}^{N} V(\lambda_i) + \sum_{i \neq j}^{N} \log|\lambda_i - \lambda_j| \right\} \end{aligned}$$

Thus if we interpret N eigenvalues $\lambda_1, \cdots, \lambda_N$ as the positions of N 'particles', the matrix dynamics is reduced to that of N particles moving on a real line, interacting with each other as well as external potential $V(\lambda)$. Since permutations of eigenvalues are inner automorphisms of $U(N)$ gauge symmetry, these 'particles' are indistinguishable. The large N asymptotics of the partition function (1) can be estimated using various techniques. A similar model is also used to study the intersection theory of moduli space of Riemann surfaces. [10], [11].

The 'new' matrix model, which is our current concern, includes multiple hermitian matrices $X^1, \ldots, X^d$. It is defined by the partition function

$$Z_N = \int_{\underbrace{\mathcal{H}_N \times \cdots \times \mathcal{H}_N}_{d}} dX^1 \cdots dX^d \exp S(X^1, \cdots, X^d)$$

where the action $S(X^1, \cdots, X^d)$ is typically given by

$$S(X^1, \cdots, X^d) = \frac{1}{4g^2} \sum_{\mu,\nu=1}^{d} \mathrm{Tr}\, [X^\mu, X^\nu]^2 + \cdots. \tag{2}$$

The new matrix model also enjoys the (diagonal) $U(N)$ gauge invariance

$$X^\mu \mapsto (X^\mu)^U := U^{-1} X^\mu U, \qquad (\mu = 1, 2, \ldots, D), \tag{3}$$

and we are, as usual, interested in the gauge invariant correlation functions. In contrast to the old matrix model, all the matrices cannot be simultaneously diagonalized by the $U(N)$ gauge action, and this makes it more difficult and interesting to analyze the model. In a sense, the new matrix model is a *non-commutative* extension of the old one. Hereafter, when we say 'matrix model' we will refer to the new one.

Since exact results are scarce*, we begin by studying the model semi-classically. First, we need to identify the classical moduli space (or vacuum)

---

* See, however, a very interesting paper [12].

which is a gauge equivalence class of critical points of the action (2):

$$\left\{X=(X^1,\ldots,X^d)\in\mathcal{H}_N^d\,\middle|\,[X^\mu,[X^\nu,X^\mu]]=0\quad(\forall\nu)\right\}\Big/\sim.$$

Consider a point $X_0\in\mathcal{H}_N^d$ where all d matrices are simultaneously diagonal. Obviously X_0 is a point in the classical moduli space and it is a natural starting point for a semi-classical analysis. By definition the vacuum expectation values of the matrices are given by

$$\langle X^\mu\rangle=X_0^\mu=\begin{pmatrix}x_1^\mu&&&\\&x_2^\mu&&\\&&\ddots&\\&&&x_N^\mu\end{pmatrix}. \tag{4}$$

As in the old matrix models, N vectors $\vec{x}_i:=(x_i^1,\ldots,x_i^d)\in\mathbb{R}^d$ $(i=1,2,\ldots,N)$ are interpreted as positions of N indistinguishable objects sitting in $\mathbb{R}^D$. These objects will be called D-branes.

Let us study the quantum fluctuation around the classical background (4), i.e. the effective dynamics induced on N D-branes located at $\vec{x}_1,\ldots,\vec{x}_N$. We thus Taylor expand the action $S(X)$ around $X=X_0$. Among physically meaningful quantities are the masses of the quantum excitations, the eigenvalues of Hessian of S. The leading part of the Taylor expansion reads as

$$\begin{aligned}S(X)&=S(X_0+\widetilde{X})\\&=\frac{1}{2g^2}\sum_{i,j}\{(x_i-x_j)^2\widetilde{X}_{ij}^\nu\widetilde{X}_{ji}^\nu\}\\&\quad+\frac{1}{2g^2}\sum_{i,j}\{(x_i^\mu-x_j^\mu)(x_j^\nu-x_i^\nu)\widetilde{X}_{ij}^\mu\widetilde{X}_{ji}^\nu\}+\cdots.\end{aligned} \tag{5}$$

where the remaining terms $\cdots$ are at least cubic in quantum fluctuations $\widetilde{X}^\mu:=X^\mu-X_0^\mu$ and are irrelevant to our subsequent discussion.

At the critical point (4), there is a unbroken gauge symmetry $U(1)^N\subset U(N)$ which leaves invariant the classical background X_0. The quantum fluctuations corresponds to the tangent vectors to $\mathcal{H}_N^d$ at X_0, and are naturally decomposed into the irreducible representations of $\mathfrak{u}(1)\oplus\cdots\oplus\mathfrak{u}(1)$ (i.e., classified by $U(1)^N$ charges.) Usual gauge fixing prescription allows us to restrict our attention to the fluctuation $\widetilde{X}_{ij}$ which is transverse to the relative vector $\vec{x}_i-\vec{x}_j$, i.e., $\sum_\mu(x_i^\mu-x_j^\mu)\widetilde{X}_{ij}^\mu=0$. Thus we can neglect the last term of (5).

From the first term of (5), we immediately notice:

(i) In the classical background (4) there are vector-like quantum excitations $(X_{ij}^\mu)_{\mu=1}^d$ labeled by the ordered pair (ij) of D-brane indices;

(ii) The mass of the excitation X^{μ}_{ij} is proportional to the Euclidean distance between the D-branes i and j

$$d(\vec{x}_i, \vec{x}_j) = \sqrt{\sum_{\mu=1}^{d} (\vec{x}_i - \vec{x}_j)^2},$$

and is independent of the index μ;

(iii) Off diagonal excitations $X^{\mu}_{ij} (i \neq j)$ are charged under $U(1)_i \times U(1)_j$ and neutral with respect to other $U(1)$'s.

In the usual terminology of string theory, D-branes are defined to be the place where open strings can be attached. Thus, from (i) and (iii), it is natural to identify the quantum fluctuation represented by X^{μ}_{ij} with an open string[†] stretched between the D-branes i and j, polarized in μ-th direction (See Figure 1).

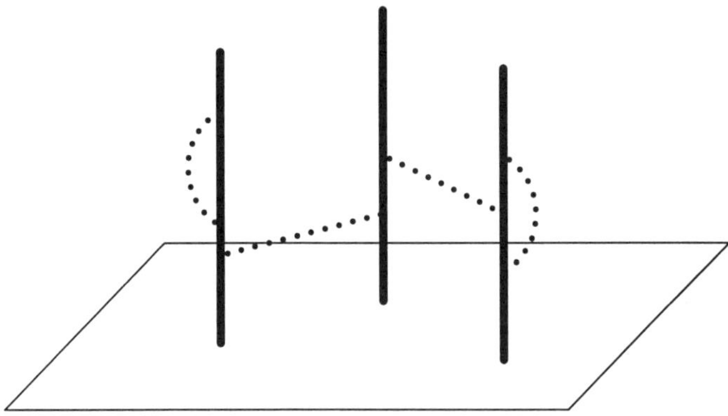

Figure 1 D-branes (thick lines) with open strings (dotted lines) stretched among them.

Property (ii) also supports the above identification, since strings have a definite tension α':

$$\begin{aligned} \text{(excitation energy)} &= \alpha' \times \text{(length of the string)} \\ &= \alpha' \times \text{(distance between the branes).} \end{aligned} \tag{6}$$

It is important to note the relation (6) is coordinate independent and makes sense on curved manifolds, too.

[†] The strings are oriented. This is because we started from Hermitian matrices rather than real symmetric ones.

It is worthwhile to mention the relationship between the Yang–Mills theory and matrix models. The action of d dimensional Yang–Mills theory with gauge group $U(N)$ is given by

$$S = \frac{1}{4g^2}\int_{\mathbb{R}^d} d^d x ||F_A||^2 \tag{7}$$

where $F_A = dA + A \wedge A$ is the curvature of the connection form $A = dx^\mu(\partial_\mu + A_\mu)$.

Locally, the gauge connection A_μ can be regarded as a d matrix valued functions $A_\mu : \mathbb{R}^d \to \mathfrak{u}(N)$. Dimensional reduction to $p+1$ dimension is a requirement that these functions depend only on the first $p+1$ coordinates x^1 $, \ldots, x^{p+1}$ but not on the others. $A_\mu(\mu = 1, \ldots, p+1)$ still have a meaning as connections on $\mathbb{R}^{p+1}$, while $A_\mu(\mu = p+2, \ldots, d)$ do not. The latter being denoted by X_μ instead, the action (7) becomes

$$\begin{aligned} S &= \frac{1}{4g^2}\int_{\mathbb{R}^{p+1}} d^{p+1}x \left\{ ||F_A||^2 + \sum_{\mu=p+2}^{d} ||d_A X^\mu||^2 \right. \\ &\qquad \left. + \sum_{\mu,\nu=p+2}^{d} \mathrm{Tr}\,[X^\mu, X^\nu]^2 + \cdots \right\}, \end{aligned} \tag{8}$$

where $d_A X^\mu = \sum_{\alpha=1}^{p+1} dx^\alpha(\partial_\alpha + [A_\alpha, X^\mu])$ is the covariant derivative. Thus, the world volume theory on $p+1$ brane is $(p+1)$-dimensional Yang–Mills theory coupled with $d-1-p$ adjoint Higgs field. The properties (ii), (iii) are nothing but the usual mass formula when Higgs fields have vacuum expectation value (4) and the $U(N)$ gauge group is broken down to $U(1)^N$.

The S given in (8) is the world volume action of D p-brane. The matrix model action (2) is nothing but the zero dimensional reduction [13] of d-dimensional Yang–Mills theory, and revived in [14]. The gauge connection becomes just d $N \times N$ matrices. In the above nomenclature, the D-branes in the matrix model (2) is called D (-1)-branes, or D-instantons.

2. D-branes on curved space

Matrix models are expected to capture non-perturbative aspects of string theory; they should reproduce the Einstein theory in the classical limit.

Gravity means space-time can be curved. Einstein's theory is formulated on any Riemannian manifold. The matrix models, however, appear to be in special affinity with flat space-time; the matrix dynamics has a natural geometrical interpretation in terms of strings and D-branes in *Euclidean* space, as we saw in section 1. Thus, first question is:

- How can we formulate matrix models on curved manifolds?

One goal of this article is to list up the minimal set of axioms which all physically sensible matrix theories should satisfy. This is first proposed by Douglas [1] under the name of D-geometry, and further studied in [2]. In the next section, we will describe what it is.

Suppose such a formulation exists. If a matrix model actually defines a consistent theory including gravity, it should be possible to derive vacuum Einstein equation $R_{\mu\nu} = 0$. In the usual formulation of perturbative string theory, this follows from the world sheet conformal invariance. Matrix models seem to require independent argument to derive Ricci flatness. Thus our second question is:

- Where is 'Einstein's equation of motion' in matrix models?

This question is discussed in section 5

Before studying multiple D-brane dynamics, let us summarize what is known for the single D-branes world volume action. Consider a D p-brane is embedded into a Riemannian manifold $\mathcal{M}$ with metric g. By choosing a local coordinate on $\mathcal{M}$, this embedding is specified by the map

$$X : \mathbb{R}^{p+1} \to \mathcal{M}.$$

The world volume action S is a functional of X, which determine the quantum measure of such configuration. Boundary CFT technique has been useful to extract the effective world volume action in the slowly varying background [15], [16],[17] . The result is given by the (supersymmetrized) Nambu–Born–Infeld action

$$S_{\text{DBI}} = T^{(p)} \int_{\mathbb{R}^{p+1}} d^{p+1}x \, \sqrt{\det(G+F)} + \cdots . \tag{9}$$

Here T^p is the p-brane tension, G and F are the pullback of the metric and the $U(1)$ gauge field, respectively, regarded as $(p+1) \times (p+1)$ matrix. We have neglected fermionic and topological terms which do not play important rôle in the following discussion.

At low energies, the action (9) reduces to two decoupled sectors, supersymmetric nonlinear σ-model and $U(1)$ gauge theory. The Lagrangian density reads

$$\begin{aligned} \mathcal{L} &= \int d^4\theta \, K_1(z,\bar{z}) + \text{Re} \int d^2\theta \, W^\alpha W_\alpha \\ &= g_{\mu\nu}(X)\partial X^\mu \partial X^\nu + \frac{1}{4}\text{Tr}\, F^2 + \cdots , \end{aligned}$$

where terms with fermions are suppressed. In particular, if the ambient space is flat $\mathbb{R}^d$, this reduces to the standard p-brane action (8).

The arguments given above also apply to the transverse dynamics of D p-branes, i.e., to the cases where the world volume of p brane is flat, but the transverse dimension is a general Kähler manifold $\mathcal{M}$.‡ Thus each D-brane is embedded in $\mathcal{M} \times \mathbb{R}^{p+1}$ as $\{\text{point}\} \times \mathbb{R}^{p+1}$ (see fig 2).

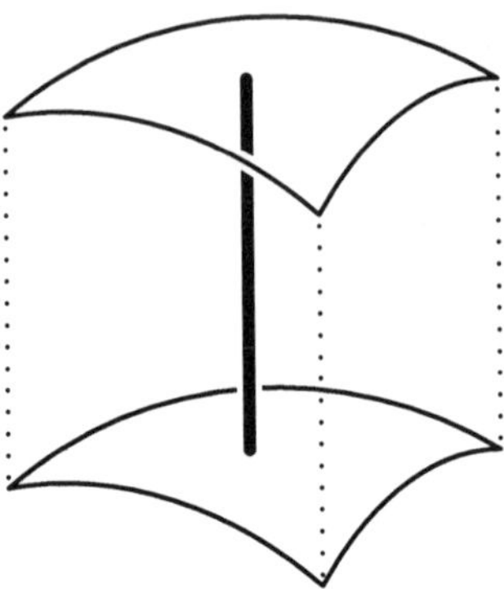

Figure 2 A D-brane sitting at a point on the curved manifold $\mathcal{M}$. The world volume of the brane (thick line) is flat while the transverse directions are curved.

A salient feature of D-brane dynamics is that coincident D-branes have extra degrees of freedom. As we saw in section 1, these come from the off-diagonal components of matrix fields. In order to clarify this and to get an insight into 'D-branes on curved manifolds', it is useful to compare the D 0-brane (D-particle) dynamics with that of the usual second quantized particles.

A quantum mechanical system is specified by the pair $(\mathcal{H}, H)$, a Hilbert space $\mathcal{H}$ and a Hamiltonian H. Let $\mathcal{M}$ be a smooth d dimensional Riemannian manifold. Quantum mechanics of a single particle moving on $\mathcal{M}$ is encoded into the following data:

$$\left(\mathcal{H}_{1\text{particle}} = L^2(\mathcal{M}),\ H = \triangle\right)$$

where, $\triangle$ is the Laplacian on $\mathcal{M}$. It is straightforward to generalize this to the system of N identical particles. We just need to replace the above data by

$$\left(\mathcal{H}_{N\text{particles}} = (L^2(\mathcal{M}) \otimes \cdots \otimes L^2(\mathcal{M}))^{S_N},\ H_{N\text{particles}} = \sum_{i=1}^{N} H_i\right),$$

where $(\cdots)^{S_N}$ is the invariant part under the permutations of particles and $H_i = 1 \otimes \cdots \otimes H \otimes \cdots \otimes 1$. Through the identification $(L^2(\mathcal{M}) \otimes \cdots \otimes L^2(\mathcal{M}))^{S_N} \simeq L^2((\mathcal{M} \times \cdots \times \mathcal{M})/S_N)$, the N particle problem is essentially equivalent to the one body problem on an orbifold

$$\mathrm{Sym}^N \mathcal{M} = (\mathcal{M} \times \cdots \times \mathcal{M})\,/S_N.$$

‡The curved world volume is also very interesting, but we need a supergravity theory on the p-brane world volume.

The degrees of freedom of the system is given by $\dim \mathrm{Sym}^N \mathcal{M} = Nd$.

The dynamics of a *single* D-particle on $\mathcal{M}$ is the same as that of a single particle:

$$\left(\mathcal{H}_{1\text{D-particle}} = L^2(\mathcal{M}),\ H = \triangle\right).$$

When multiple D-particles are put on $\mathcal{M}$, the situation is different. We inevitably have open string degrees of freedom among D-particles. The previous argument suggests that we will need a 'matrix analogue' of $\mathcal{M}$ which would look like (at least locally)

$$\mathcal{X}_N(\mathcal{M}) := \text{“}\mathrm{Mat}_N(\mathcal{M})\text{”} = \left\{ X = \begin{pmatrix} \mathcal{M} & \cdots & \mathcal{M} \\ \vdots & & \vdots \\ \mathcal{M} & \cdots & \mathcal{M} \end{pmatrix} \right\}.$$

The dimension of this space is $\dim \mathcal{X}_N(\mathcal{M}) = N^2 d$, much larger than the ordinary particle case, $\dim \mathrm{Sym}^N \mathcal{M} = Nd$. Moreover, instead of discrete the symmetry group S_N of permutations, we need to consider $U(N)$ gauge action $X \mapsto U^{-1} X U$.

Thus to describe quantum dynamics of multiple D-particles, we will need to prepare a system

$$\left(L^2(\mathcal{X}_N(\mathcal{M})/U(N)),\ H = ?\right)$$

Like $\mathrm{Sym}^N \mathcal{M}$, the configuration space $\mathcal{X}_N(\mathcal{M})/U(N)$ has singularities, but with more delicate structure. According to the standard recipe of quantum field theory, such quotient spaces are best described by introducing gauge fields A coupled with X and quantizing the whole system. This is equivalent to replacing $L^2(\mathcal{X}_N(\mathcal{M})/U(N))$ with $L^2(\mathcal{X}_N(\mathcal{M}))^{U(N)}$. Of course, we need to specify an appropriate Hamiltonian H which correctly describe the open string dynamics as well. This is the main subject of our work.

In summary, to describe multiple D p-brane dynamics a natural starting point seems to be a $U(N)$ gauged nonlinear σ model in $p+1$ dimensions with target space $\mathcal{X}_N(\mathcal{M})$.

We have stressed the difference between ordinary and D-particle dynamics. But we also expect a certain 'correspondence principle' should hold: when the stringy effects are turned off, D-brane dynamics should tend to that of the ordinary particles.

Few remarks are in order. First, the action we will construct should be regarded as an 'effective' matrix theory for N D-branes. The curvature of the background manifold $\mathcal{M}$ is treated as a given classical data. In the ultimate formulation of the Matrix conjecture, background would be determined by matrix dynamics itself. Second, we will be interested only in the 'local' geometry of $\mathcal{M}$, such as metric and Riemann curvature. The effect of wrapped

D-branes or strings around nontrivial cycles of $\mathcal{M}$ is of course very interesting, but is not pursued here.

3. D-geometry axioms

Now let us formulate the most general form of the effective action which governs the many-body dynamics of D-branes on general curved background. It must correctly reproduce physics such as gauge symmetry enhancement, BPS mass spectra etc.

Supersymmetry. We want to make a quantum theory whose classical moduli space is given by the configuration space of N points on $\mathcal{M}$. In general, classical moduli spaces do not survive quantum fluctuation. For example, if we take a bosonic nonlinear σ model, bosonic open strings between the branes exert purely attractive force among D-branes: D-branes coalesce, and the moduli space will shrink to a point.

Inclusion of supersymmetry is the most natural way to avoid such instability and to protect the classical moduli space. In particular, one can expect cancellation of force from bosonic and fermionic contribution for BPS states. For the general story of supersymmetries, the reader should consult [18],[19].

Supersymmetric nonlinear σ models are classified by the number $\mathcal{N}$ of supersymmetries. Possible values are $\mathcal{N} = 1, 2, 4$ and 8. Depending on the number of supersymmetry, the geometric structure of the target space is restricted.

$\mathcal{N} = 8$ theory can be formulated only when $\mathcal{M}$ is hyperkähler. Its maximal world volume dimension is 6, i.e. five-branes. This case is interesting in its own right because five-branes are dual to strings in ten dimensions. But a hyperkähler manifold is automatically Ricci flat. Moreover, the $\mathcal{N} = 8$ theory in six dimensions is chiral, which prohibits covariant Lagrangian formulation of the world volume theory. Thus we will not consider this case. $\mathcal{N} = 4$ theory can be formulated on general Kähler manifold $\mathcal{M}$; the maximal world volume dimension is four, i.e., D 3-branes. By dimensional reduction, it can describe p-branes ($p \leq 3$) as well. In the following we shall specialize to $\mathcal{N} = 4$ supersymmetric models.

In principle, we can reduce the number of supersymmetry down to $\mathcal{N} = 2$, general Riemannian manifold. But less supersymmetry means less control over the form of world volume Lagrangian and, naturally, more complexity. This is an interesting problem for future research.

$\mathcal{N} = 4$ nonlinear σ model. A complex manifold $\mathcal{X}$ with Hermitian metric g is called kähler manifold if the fundamental two form $\omega(x, y) := g(Jx, y)$ is closed. Let $(z^1, \ldots, z^n)$ be a local complex coordinate, and express the metric as

$$g = \sum_{i,j=1}^{n} g_{i\bar{j}}(z,\bar{z}) dz^i d\bar{z}^{\bar{j}}.$$

On a Kähler manifold there exists locally a real valued function $K(z,\bar{z})$ whose second derivative gives the metric:

$$g_{i\bar{j}}(z,\bar{z}) = \frac{\partial^2}{\partial z^i \partial \bar{z}^{\bar{j}}} K(z,\bar{z}).$$

$K(z,\bar{z})$ is called the Kähler potential. For example, the standard flat metric on $\mathbb{C}^n$ corresponds to $K(z,\bar{z}) = \sum_\mu |z^\mu|^2$.

$\mathcal{N}=4$ nonlinear σ models are best described in superfield formalism, i.e., as a geometry on the superspace $\mathbb{R}^{4|4}$. Let $(u^i;\theta_\alpha)$ be the standard even|odd coordinates on $\mathbb{R}^{4|4}$. The chiral field z is a map from superspace $\mathbb{R}^{4|4}$ to the Kähler manifold

$$z : \mathbb{R}^{4|4} \to \mathcal{X}$$

subject to a chiral condition $\bar{D}_\alpha z^i = 0$. Choosing a local complex coordinate $(z^1,\ldots,z^n)$ on $\mathcal{X}$, we can regard chiral superfields as the functions $z^i : \mathbb{R}^{4|4} \to \mathbb{C}$, $(i=1,\ldots,n)$.

In terms of chiral superfields and a Kähler potential, the action of $\mathcal{N}=4$ nonlinear σ model is neatly expressed as

$$S = \int_{\mathbb{R}^{4|4}} d^4u \, d^4\theta \, K(z,\bar{z}).$$

Actually, we need a gauged version of it. The most general from needs the following data. Let G be a compact Lie group acting on $\mathcal{X}$ by isometries. Let Tr be a bi-invariant scalar product in $\mathfrak{g}$. A gauge field $\mathcal{A}$ is a connection on a principal G bundle $\mathcal{P} \to \mathbb{R}^{4|4}$. Then, chiral superfields should be viewed as a chiral section of a bundle over $\mathbb{R}^{4|4}$ with typical fiber $\mathcal{X}$. Finally we can introduce a superpotential W, which is a gauge invariant holomorphic function on $\mathcal{X}$.

Given all these, one can construct a supersymmetric gauged nonlinear σ model. We need only the bosonic parts of the full action:

$$\begin{aligned}\mathcal{L}_{\text{bosonic}} &= \left\{\left(\frac{1}{2g^2}|F_A|^2 + ||d_A z||^2\right) - z^*(2g^2|\mu|^2 + ||\text{grad}W||^2)\right\} d^4u \\ &\quad + \frac{\theta}{16\pi^2}\text{Tr}\,(F_A \wedge F_A)\end{aligned}$$

The last term is the θ term: this does not affect the equation of motion and will be ignored. μ is the moment map for the $U(N)$ action

$$\mu : \mathcal{X} \longrightarrow \mathfrak{u}(N)^*.$$

The potential energy

$$V = 2g^2|\mu|^2 + ||\mathrm{grad}W||^2 \ : \ \mathcal{X} \longrightarrow \mathbb{R}$$

is a sum of contributions from the moment map (D-term) and the superpotential (F-term). As expected in a supersymmetric theory, V is non-negative. The classical moduli space is defined to be $V^{-1}(0)/G$. From the above structure of V, it is given by $\mu^{-1}(0) \cap \mathrm{Crit}(W)$.

D-geometry axioms. Now we come back to the multiple D-brane systems.

Let $\mathcal{M}$ be a complex d dimensional Kähler manifold. Choose a complex analytic local coordinate $(z^1, \ldots, z^d)$ and a Kähler potential $K_1(z, \bar{z})$. As we argued in the previous section, chiral superfields (σ model coordinates) are promoted to $N \times N$ complex valued matrices[§]

$$z^\mu \in \mathcal{M} \dashrightarrow Z^\mu = \begin{pmatrix} Z^\mu_{11} & \cdots & Z^\mu_{1N} \\ \vdots & & \vdots \\ \vdots & & \vdots \\ Z^\mu_{N1} & \cdots & Z^\mu_{NN} \end{pmatrix} \in \mathcal{X}_N(\mathcal{M})$$

which are regarded as the local coordinates on a new target space $\mathcal{X}_N(\mathcal{M})$ with $U(N)$ isometries $Z^\mu \to U^{-1} Z^\mu U$. As usual, the complex eigenvalues of $\{Z^\mu\}$ are interpreted as the positions of D-branes. Our problem is to find an appropriate σ model action which describes N D-branes sitting on $\mathcal{M}$. Let us state the requirements more precisely.

For a d-dimensional Kähler manifold $\mathcal{M}$ with Kähler potential K_1, the problem will be to find $U(N)$ gauged nonlinear σ-model with $\mathcal{N} = 4$ supercharges satisfying the axioms below. The low energy effective action will be determined by a configuration space $\mathcal{X}_N(\mathcal{M})$, a $N^2 d$-dimensional Kähler manifold with potential K_N, action $U(N) \curvearrowright \mathcal{X}_N(\mathcal{M})$ by holomorphic isometries, and a superpotential $W_N : \mathcal{X}_N(\mathcal{M}) \to \mathbb{C}$.

DG1 The classical moduli space (vacuum) is isomorphic to the symmetric product $\mathcal{M}^N/S_N$ (the set of unordered N points on $\mathcal{M}$).

DG2 The generic unbroken gauge symmetry is $U(1)^N$, while if two branes coincide the unbroken symmetry is $U(2) \times U(1)^{N-2}$, and so on.

DG3 Given two non-coincident branes at points $P_i \neq P_j$, all states charged under $U(1)_i \times U(1)_j$ have mass $m_{ij} = d(P_i, P_j)$, the distance along the shortest geodesic between the two points.

[§] Just as real and imaginary parts of complex coordinate gives two real coordinates $z = x + \sqrt{-1}y$, matrix chiral field $Z = X + \sqrt{-1}Y$ is a combination of two Hermitian matrices X and Y.

DG4 The action is a single trace (in terms of matrix coordinates),

$$S = \mathrm{Tr}\,(\cdots).$$

In particular, the action has no explicit N dependence.

Here some remarks are in order. DG1 implies the background manifold $\mathcal{M}$ must be reproduced for $N = 1$, single D-brane case. For $N > 1$, the dimension $N^2 d$ of the target space $\mathcal{X}_N(\mathcal{M})$ exceeds the dimension Nd of classical moduli space.

Axiom DG4 looks somewhat technical. This condition follows from the factorization property: when D-branes are moved very far from each other, open strings become extremely heavy. We should be able to ignore such objects. In matrix terms, off-diagonal components can be neglected and the eigenvalues $\{z_i\}$ becomes the only relevant degrees of freedom. In this limit the action should tend to the decoupled form

$$S(Z,\bar{Z}) \longrightarrow \sum_{i=1}^{N} S(z_i,\bar{z}_i).$$

This is true if and only if the condition DG4 is satisfied.

Let us see more explicitly how D-geometry axiom restricts the form of effective actions. For simplicity, we consider dimensional reduction to $(1+0)$ dimension where gauge sector is trivial. From the supersymmetry and axiom DG4, the most general $U(N)$ invariant Lagrangian is

$$\mathcal{L}_N = \int d^4\theta \,\mathrm{Tr}\,\mathcal{K}_N(Z,\bar{Z}) \;+ \mathrm{Re}\int d^2\theta\,\mathrm{Tr}\,\mathcal{W}_N(Z)$$

Here we use the notation $\bar{Z}^\mu := (Z^\mu)^\dagger$. $\mathcal{K}_N(Z,\bar{Z})$ is a polynomial (power series) in Z^μ and $\bar{Z}^\mu$ while $\mathcal{W}_N(Z)$ is a polynomial (power series) in Z^μ only.

As discussed before, the classical moduli space is given by the common locus of two types of constraints:

$$D\text{-term constraint}: \qquad \sum_\mu\,[Z^\mu, \frac{\partial}{\partial Z^\mu}\mathrm{Tr}\,\mathcal{K}_N] = 0,$$

$$F\text{-term constraint}: \qquad \frac{\partial}{\partial Z^\mu}\,\mathrm{Tr}\,\mathcal{W}_N = 0.$$

Here $\partial f/\partial Z^\mu$ stands for the matrix-valued gradient

$$\left(\frac{\partial f}{\partial Z}\right)_{ij} := \frac{\partial f}{\partial Z_{ij}}.$$

The first nontrivial check will be whether the dimension of the classical moduli space is correctly reproduced or not.

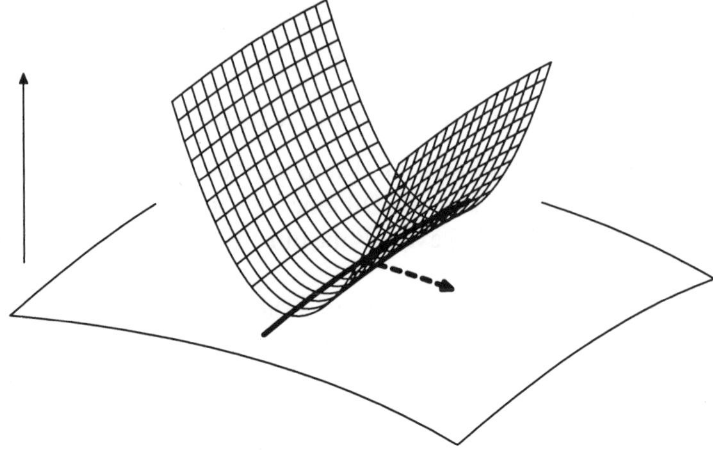

Figure 3 Potential energy V of the nonlinear σ model as a function on $\mathcal{X}_N(\mathcal{M})$. Classical moduli space is the zero locus of the potential energy (thick curve). The dashed arrow represents an off-diagonal quantum fluctuation Z^{μ}_{ij}.

The most important axiom is DG3, the mass condition for the off-diagonal excitations. On curved space, we need to be careful about the definition of mass. Suppose the Lagrangian is given by

$$\mathcal{L} = g_{\mu\nu}(x)\dot{x}^{\mu}\dot{x}^{\nu} + V(x),$$

and $x = x_0$ is the critical point of $V(x)$. The mass is determined by the ratio of quadratic part of potential energy to the kinetic energy. More precisely, it is expressed by the 'mass matrix'

$$m^2_{\mu\nu}(x_0) = (g(x_0)^{-1/2})_{\mu\rho}\left(\left.\frac{\partial^2 V(x)}{\partial x^{\rho}\partial x^{\sigma}}\right|_{x=x_0}\right)(g(x_0)^{-1/2})_{\sigma\nu}. \tag{10}$$

Having stated the axiom, our task is to find a matrix action satisfying all the axioms. It is difficult to tackle this problem in full generality, we will try to construct such an action perturbatively — treating the curvature effect as a perturbation of the flat metric. For our purpose it is useful to work in the normal coordinate system.

4. Complex dimension one

Let us study the case where branes are moving on a Riemann surface $\mathcal{M}$. Since $\mathcal{M}$ is automatically Kähler, we can, without loss of generality, choose a holomorphic normal coordinate z on a local patch $\mathcal{U} \subset \mathcal{M}$. Kähler potential is given by

$$K_1(z,\bar{z}) = |z|^2 + \sum_{p,q=2}^{\infty} K_1^{(p,q)} z^p \bar{z}^q.$$

First we replace z by $N \times N$ matrix chiral superfield Z. Accordingly, we need to find *non-commutative* polynomial in Z and $\bar{Z}$:

$$\begin{aligned} K_N(Z,\bar{Z}) &= \mathrm{Tr}\,\mathcal{K}_N(Z,\bar{Z}) \\ &= \mathrm{Tr}\,Z\bar{Z} + \sum_{p,q=2}^{\infty} \sum_{w\in W^{(p,q)}} \mathcal{K}^w\, \mathrm{Tr}\,(\cdots Z \cdots \bar{Z} \cdots Z \cdots \bar{Z} \cdots). \end{aligned}$$

Here $W^{(p,q)}$ is the set of words made of p Z's and q $\bar{Z}$'s modulo cyclic permutations. According to the axiom DG1, this K_N must reproduce the metric on $\mathcal{M}$ for a single D-brane, or equivalently, $K_N(z\,\mathbf{1}_N, \bar{z}\,\mathbf{1}_N) = NK_1(z,\bar{z})$. This leads to a sum rule for the coefficients of K_N:

$$K_1^{(p,q)} = \sum_{w\in W^{(p,q)}} \mathcal{K}^w. \tag{11}$$

As we will see, it is highly nontrivial how to distribute $K_1^{(p,q)}$ into $\mathcal{K}^w$'s, each corresponding to different ordering of Z and $\bar{Z}$.

Before discussing this problem consider the condition on the superpotential $\mathcal{W}_N$. It turns out that:

Proposition 1 *We cannot introduce nontrivial superpotential.*

The reason is very simple. If we had a nontrivial superpotential $\mathcal{W}_N(Z)$, F-flatness condition $\partial \mathrm{Tr}\,\mathcal{W}_N(Z)/\partial Z = 0$ would put N^2 constraints on N^2 variables $\{Z_{ij}\}$, thus we would get only isolated solutions; D-branes could be located only at particular places, and we cannot have the classical moduli space $\mathrm{Sym}^N\mathcal{M}$.

Since there is no superpotential the potential energy entirely comes from D-terms. Written out explicitly,

$$V = \mathrm{Tr}\,D^2, \qquad D = \left[Z, \frac{\partial K_N}{\partial Z}\right] = -\left[\bar{Z}, \frac{\partial K_N}{\partial \bar{Z}}\right].$$

We are now in a position to discuss axiom DG3. This is concerned with the mass matrix (see eq. (10)) at the classical background where Z takes diagonal form

$$\langle Z \rangle = Z_0 = \begin{pmatrix} z_1 & & \\ & \ddots & \\ & & z_N \end{pmatrix}. \tag{12}$$

Eigenvalues $z_1, \ldots, z_N$ are, of course, the positions of D-branes. The mass matrix for the quantum fluctuation[¶] Z_{ij} is

$$m_{ij}^2 = \frac{\dfrac{\partial^2 V}{\partial Z_{ij} \partial \bar{Z}_{ji}}}{\dfrac{\partial^2 K_N}{\partial Z_{ij} \partial \bar{Z}_{ji}}} = \frac{\dfrac{\partial D}{\partial Z_{ij}} \dfrac{\partial D}{\partial \bar{Z}_{ji}}}{\dfrac{\partial^2 K_N}{\partial Z_{ij} \partial \bar{Z}_{ji}}} . \tag{13}$$

In the last equality we have used $D = 0$ on the classical moduli space. For arbitrary K_N one can show

$$\left. \frac{\partial D}{\partial Z_{ij}} \right|_{Z_0} = (z_j - z_i) \left. \frac{\partial^2 K_N}{\partial Z_{ij} \partial \bar{Z}_{ji}} \right|_{Z_0} .$$

Thus (13) can be further simplified to

$$m_{ij}^2(Z_0) = |z_i - z_j|^2 \left. \frac{\partial^2 K_N}{\partial Z_{ij} \partial \bar{Z}_{ji}} \right|_{Z_0} . \tag{14}$$

The axiom DG3 demands that this is given by the geodesic distance between the branes:

$$m_{ij}^2(Z_0) = d^2(z_i, z_j), \qquad (i, j = 1, 2, \ldots, N). \tag{15}$$

Note that the left hand side is determined from the 'matrix' Kähler potential $K_N(Z, \bar{Z})$ while the right hand side is completely fixed by the original Kähler potential $K_1(z, \bar{z})$.

Let us see how the ordering of noncommutative polynomial $\mathcal{K}_N(Z, \bar{Z})$ affects the mass matrix. Consider, for example, two possible orderings $ZZ\bar{Z}\bar{Z}$ and $Z\bar{Z}Z\bar{Z}$ at $(p, q) = (2, 2)$. They would give different contribution to mass matrix (14):

$$K_N = \mathrm{Tr}\, ZZ\bar{Z}\bar{Z} \quad \longrightarrow \quad \left. \frac{\partial^2 K_N}{\partial Z_{ij} \partial \bar{Z}_{ji}} \right|_{Z_0} = (z_i + z_j)(\bar{z}_i + \bar{z}_j),$$

$$K_N = \mathrm{Tr}\, Z\bar{Z}Z\bar{Z} \quad \longrightarrow \quad \left. \frac{\partial^2 K_N}{\partial Z_{ij} \partial \bar{Z}_{ji}} \right|_{Z_0} = 2(z_i \bar{z}_j + z_j \bar{z}_i).$$

Conversely, by carefully analyzing the geodesic distance $d^2(z_i, z_j)$ as a function of z_i and z_j, it is possible to some extent to determine the noncommutative polynomial $\mathcal{K}_N(Z, \bar{Z})$ from the 'commutative' or 'diagonal' data $K_1(z, \bar{z})$.

[¶] It might be better to denote quantum fluctuation by $\widetilde{Z}$. Since we choose the vacuum $\langle Z_{ij} \rangle = 0$ $(i \neq j)$, we drop '~' to avoid heavy notations.

Note that the D-geometry axiom will only determine $K_N(Z,\bar{Z})$ up to terms involving two commutators. This is because the mass condition needs only the second derivative of $K_N(Z,\bar{Z})$ evaluated at the diagonal Z_0. For example, consider a term with triple commutators:

$$\Delta K = \mathrm{Tr}\, f_1(Z,\bar{Z})[Z,\bar{Z}]f_2(Z,\bar{Z})[Z,\bar{Z}]f_3(Z,\bar{Z})[Z,\bar{Z}].$$

Its second derivative vanishes on the diagonal vacuum:

$$\left.\frac{\partial^2 \Delta K}{\partial Z_{ij}\partial \bar{Z}_{ji}}\right|_{Z_0} = 0.$$

The question becomes 'Can we always find a suitable ordering prescription for $\mathcal{K}_N$?' The answer is yes.

Theorem 1 (Douglas [1]) *For any Kähler potential $K_1(z,\bar{z})$, one can find $K_N(Z,\bar{Z})$ which satisfies the mass condition (15). This is unique up to higher commutator terms.*

The idea of the proof is as follows. We want to show that there exists an ordering for K_N which will reproduce any desired $\delta\bar{\delta}K_N$. It is sufficient to consider an expansion with terms

$$K_N = \mathrm{Tr}\, K_1(Z,\bar{Z}) + \sum_a \mathrm{Tr}\, f_a(Z,\bar{Z})[Z,\bar{Z}]\, g_a(Z,\bar{Z})[Z,\bar{Z}], \tag{16}$$

where we take the convention that functions of two variables are ordered with all Z's before all $\bar{Z}$'s, for example

$$K_1(Z,\bar{Z}) = \sum_{p,q} K_1^{(p,q)} Z^p \bar{Z}^q.$$

The ansatz (16) automatically satisfies the sum rule (11). Putting (16) into (15) and expanding as a power series in $\{z_i,\bar{z}_i\}_{i=1}^N$, we can recursively determine the polynomials f_a and g_a.

5. Complex dimension three

Let us consider a more 'realistic' situation where D-branes are moving on a higher dimensional Kähler manifold. In the context of superstring compactification, the most interesting case is when D 3-branes sit at points on $\mathcal{M}$ with $\dim_{\mathbb{C}} = 3$. Let us study this model.

New features. In contrast to the case $\dim_{\mathbb{C}} \mathcal{M} = 1$ there are some new features. First, of course, we have to tackle with a plethora of non-commutativity; an ordering among the holomorphic coordinates matters, too. More importantly, we need a non-trivial superpotential $\mathcal{W}_N(Z)$ to correctly reproduce the classical moduli space. To see this, let us estimate its dimensions assuming we had only D-term potential V_D:

$$\begin{aligned}
&\dim_{\mathbb{C}} (\{Z^\mu | V_D = 0\}/\sim) \\
&\quad = \text{(\# of matrix elements)} \\
&\qquad -(D\text{-term const. \& } U(N) \text{ gauge equiv.}) \\
&\quad = dN^2 - (N^2 - N) \\
&\quad = (d-1)N^2 + N.
\end{aligned}$$

This grows quadratically with N while the expected dimension is Nd. We see that $\dim_C \mathcal{M} = 1$ is the only dimension where no superpotential is needed.

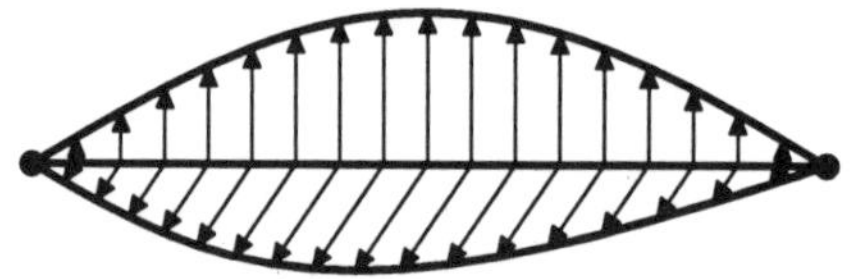

Figure 4 Two polarization of string stretched between branes (blobs).

Finally, the open strings stretched between the branes can oscillate in more than one directions — they can have different polarizations. Axiom DG3 demands, in particular:

- The mass of a stretched open string is independent of its polarization.

This will be referred to as *isotropic mass condition.* At first sight this condition appears to be too restrictive. But a little thought reveals that there is no invariant way to write down a mass splitting formula to the leading order in off-diagonal excitations.

Gauge boson mass and Higgs mass. Now let us study the mass of various modes on the p-brane world volume. Relevant fields are off-diagonal parts of the gauge fields (W-bosons) A^μ_{ij} and chiral fields (Higgs bosons) Z^μ_{ij}. They have different dependence on the Kähler potential $K_N(Z, \bar{Z})$ and the superpotential $W_N(Z) = \mathrm{Tr}\, \mathcal{W}_N(Z)$. Nevertheless, axiom DG3 demands that both masses agree and are proportional to the geodesic distance between the branes.

In component fields, the mass term for the $U(N)$ gauge boson is

$$\frac{\partial^2 K_N}{\partial Z^i_{ab} \partial \bar{Z}^{\bar{j}}_{cd}} [A_\mu, Z^i]_{ab} [A^\mu, \bar{Z}^{\bar{j}}]_{cd}.$$

Expanding about the diagonal matrices $Z^i = Z_0^i = \mathrm{diag}(z_1^i, \ldots, z_N^i)$, we obtain the *gauge boson mass condition*

$$\frac{\partial^2 K_N}{\partial Z_{12}^i \partial \bar{Z}_{21}^{\bar{j}}} (z_1 - z_2)^i (\bar{z}_1 - \bar{z}_2)^{\bar{j}} \equiv d(z_1, z_2)^2. \tag{17}$$

The computation and the result is almost the same as in $\dim_{\mathbb{C}} \mathcal{M} = 1$ (cf. eq. (14), (15)). The gauge boson mass condition is independent of the superpotential W_N.

The mass of Higgs fields Z_{ij}^μ, on the other hand, depends on both $\mathcal{K}_N$ and $\mathcal{W}_N$. Recall that the potential energy is given as the sum of so called 'D-term' and 'F-term' contribution

$$V(Z, \bar{Z}) = V_D + V_F.$$

Both terms are positive definite:

$$V_D = -\frac{1}{2} \sum_{i,\bar{j}} \mathrm{Tr} \left[Z^i, \frac{\partial K_N}{\partial Z^i} \right] \left[\bar{Z}^{\bar{j}}, \frac{\partial K_N}{\partial \bar{Z}^{\bar{j}}} \right], \tag{18}$$

$$V_F = \left(\frac{\partial^2 K_N}{\partial Z_{ab}^i \partial \bar{Z}_{ba}^{\bar{j}}} \right)^{-1} \frac{\partial W_N}{\partial Z_{ab}^i} \frac{\partial \bar{W}_N}{\partial \bar{Z}_{ba}^{\bar{j}}}. \tag{19}$$

Using the shorthand notation $\delta_i = \partial / \partial Z_{12}^i$, $\bar{\delta}_{\bar{j}} = \partial / \partial \bar{Z}_{21}^{\bar{j}}$, the isotropic mass condition is expressed as

$$\delta_i \bar{\delta}_{\bar{j}} V = \delta_i \bar{\delta}_{\bar{j}} V_D + \delta_i \bar{\delta}_{\bar{j}} V_F \equiv d^2(z_1, z_2) \delta_i \bar{\delta}_{\bar{j}} \mathrm{Tr}\, K_N \tag{20}$$

If we denote the second derivatives as

$$\begin{aligned} \hat{g}_{i\bar{j}}(z_1, \bar{z}_1, z_2, \bar{z}_2) &\equiv \delta_i \bar{\delta}_{\bar{j}} \, \mathrm{Tr}\, \mathcal{K}_N(Z, \bar{Z}) \Big|_{Z_0}, \\ \hat{\Omega}_{ijk}(z_1, z_2)(z_1 - z_2)^k &\equiv \delta_i \bar{\delta}_{\bar{j}} \, \mathrm{Tr}\, \mathcal{W}_N(Z) \Big|_{Z_0}, \end{aligned}$$

the gauge boson mass condition (17) reads

$$\hat{g}_{i\bar{j}} (z_1 - z_2)^i (\bar{z}_1 - \bar{z}_2)^{\bar{j}} \equiv d(z_1, z_2)^2, \tag{21}$$

whereas the isotropic mass condition (20) becomes

$$\begin{aligned} &\hat{g}_{i\bar{n}} \, (\bar{z}_1^{\bar{n}} - \bar{z}_2^{\bar{n}}) \, \hat{g}_{m\bar{j}} \, (z_1^m - z_2^m) + \, \hat{g}^{k\bar{l}} \, \hat{\Omega}_{ikm} \, (z_1^m - z_2^m) \, \hat{\Omega}_{\bar{j}\bar{l}\bar{n}} \, (\bar{z}_1^{\bar{n}} - \bar{z}_2^{\bar{n}}) \\ &\quad \equiv \hat{g}_{i\bar{j}} \, \hat{g}_{m\bar{n}} \, (z_1^m - z_2^m) \, (\bar{z}_1^{\bar{n}} - \bar{z}_2^{\bar{n}}). \end{aligned} \tag{22}$$

Constraints from D-geometry axioms. Functional equations (21), (22) together with the holomorphy of $\hat{\Omega}_{ijk}$ tightly constrain the form of $\hat{g}_{i\bar{j}}(z_1, \bar{z}_1, z_2, \bar{z}_2)$ and $\hat{\Omega}_{ijk}(z_1, z_2)$. One can show following statements [2].

Proposition 2 *In a holomorphic normal coordinate system, (21) and (22) are satisfied if and only if*

$$\det \hat{g}_{i\bar{j}}(z_1, \bar{z}_1, z_2, \bar{z}_2) \equiv 1,$$

and

$$\hat{\Omega}_{ijk}(z_1, z_2)(z_1 - z_2)^k \equiv \epsilon_{ijk}(z_1 - z_2)^k.$$

This constrains the background geometry:

Proposition 3 *$\mathcal{M}$ must be Ricci flat.*

Proposition 3 is a corollary of Proposition 2. From (11) we can show that the original Kähler metric $g_{i\bar{j}}(z, \bar{z})$ is recovered from $\hat{g}_{i\bar{j}}(z_1, \bar{z}_1, z_2, \bar{z}_2)$ as follows:

$$\lim_{z_1 \to z_2} \hat{g}_{i\bar{j}}(z_1, \bar{z}_1, z_2, \bar{z}_2) = g_{i\bar{j}}(z_2, \bar{z}_2).$$

Therefore, combined with Proposition 2, the Ricci tensor on $\mathcal{M}$ can be computed as

$$\begin{aligned} R_{i\bar{j}} &= \frac{\partial^2}{\partial z^i \partial \bar{z}^{\bar{j}}} \log \det g_{m\bar{n}}(z, \bar{z}) \\ &= \frac{\partial^2}{\partial z^i \partial \bar{z}^{\bar{j}}} \log \det \lim_{w \to z} \hat{g}_{m\bar{n}}(w, \bar{w}, z, \bar{z}) = 0. \end{aligned}$$

The superpotential is also fixed by the axioms:

Proposition 4 *In a holomorphic normal coordinate system the superpotential is given by*

$$\operatorname{Tr} \mathcal{W}_N(Z) = \operatorname{Tr} Z^1[Z^2, Z^3] = \tfrac{1}{3}\epsilon_{ijk} \operatorname{Tr} Z^i Z^j Z^k \tag{23}$$

to all orders in diagonal coordinates and to the second order in off-diagonal coordinates.

Existence of D-brane action. We have seen that D-geometry axioms are satisfied only if the background geometry is Ricci flat and the superpotential has a special form (23). A natural question arises: 'Can we actually find $K_N(Z, \bar{Z})$ which satisfy all these properties?'

We have no general answer to this, but we conjecture that it is true. As a support for this conjecture let us construct $K_N(Z, \bar{Z})$ order by order in Z and $\bar{Z}$.

Suppose we are given a Kähler potential to quartic order in a normal coordinate patch $\mathcal{U} \subset \mathcal{M}$:

$$K_1(z, \bar{z}) = z^i \bar{z}^i - \tfrac{1}{4} R_{i\bar{j}k\bar{l}}\, z^i \bar{z}^{\bar{j}} z^k \bar{z}^{\bar{l}} + \cdots .$$

The coefficients $R_{i\bar{j}k\bar{l}}\, z^i \bar{z}^{\bar{j}}$ are Riemann curvatures at $z = 0$. By definition the metric on $\mathcal{M}$ is given by the second derivative of K_1:

$$g_{i\bar{j}}(z, \bar{z}) = \delta_{i\bar{j}} - R_{i\bar{j}k\bar{l}}\, z^k \bar{z}^{\bar{l}} + \cdots .$$

The geodesic distance between two points $x, z \in \mathcal{U}$ is then given by

$$\begin{aligned} d^2(z, \bar{z}; x, \bar{x}) \;&=\; (z - x)^i (\bar{z} - \bar{x})^i \\ &\quad -\tfrac{1}{4} R_{i\bar{j}k\bar{l}}\,(x - z)^i (\bar{x} - \bar{z})^{\bar{j}} (x + z)^k (\bar{x} + \bar{z})^{\bar{l}} \\ &\quad -\tfrac{1}{12} R_{i\bar{j}k\bar{l}}\,(x - z)^i (\bar{x} - \bar{z})^{\bar{j}} (x - z)^k (\bar{x} - \bar{z})^{\bar{l}} + \cdots . \end{aligned}$$

The quartic terms in $d^2(z, \bar{z}; x, \bar{x})$ are related to quartic terms in $K_N(Z, \bar{Z})$, as we discussed before. The geodesic distance is correctly reproduced to this order by

$$K_N(Z, \bar{Z}) = \mathrm{Tr}\, Z^i \bar{Z}^{\bar{i}} - \tfrac{1}{4} R_{i\bar{j}k\bar{l}}\, \mathrm{STr}\, Z^i \bar{Z}^{\bar{j}} Z^k \bar{Z}^{\bar{l}} + \cdots$$

where STr is the symmetrized trace

$$\mathrm{STr} A_1 A_2 \cdots A_n = \frac{1}{n!} \mathrm{Tr}\, A_{\sigma(1)} A_{\sigma(2)} \cdots A_{\sigma(n)} .$$

Higher order terms can be computed in a similar fashion, though very cumbersome. For the Kähler potential

$$K_1(z, \bar{z}) = |z|^2 + \sum_{p,q \geq 2} K^{(p,q)}_{i_1 \ldots i_p; \bar{j}_1 \ldots \bar{j}_q} z^{i_1} \cdots z^{i_p} \bar{z}^{\bar{j}_1} \cdots \bar{z}^{\bar{j}_q} ,$$

one can check that the following $K_N(Z, \bar{Z})$ satisfies all the requirements (to the sixth order):

$$\begin{aligned} K_N(Z, \bar{Z}) \;&=\; \mathrm{Tr}\, Z\bar{Z} + \sum K^{(p,q)}_{i_1, \ldots, i_p; \bar{j}_1, \ldots, \bar{j}_q} \mathrm{Str}\, Z^{i_1} \cdots Z^{i_p} \bar{Z}^{\bar{j}_1} \cdots \bar{Z}^{\bar{j}_q} \\ &\quad + \frac{1}{1440} R_{i\bar{q}m\bar{n}} \delta^{\bar{q}p} R_{p\bar{l}k\bar{j}} \\ &\quad \times \big(-15\, \mathrm{Tr}\, \{Z^k, [Z^i, \bar{Z}^{\bar{l}}]\}\{\bar{Z}^{\bar{n}}, [Z^m, \bar{Z}^{\bar{j}}]\} \\ &\qquad + \mathrm{Tr}\, [Z^k, [Z^i, \bar{Z}^{\bar{l}}]]\, [\bar{Z}^{\bar{n}}, [Z^m, \bar{Z}^{\bar{j}}]] \\ &\qquad + 15\, \mathrm{Tr}\, \{Z^k, [Z^i, \bar{Z}^{\bar{l}}]\}\{Z^m, [\bar{Z}^{\bar{j}}, \bar{Z}^{\bar{n}}]\} + \text{c.c.} \\ &\qquad - \mathrm{Tr}\, [Z^k, [Z^i, \bar{Z}^{\bar{l}}]]\, [Z^m, [\bar{Z}^{\bar{j}}, \bar{Z}^{\bar{n}}]] + \text{c.c.} \big) . \end{aligned}$$

Note that a naive STr prescription works only up to the quartic order, and that the 'correction terms' are needed when we go to higher order. At the sixth order these correction terms are proportional to the square of the Riemann curvature tensor and contain at least two commutators.

In fact, the form is not unique even if the ambiguity discussed in section 4 is taken into account.

6. Conclusions

We considered actions for N D-branes at points in a general Kähler manifold, which satisfy the axioms of D-geometry, and could be used as starting points for defining matrix theory in curved space.

We showed that the axioms cannot be satisfied unless the metric is Ricci flat, and argue that such actions do exist when the metric is Ricci flat. This may provide an argument for Ricci flatness in matrix theory.

Acknowledgments

It is a pleasure to acknowledge fruitful collaborations on these topics with M. Douglas and H. Ooguri. I am also very grateful to the organizers and the participants of this conference for giving me a opportunity to convey ideas behind D-branes and noncommutative geometry.

This work is supported in part by the Grant-in-Aid for Scientific Research on Priority Area 707 'Supersymmetry and Unified Theory of Elementary Particles', Japan Ministry of Education.

References

[1] M.R. Douglas, D-branes in curved space, *Adv. Theor. Math. Phys.* **1**, 198 (1998) hep-th/9703056.

[2] M.R. Douglas, A. Kato and H. Ooguri, D-brane actions on Kähler manifolds, *Adv. Theor. Math. Phys.* **1**, 237 (1998) hep-th/9708012.

[3] J. Polchinski, TASI lectures on D-branes, hep-th/9611050.

[4] E. Witten, Bound States Of Strings And p-Branes, *Nucl. Phys.* **B460**, 335 (1996) hep-th/9510135.

[5] T. Banks, W. Fischler, S.H. Shenker and L. Susskind, M theory as a matrix model: A conjecture, *Phys. Rev.* **D55**, 5112 (1997) hep-th/9610043.

[6] K. Becker and M. Becker, A two-loop test of M(atrix) theory, *Nucl. Phys.* **B506**, 48 (1997) hep-th/9705091.

[7] K. Becker, M. Becker, J. Polchinski and A. Tseytlin, Higher order graviton scattering in M(atrix) theory, *Phys. Rev.* **D56**, 3174 (1997) hep-th/9706072.

[8] Y. Okawa and T. Yoneya, Multi-body interactions of D-particles in supergravity and matrix theory, *Nucl. Phys.* **B538**, 67 (1999) hep-th/9806108.

[9] For a review, see P. Ginsparg and G. Moore, Lectures on 2-D gravity and 2-D string theory, hep-th/9304011; P. Di Francesco, P. Ginsparg and J. Zinn-Justin, 2-D Gravity and random matrices, *Phys. Rept.* **254**, 1 (1995) hep-th/9306153.

[10] E. Witten, Two-dimensional gravity and intersection theory on moduli space, *Surveys Diff. Geom.* **1** (1991) 243.

[11] M. Kontsevich, Intersection theory on the moduli space of curves and the matrix Airy function, *Comm. Math. Phys.* **147**,1 (1992).

[12] G. Moore, N. Nekrasov and S. Shatashvili, D-particle bound states and generalized instantons, hep-th/9803265.

[13] T. Eguchi and H. Kawai, Reduction Of Dynamical Degrees Of Freedom In The Large N Gauge Theory, *Phys. Rev.* Lett. **48**, 1063 (1982).

[14] N. Ishibashi, H. Kawai, Y. Kitazawa and A. Tsuchiya, A Large-N Reduced Model as Superstring, *Nucl. Phys.* **B498**, 467 (1997) hep-th/9612115.

[15] R.G. Leigh, Dirac-Born-Infeld Action From Dirichlet Sigma Model, *Mod. Phys. Lett.* **A4**, 2767 (1989).

[16] J. Polchinski and Y. Cai, Consistency Of Open Superstring Theories, *Nucl. Phys.* **B296**, 91 (1988).

[17] C.G. Callan, C. Lovelace, C.R. Nappi and S.A. Yost, Loop Corrections To Superstring Equations Of Motion, *Nucl. Phys.* **B308**, 221 (1988).

[18] J. Wess, J. Bagger, *Supersymmetry and Supergravity* (Princeton Series in Physics) 2nd Rev. edition, 1992, Princeton Univ Press.

[19] D.S. Freed, *Five Lectures on Supersymmetry* American Mathematical Society, 1999.

ON THE PROJECTIVE CLASSIFICATION OF THE MODULES OF DIFFERENTIAL OPERATORS ON $\mathbb{R}^m$

Pierre B. A. Lecomte

Institut de Mathématiques, Université de Liège

Sart Tilman, Grande Traverse, 12 (B 37), B–4000 Liège, BELGIUM

plecomte@ulg.ac.be

Abstract The Lie algebra $\mathfrak{sl}_{m+1}$ of infinitesimal projective linear transformations acts via Lie derivatives on the space $\mathcal{D}_{\lambda,\mu}$ of differential operators between densities on $\mathbb{R}^m$ of weights λ and μ. In most of the cases this module is isomorphic to the graded module associated to the filtration by the order of differentiation. This is, in particular, the case when λ equals μ and this leads to a $\mathfrak{sl}_{m+1}$-equivariant quantization. The modules $\mathcal{D}_{\lambda,\mu}$ are more generally classified by two sets of integers. For a given difference $\delta = \mu - \lambda$, there are finitely many isomorphisms classes.

Mathematics Subject Classification (2000): 58B99, 17B56, 17B65

Keywords: Lie algebra modules, vector fields, differential operators, densities, projective structure

1. Position of the problem

Denote by $\mathcal{D}_{\lambda\mu}$ the space of differential operators on $\mathbb{R}^m$ whose arguments are λ-densities and whose values are μ-densities, where λ, μ are given real numbers.

The Lie derivatives acting on densities lift on $\mathcal{D}_{\lambda\mu}$ and turn it into a representation of the Lie algebra $\mathrm{Vect}(\mathbb{R}^m)$ of vector fields of $\mathbb{R}^m$.

Y. Maeda, et al. (eds.), Noncommutative Differential Geometry and Its Applications to Physics, 123–129.

Seeking equivariant quantization leads us to restrict this representation to some preferred Lie subalgebras of $\mathrm{Vect}(\mathbb{R}^m)$, as explained in [1], where the reader will find the motivations and origin of the present paper.

Here we restrict the representation to the linear span $\mathfrak{sl}_{m+1}$ of the vector fields

$$\frac{\partial}{\partial x^i},\ x^j\frac{\partial}{\partial x^i},\ x^i\left(x^1\frac{\partial}{\partial x^1}+\cdots+x^m\frac{\partial}{\partial x^m}\right),\quad i,j\in\{1,\dots,m\}.$$

It is isomorphic to $\mathfrak{sl}(m+1,\mathbb{R})$. It is called the *projective* embedding of $\mathfrak{sl}(m+1,\mathbb{R})$ because it generates the homographies of $\mathbb{R}^m$ [2]. It is a maximal Lie subalgebra of the Lie algebra of vector fields that have polynomial coefficients.

Our purpose is to present and to explain the classification of the $\mathfrak{sl}_{m+1}$-modules $\mathcal{D}_{\lambda\mu}$. We assume $m > 1$. The results have been announced in [3].

2. The main result

Let us describe the two invariants that characterize $\mathcal{D}_{\lambda\mu}$ as a representation of $\mathfrak{sl}(m+1,\mathbb{R})$.

Set $\delta = \mu - \lambda$ and define I_δ to be the set of positive integer k for which there exists $n \in \{1,\dots,k\}$ such that $(m+1)\delta - m = 2k - n$. If $(m+1)\delta - m$ is not a positive integer, then I_δ is empty.

Given a positive integer u, if $(m+1)\delta - m = 2u$, then $I_\delta = \{u+1,\dots,2u\}$ and if $(m+1)\delta - m = 2u-1$, then $I_\delta = \{u,\dots,2u-1\}$. The number δ or, equivalently, the possibly empty interval I_δ is the first invariant.

Define now $I_{\lambda\mu}$ to be the set of $k \in I_\delta$ such that

$$\prod_{k-n\leq i<k}[(m+1)\lambda+i]\neq 0$$

where $n = 2k + m - (m+1)\delta$. As easily seen, if it is not empty $I_{\lambda\mu}$ is a subinterval of I_δ sharing with it the same infimum.

Theorem 2.1. *The $\mathfrak{sl}_{m+1}$-modules $\mathcal{D}_{\lambda\mu}$ and $\mathcal{D}_{\lambda'\mu'}$ are isomorphic if and only if $\delta = \delta'$ and $I_{\lambda\mu} = I_{\lambda'\mu'}$. In particular, the $\mathfrak{sl}_{m+1}$-modules $\mathcal{D}_{\lambda\lambda}$, $\lambda \in \mathbb{R}$, are isomorphic.*

For the particular case, one notes that if $\delta = 0$, then I_δ and $I_{\lambda\mu}$ are empty.

3. The geometrical meaning of the invariants

Let us denote by $\mathcal{D}^k_{\lambda\mu}$ the space of operators $A \in \mathcal{D}_{\lambda\mu}$ of order less or equal to k. Since $\mathcal{D}^{k-1}_{\lambda\mu}$ is a submodule of $\mathcal{D}^k_{\lambda\mu}$, one has a short exact sequence of $\mathfrak{sl}(m+1)$-modules

$$0\to\mathcal{D}^{k-1}_{\lambda\mu}\to\mathcal{D}^k_{\lambda\mu}\to\mathcal{S}_{k,\delta}\to 0 \tag{1}$$

where the quotient is known to be the space of symbols of the operators A of order k. It is the space of symmetric k-contravariant tensor fields on $\mathbb{R}^m$ with coefficients in the δ-densities. Equivalently, it is the space of smooth functions on $T^*\mathbb{R}^m$, which are homogeneous polynomials on each cotangent space and which are valued in the space of δ-densities of $\mathbb{R}^m$. The action of $\mathrm{Vect}(\mathbb{R}^m)$ is the Lie derivative of tensor fields in the first interpretation. It is the usual derivative of function in the direction of the canonical lift of vector fields on $T^*\mathbb{R}^m$ in the second one.

Theorem 3.1. *Let k be a positive integer. Then $k \in I_\delta$ if and only if there exists $n \in \{1, \dots, k\}$ such that*

$$\mathrm{Hom}_{\mathfrak{sl}_{m+1}}(\mathcal{S}_{k,\delta}, \mathcal{S}_{k-n,\delta}) \neq 0.$$

Theorem 3.2. *Let $k \in I_\delta$. Then $k \in I_{\lambda\mu}$ if and only if the sequence (1) of $\mathfrak{sl}_{m+1}$-modules is not split.*

This theorem have been shown in [4].

Assume that $I_{\lambda\mu}$ is empty. Then, by theorem 3.2, for each positive integer k, there is a splitting $\tau_k : \mathcal{S}_{k,\delta} \to \mathcal{D}^k_{\lambda,\mu}$ of the sequence of $\mathfrak{sl}_{m+1}$-modules (1). We get thus a $\mathfrak{sl}_{m+1}$-equivariant isomorphism $\tau : \mathcal{S}_\delta = \oplus_{k\in\mathbb{N}} \mathcal{S}_{k,\delta} \to \mathcal{D}_{\lambda\mu}$ by gluing the τ_k's. Assume that θ is some automorphism of $\mathfrak{sl}_{m+1}$-modules of $\mathcal{D}_{\lambda\mu} \simeq \mathcal{S}_\delta$. By restriction to the homogeneous components $\mathcal{S}_{k,\delta}$, it — or its inverse — induces between these components. If I_δ is empty then the only nontrivial components of θ are these from each $\mathcal{S}_{k,\delta}$ into itself. It can then be easily shown that they are constant multiples of the identity. In conclusion, if I_δ is empty then there is an unique isomorphism of $\mathfrak{sl}_{m+1}$ modules from $\mathcal{D}_{\lambda\mu}$ into $\mathcal{S}_\delta$ that preserves the symbol of the differential operators.

4. The Casimir operator of $\mathcal{D}_{\lambda\mu}$

As vector spaces, $\mathcal{D}_{\lambda\mu}$ and $\mathcal{S}_\delta$ are isomorphic.

For instance, expand each $A \in \mathcal{D}_{\lambda\mu}$ in terms of partial derivatives

$$A = \sum_\alpha A_\alpha\, D^\alpha_x,$$

where $\alpha \in \mathbb{N}^m$ and

$$D^\alpha_x = \left(\frac{\partial}{\partial x^1}\right)^{\alpha^1} \cdots \left(\frac{\partial}{\partial x^m}\right)^{\alpha^m}.$$

Then mapping D^α_x onto the monomial:

$$\xi \to \xi^\alpha = \xi_1^{\alpha^1} \dots \xi_m^{\alpha^m}$$

extends to an isomorphism $\mathcal{D}_{\lambda\mu} \to \mathcal{S}_\delta$ of vector spaces. We will use it from now on to identify $\mathcal{D}_{\lambda\mu}$ and $\mathcal{S}_\delta$ as vector spaces. Of course, the $\mathfrak{sl}(m+1)$-module structures are not the same, although they could be isomorphic. If we want to stress the difference, we will denote L_X^{op} and L_X^{po} the action of $X \in \mathfrak{sl}(m+1)$ on $\mathcal{D}_{\lambda\mu}$ and $\mathcal{S}_\delta$ respectively.

Theorem 4.1. *The Casimir Operator $\mathcal{C}_{\lambda\mu}$ of the $\mathfrak{sl}(m+1)$-module $\mathcal{D}_{\lambda\mu}$ is*

$$\mathcal{C}_{\lambda\mu} = 2E^2 - 2[(m+1)\delta - m]E + m(m+1)\delta(\delta-1) + 2[E + (m+1)\lambda]T$$

where the linear maps $E, T : \mathcal{S}_\delta \to \mathcal{S}_\delta$ are given by

$$E = \sum_i \xi_i \frac{\partial}{\partial \xi_i}, \qquad T = \sum_i \frac{\partial}{\partial x^i} \frac{\partial}{\partial \xi_i}.$$

The proof is a simple matter of computation.

Let us seek for the eigenvalues and the eigenvectors of $\mathcal{C}_{\lambda\mu}$. Let

$$A = A_k + \cdots + A_0, \qquad A_i \in \mathcal{S}_{i,\delta}, \qquad A_k \neq 0,$$

be the decomposition of $A \in \mathcal{S}_\delta$ into homogeneous components. Then A is an eigenvector of $\mathcal{C}_{\lambda\mu}$ of eigenvalues α if and only if

$$\begin{cases} (\alpha_k - \alpha)A_k = 0, \\ (\alpha_i - \alpha)A_i + \beta_i\, T\, A_{i+1} = 0, \quad i = 1, \ldots, k-1, \end{cases} \tag{2}$$

where

$$\alpha_i = 2i^2 - 2[(m+1)\delta - m]i + m(m+1)\delta(\delta-1)$$

and

$$\beta_i = 2[(m+1)\lambda + i].$$

Since $A_k \neq 0$, it follows that $\alpha = \alpha_k$.

Assume that $\alpha_k \notin \{\alpha_0, \ldots, \alpha_{k-1}\}$. Then for any prescribed $A_k \in \mathcal{S}_{k,\delta}$ the above system has an unique solution, namely

$$A_i = \frac{\beta_i \ldots \beta_{k-1}}{(\alpha_k - \alpha_i) \ldots (\alpha_k - \alpha_{k-1})}\, T^{k-i}\, A_k \tag{3}$$

for $i = 0, \ldots, k-1$.

Suppose now that $\alpha_k = \alpha_j$ for some $j < k$. Then

$$(m+1)\delta_m = k + j.$$

This shows that there is only one such j so that $\alpha_k \notin \{\alpha_0, \ldots, \alpha_{j-1}\}$, and for a given $P \in \mathcal{S}_{j,\delta}$, there is an unique eigenvector of $\mathcal{C}_{\lambda\mu}$ of eigenvalue

α_k having the symbol P. This proves that the spectrum of $\mathcal{C}_{\lambda\mu}$ is the set $\{\alpha_k \mid k \in \mathbb{N}\}$.

It is then easy to show that

Proposition 4.2. *If $\mathcal{C}_{\lambda\mu}$ and $\mathcal{C}_{\lambda'\mu'}$ have the same spectrum, then $\delta = \delta'$.*

Suppose again that $\alpha_k = \alpha_j$ for some $j < k$ and set $j = k - n$.

Then the first equations in (2) still have an unique solution $A_k + \cdots + A_{j+1}$ for a given $A_k \in \mathcal{S}_{k,\delta}$. It is given by (3) for $i \in \{j+1, \ldots, k-1\}$.

If, in addition,

$$\beta_j \ldots \beta_{k-1} = 2^{k-j}[(m+1)\lambda + k - n] \ldots [(m+1)\lambda + k - 1] = 0,$$

then the complete system (2) has at least one solution for a given $A_k \in \mathcal{S}_{k,\delta}$. It suffices indeed to take $A_j = 0, \ldots, A_0 = 0$. It is of course no longer unique, since, in fact, A_j may be chosen arbitrarily.

This shows that if $I_{\lambda\mu}$ is empty, then $\mathcal{C}_{\lambda\mu}$ is semi-simple since each $P \in \mathcal{S}_{k,\delta}$, $k \in \mathbb{N}$, is the symbol of an eigenvector of $\mathcal{C}_{\lambda\mu}$.

Thus the $k \in I_{\lambda\mu}$ are, roughly speaking, those which prevent $\mathcal{C}_{\lambda\mu}$ from being semi-simple. We can be more precise.

Proposition 4.3. *Let $k \in I_\delta$ be given. Then $k \in I_{\lambda\mu}$ if and only if there is an $A \in \mathcal{D}_{\lambda\mu}$ such that the characteristic polynomial of the restriction of $\mathcal{C}_{\lambda\mu}$ to the subspace $\rangle \mathcal{C}^i_{\lambda\mu}\ A,\ i \in \mathbb{N} \langle$ of $\mathcal{D}_{\lambda\mu}$ is its minimum polynomial and is $\prod_{0 \le i \le k} (t - \alpha_i)$.*

The idea behind this proposition is to reduce the situation to a finite-dimensional subspace of $\mathcal{D}_{\lambda\mu}$ on which $\mathcal{C}$ is not semi-simple because its minimal polynomial has a root $\alpha_k = \alpha_{k-n}$ $(1 \le n \le k)$ of multiplicity two. The proof is rather technical and we only sketch it.

Step one: $k \in I_{\lambda\mu}$

Let $P \in \mathcal{S}_{k,\delta}$ be such that $T^k\ P \neq 0$. It is then clear that the linear span ξ of the $\mathcal{C}^i_{\lambda\mu}(P)$, $i \in \mathbb{N}$, has the basis $T^j\ P$, $j \in \{0, \ldots, k\}$. In this basis, $\mathcal{C}_{\lambda\mu}|_\xi$ is represented by the matrix

$$\begin{bmatrix} \alpha_k & & & & & \\ \beta_k & \alpha_{k-1} & & & & \\ & \beta_{k-1} & & & & \\ & & \ddots & & & \\ & & & & \alpha_1 & \\ & & & & \beta_1 & \alpha_0 \end{bmatrix},$$

where only the main diagonal and the subdiagonal just below it are possibly non-vanishing. It is then clear that the minimal polynomial of $\mathcal{C}_{\lambda\mu}|_\xi$ is $(t -$

$\alpha_0)\dots(t-\alpha_k)$ provided that $\beta_1\dots\beta_k\neq 0$. This may be assumed to be the case. Indeed, using conjugacy, we can replace $\mathcal{D}_{\lambda\mu}$ by the isomorphic module $\mathcal{D}_{1-\mu,\,1-\lambda}$ ([1]). Moreover, it is easy to see that the product $\beta_1\dots\beta_k$ is non-vanishing for one of the modules $\mathcal{D}_{\lambda\mu}$ and $\mathcal{D}_{1-\mu,\,1-\lambda}$.

Step two: the restriction of $\mathcal{C}_{\lambda\mu}$ to $\rangle\mathcal{C}^i_{\lambda\mu}\,A,\ i\in\mathbb{N}\langle$ has the prescribed minimal polynomial

We explain how to see that $k\notin I_{\lambda\mu}$ leads to a contradiction. One can check that the order a of A is greater than k and that there is $j\in(I_\delta\setminus I_{\lambda\mu})\cap\{k-n+1,\dots,k\}$ such that $\beta_j=0$, where $n=2k-(m+1)\delta+m$ (this may use the fact that $\mathcal{D}_{1-\mu,\,1-\lambda}$ is isomorphic to $\mathcal{D}_{\lambda\mu}$). Then

$$A=\oplus_{i\in I}\,V_i+A'$$

where $I\subset[j,a]$, V_i is an eigenvector of $\mathcal{C}_{\lambda\mu}$ and $A'\in\mathcal{D}^{j-1}_{\lambda\mu}$. One has

$$\xi=\rangle\mathcal{C}^i_{\lambda\mu}\,A,\ i\in\mathbb{N}\langle\subset\xi'=\oplus_{i\in I}\,\mathbb{R}\,V_i\oplus\rangle\,\mathcal{C}^i_{\lambda\mu}\,A',\ i\in\mathbb{N}\langle$$

and $C':=\mathcal{C}_{\lambda\mu}|_{\xi'}$ is of the form

$$\begin{bmatrix}\operatorname{diag}(\alpha_i,\ i\in I) & 0\\ 0 & C''\end{bmatrix}$$

where C'' is the restriction of $\mathcal{C}_{\lambda\mu}$ to $\rangle\mathcal{C}^i_{\lambda\mu}\,A',\ i\in\mathbb{N}\langle$. It can be seen that $\alpha_j,\dots,\alpha_a\in\{\alpha_0,\dots,\alpha_k\}$ and thus that the minimal polynomial of C' is $(t-\alpha_0)\dots(t-\alpha_k)$. Moreover, the roots of the minimal polynomial $\mathcal{M}$ of C'' belong to $\{\alpha_0,\dots,\alpha_{j-1}\}$. In particular, α_k is a root of $\mathcal{M}$ of multiplicity at most one and $k\in I$ because $\prod_{i\in I}(C-\alpha_i)\mathcal{M}(C)=0$. The contradiction follows then from the property $\prod_{i\in I\setminus k}(C-\alpha_i)\mathcal{M}(C)=0$ although α_k has multiplicity at most one as a root of $\mathcal{M}(t)\prod_{i\in I\setminus k}(t-\alpha_i)$.

The following properties are now obvious.

Corollary 4.4. *The Casimir operator $\mathcal{C}_{\lambda\mu}$ is semi-simple if and only if $I_{\lambda\mu}$ is empty.*

Corollary 4.5. *If the $\mathfrak{sl}(m+1)$-modules $\mathcal{D}_{\lambda\mu}$ and $\mathcal{D}_{\lambda'\mu'}$ are isomorphic, then $I_{\lambda\mu}=I_{\lambda'\mu'}$.*

5. A model for $\mathcal{D}_{\lambda\mu}$

We recall here the construction of a model of $\mathcal{D}_{\lambda\mu}$ given in [3].

It shows that if $\delta=\delta'$ and $I_{\lambda\mu}=I_{\lambda'\mu'}$, then the $\mathfrak{sl}_{m+1}$-modules $\mathcal{D}_{\lambda\mu}$ and $\mathcal{D}_{\lambda'\mu'}$ are isomorphic.

We denote as $\mathcal{D}^k_m$ the space $\mathcal{S}_{0,\delta} \oplus \cdots \oplus \mathcal{S}_{k,\delta}$ equipped with the $\mathfrak{sl}_{m+1}$-module structure given by

$$X.(P_0, \ldots, P_k) = \left(L^{po}_X P_0 + \gamma_X P_1, \ldots, L^{po}_X P_i + \gamma_X P_{i+1}, \ldots, L^{po}_X P_k\right),$$

where $P_i \in \mathcal{S}_{i,\delta}$ and where the cocycle γ of $\mathfrak{sl}(m+1)$ with values in $\mathrm{Hom}(\mathcal{S}_\delta, \mathcal{S}_\delta)$ is defined by

$$\gamma_X(P) = \frac{1}{m+1} \sum_{i,j} \frac{\partial^2 X^j}{\partial x^i \partial x^j} \frac{\partial P}{\partial \xi_i}.$$

Proposition 5.1. *The* $\mathfrak{sl}(m+1)$ *module* $\mathcal{D}_{\lambda\mu}$ *is isomorphic to*

$$\mathcal{D}^{\delta,s}_m = \oplus_{i>s} \mathcal{S}_{i,\delta} \oplus \mathcal{D}^s_m,$$

where $s = \sup\ I_{\lambda\mu} \quad (\sup \emptyset = -1$ *and* $\mathcal{D}^{-1}_m = 0)$.

Acknowledgments

This is the content of a lecture which I gave at the workshop 'Noncommutative Differential Geometry and its applications to Physics'.

I would like to thank very much the organizers Y. Maeda, S. Watamura, H. Morijoshi and H. Omori for their invitation and to congratulate them for this enjoyable meeting.

References

[1] C. Duval, V. Ovsienko, P. Lecomte. Method of Equivariant Quantization. In the present proceedings.

[2] D.B. Fucks. *Cohomology of infinite-dimensional Lie algebras*, Consultants Bureau, New York, London, 1986.

[3] P. Lecomte. Classification projective des espaces d'opérateurs différentiels agissant sur les densités. *CRAS* t.328, tome I, p. 287-290, 1999.

[4] P. Lecomte. *On the Cohomology of* $\mathfrak{sl}(m+1, \mathbb{R})$ *acting on differential operators and* $\mathfrak{sl}(m+1, \mathbb{R})$*-equivariant symbol.* To appear in *Indagationes Math.*

AN INTERPRETATION OF THE SCHOUTEN–NIJENHUIS BRACKET

Kentaro Mikami*
Department of Computer Science and Engineering
Akita University, Akita, 010–8502, JAPAN
mikami@math.akita-u.ac.jp

Abstract The Schouten–Nijenhuis bracket ([4],[5]) is used to describe whether a 2-vector field becomes a Poisson tensor field or not, due to Lichnerowicz, and it is a very popular and useful tool in Poisson geometry. In this note we observe that the bracket has a very natural and simple notion in its root, namely, skew-symmetry and the Leibniz' rule. We then give an easy proof for the super Jacobi identity of the Schouten–Nijenhuis bracket.

Mathematics Subject Classification (2000): 17B70, 53D17

Keywords: Schouten–Nijenhuis bracket, skew-symmetry, graded algebras

1. Introduction

Let M be a smooth manifold with the function algebra $\mathfrak{F}(M) := C^\infty(M)$.

A Poisson bracket is a binary operation on $\mathfrak{F}(M)$ with two properties: $\mathfrak{F}(M)$ becomes a Lie algebra and the binary operation satisfies the Leibniz' rule. The bracket is given by a 2-vector field, say π as $\{f, g\} = \pi(df, dg)$. It is known that $\{\cdot, \cdot\}$ satisfies the Jacobi identity if and only if the Schouten–Nijenhuis bracket $[\pi, \pi]_{\mathrm{SN}} = 0$.

There is a notion of Poisson cohomology, where the cochain space is $\Gamma(\wedge^k \mathrm{T}(M))$ $(k = 0, 1, \dots)$ and the coboundary operator is $U \mapsto [\pi, U]_{\mathrm{SN}}$.

* Partially supported by Grant-in-Aid for Scientific Research (C) (No. 10640057), The Ministry of Education, Science, Sports and Culture, Japan.

Y. Maeda, et al. (eds.), Noncommutative Differential Geometry and Its Applications to Physics, 131–143.

Those two examples certainly show that the Schouten–Nijenhuis bracket is a useful tool in Poisson geometry.

The original definition of the Schouten–Nijenhuis bracket ([4],[5]) is written in an old fashion and there are several mistakes concerning parities in their formulae. Also, it is hard to see and understand it at first. There are modern ways, say [1] or [2], of describing properties of the Schouten–Nijenhuis bracket. Namely, it is a prototype of a graded algebra structure. But those seem to be not enough to explain the reason why [4] or [5] defined the bracket as we see it today. In this paper we claim that the bracket is based on natural and simple notions, namely skew-symmetry and the Leibniz' rule, and give an easy proof for the super Jacobi identity of the Schouten–Nijenhuis bracket.

Let us consider $\mathfrak{F}(M)$-valued $\mathbf{R}$-multilinear skew-symmetric maps on $\mathfrak{F}(M)$,

$$\mathcal{A} : \underbrace{\mathfrak{F}(M) \times \cdots \times \mathfrak{F}(M)}_{a\text{-times}} \longrightarrow \mathfrak{F}(M) \,.$$

We call the map $\mathcal{A}$ as a skew product of degree a. When $a = 0$, we consider the $\mathbf{R}$-linear map $\mathcal{A} : \mathbf{R} \longrightarrow \mathfrak{F}(M)$ identified by $\mathcal{A}(1) \in \mathfrak{F}(M)$.

Naturally, we can construct two new kinds of brackets for each pair of skew products $\mathcal{A}$ and $\mathcal{B}$ of degree a and b respectively.

One is a degree $(a + b)$-skew product valued bracket which is the direct analogy of the wedge product:

$$\begin{aligned}
&\mathcal{A} \wedge \mathcal{B}(f_1, \dots, f_{a+b}) \\
&\quad := \frac{1}{a!b!} \sum_{\sigma} \operatorname{sgn}(\sigma) \mathcal{A}(f_{\sigma(1)}, \dots, f_{\sigma(a)}) \mathcal{B}(f_{\sigma(a+1)}, \dots, f_{\sigma(a+b)}) \\
&\quad = \sum_{j_1 < \cdots < j_a}{}' \operatorname{sgn}(j_1 < \cdots < j_a; k_1 < \cdots < k_b) \\
&\qquad\qquad \times \mathcal{A}(f_{j_1}, \dots, f_{j_a}) \mathcal{B}(f_{k_1}, \dots, f_{k_b}) \\
&\quad = \sum_{J}{}' \operatorname{sgn}(J; K) \mathcal{A}(f_J) \mathcal{B}(f_K) \quad \text{as abbreviation} \,.
\end{aligned}$$

Here we remark that $\sum_J{}' \operatorname{sgn}(J; K)$ and $\sum_K{}' \operatorname{sgn}(J; K)$ mean the same summation because $j_1 < \cdots < j_a$ and $k_1 < \cdots < k_b$ are complementary to each other in the set $\{1, \dots, a + b\}$.

The other is a bracket with values of degree $(a + b - 1)$-skew multilinearity:

$$\begin{aligned}
&(\mathcal{A} \triangleleft \mathcal{B})(f_1, \dots, f_{a+b-1}) \\
&\quad := \frac{1}{(a-1)!b!} \sum_{\sigma} \operatorname{sgn}(\sigma) \mathcal{A}(f_{\sigma(1)}, \dots, f_{\sigma(a-1)}, \mathcal{B}(f_{\sigma(a)}, \dots, f_{\sigma(a+b-1)}))
\end{aligned}$$

$$= \sum_K{}' \operatorname{sgn}(J;K)\mathcal{A}(f_J, \mathcal{B}(f_K)) .$$

2. Leibniz' rule and super skewness

If $\mathbf{A}$ is a multi-vector field of degree a, i.e., a contravariant skew-symmetric $(a, 0)$-tensor field, then

$$\mathcal{A}(f_1, \dots, f_a) := \mathbf{A}(df_1, \dots, df_a)$$

is a skew product and satisfies Leibniz' rule for each argument. Conversely, if a skew product satisfies Leibniz' rule for each argument (in fact, it is enough to satisfy Leibniz' rule for one argument), then the skew product is given by a multi-vector field as above. We sometimes call a skew product satisfying Leibniz' rule as a skew L-product in short. From now on, we use a boldface letter to express the corresponding multi-vector field of a skew L-product, such as $\mathbf{A}$ for $\mathcal{A}$.

If $\mathcal{A}$ and $\mathcal{B}$ are skew L-products, then $\mathcal{A} \wedge \mathcal{B}$ is also a skew L-product and the corresponding multi-vector field is $\mathbf{A} \wedge \mathbf{B}$. However, $\mathcal{A} \lhd \mathcal{B}$ does not satisfy Leibniz' rule in general. The simple situation is: let X and Y be vector fields, then $X \lhd Y$ is the same with $X \circ Y$ and does not satisfy Leibniz' rule, but it is well-known that $X \lhd Y - Y \lhd X$ satisfies Leibniz' rule and is known as Jacobi–Lie bracket $[X, Y]$.

A classical linear example of super Lie algebra, denoted by $\mathfrak{gl}(m \mid n)$, is given on $\mathfrak{gl}(m + n, \mathbf{R})$ by

$$[X, Y] = XY - (-1)^{p(X)p(Y)} YX$$

where

$$\mathfrak{gl}(m+n, \mathbf{R}) = \left\{ \left(\begin{array}{c|c} A & O \\ \hline O & D \end{array} \right) \mid A \in \mathfrak{gl}_m,\ D \in \mathfrak{gl}_n \right\} \oplus \left\{ \left(\begin{array}{c|c} O & B \\ \hline C & O \end{array} \right) \mid B, {}^tC \in \mathrm{Mat}(m, n) \right\}$$

and the parity of $\left(\begin{array}{c|c} A & O \\ \hline O & D \end{array} \right)$ and $\left(\begin{array}{c|c} O & B \\ \hline C & O \end{array} \right)$ are 0 and 1 respectively.

We claim that for each skew L-products $\mathcal{A}$ and $\mathcal{B}$,

$$\mathcal{A} \lhd \mathcal{B} - (-1)^{(a-1)(b-1)} \mathcal{B} \lhd \mathcal{A}$$

also satisfies Leibniz' rule and its multi-vector field is the Schouten–Nijenhuis bracket of $\mathbf{A}$ and $\mathbf{B}$ and stress that our interpretation is like a linear super example.

Lemma 1 Let $\mathcal{A}$ and $\mathcal{B}$ be skew L-products of degree a and b respectively. Then

$$
\begin{aligned}
&(\mathcal{A} \lhd \mathcal{B})(f_1, \dots, f_{a+b-2}, \phi\psi) \\
&\quad = \underset{\phi,\psi}{\mathfrak{S}} (\mathcal{A} \lhd \mathcal{B})(f_1, \dots, f_{a+b-2}, \phi)\psi \\
&\qquad + \sum_{j_1<\cdots<j_{a-1}}{}' \operatorname{sgn}(j_1 < \cdots < j_{a-1}; k_1 < \cdots < k_{b-1}) \\
&\qquad\qquad \times \underset{\phi,\psi}{\mathfrak{S}} \mathcal{A}(f_{j_1}, \dots, f_{j_{a-1}}, \phi)\mathcal{B}(f_{k_1}, \dots, f_{k_{b-1}}, \psi),
\end{aligned}
$$

where $\underset{\phi,\psi}{\mathfrak{S}}$ means the symmetric sum by ϕ and ψ.

Proof of Lemma 1: Putting $f_{a+b-1} = \phi\psi$, we have

$$
\begin{aligned}
&(\mathcal{A} \lhd \mathcal{B})(f_1, \dots, f_{a+b-2}, f_{a+b-1}) \\
&= \sum_{j_{a-1}=a+b-1}{}' \operatorname{sgn}(j_1 < \cdots < j_{a-1}; K)\mathcal{A}(f_{j_1}, \dots, f_{j_{a-2}}, \phi\psi, \mathcal{B}(f_K)) \\
&\quad + \sum_{k_b=a+b-1}{}' \operatorname{sgn}(J; k_1 < \cdots < k_b)\mathcal{A}(f_J, \mathcal{B}(f_{k_1}, \dots, f_{k_{b-1}}, \phi\psi)) \\
&= \sum_{J'}{}' \operatorname{sgn}(J', a+b-1; K)(\mathcal{A}(f_{J'}, \phi, \mathcal{B}(f_K))\psi + \mathcal{A}(f_{J'}, \psi, \mathcal{B}(f_K))\phi) \\
&\quad + \sum_{K'}{}' \operatorname{sgn}(J; K', a+b-1)\mathcal{A}(f_J, \mathcal{B}(f_{K'}, \phi)\psi + \mathcal{B}(f_{K'}, \psi)\phi) \\
&= \sum_{J'}{}' \operatorname{sgn}(J', a+b-1; K)(\mathcal{A}(f_{J'}, \phi, \mathcal{B}(f_K))\psi + \mathcal{A}(f_{J'}, \psi, \mathcal{B}(f_K))\phi) \\
&\quad + \sum_{K'}{}' \operatorname{sgn}(J; K', a+b-1)(\mathcal{A}(f_J, \mathcal{B}(f_{K'}, \phi))\psi + \mathcal{A}(f_J, \psi)\mathcal{B}(f_{K'}, \phi) \\
&\qquad\qquad + \mathcal{A}(f_J, \mathcal{B}(f_{K'}, \psi))\phi + \mathcal{A}(f_J, \phi)\mathcal{B}(f_{K'}, \psi)) \\
&= \underset{\phi,\psi}{\mathfrak{S}} (\mathcal{A} \lhd \mathcal{B})(f_1, \dots, f_{a+b-2}, \phi)\psi \\
&\quad + \sum_{K'}{}' \operatorname{sgn}(J; K', a+b-1)(\mathcal{A}(f_J, \psi)\mathcal{B}(f_{K'}, \phi) + \mathcal{A}(f_J, \phi)\mathcal{B}(f_{K'}, \psi)) \\
&= \underset{\phi,\psi}{\mathfrak{S}} (\mathcal{A} \lhd \mathcal{B})(f_1, \dots, f_{a+b-2}, \phi)\psi \\
&\quad + \sum{}' \operatorname{sgn}(J; K') \underset{\phi,\psi}{\mathfrak{S}} \mathcal{A}(f_J, \phi)\mathcal{B}(f_{K'}, \psi) .
\end{aligned}
$$

∎

By Lemma 1, we have the theorem below directly.

Theorem 1 ([4],[5]) Let $\mathcal{A}$ and $\mathcal{B}$ be skew products satisfying Leibniz' rule of degree a and b respectively. Then

$$\mathcal{A} \lhd \mathcal{B} - (-1)^{(a-1)(b-1)} \mathcal{B} \lhd \mathcal{A}$$

is a skew product satisfying Leibniz' rule.

Definition 1 We define the skew L-product of degree $(a+b-1)$

$$[[\mathcal{A},\mathcal{B}]] := \mathcal{A} \lhd \mathcal{B} - (-1)^{(a-1)(b-1)} \mathcal{B} \lhd \mathcal{A}$$

for each skew L-products $\mathcal{A}$ and $\mathcal{B}$ of degree a and b respectively.

From the definition, we see that:

Theorem 2

$$[[\mathcal{A},\mathcal{B}]] = -(-1)^{(a-1)(b-1)} [[\mathcal{B},\mathcal{A}]] \qquad \text{(super skewness)} .$$

We can write the bracket as

$$\begin{aligned}
&[[\mathcal{A},\mathcal{B}]](f_1,\cdots,f_{a+b-1}) \\
&\quad = \sum{}' \operatorname{sgn}(j_1 < \cdots < j_{a-1}; k_1 < \cdots < k_b) \\
&\qquad\qquad \times \mathcal{A}(f_{j_1},\cdots,f_{j_{a-1}},\mathcal{B}(f_{k_1},\cdots,f_{k_b})) \\
&\qquad - (-1)^{(a-1)(b-1)} \sum{}' \operatorname{sgn}(j_1 < \cdots < j_{b-1}; k_1 < \cdots < k_a) \\
&\qquad\qquad \times \mathcal{B}(f_{j_1},\cdots,f_{j_{b-1}},\mathcal{A}(f_{k_1},\cdots,f_{k_a})) \\
&\quad = \sum{}' \operatorname{sgn}(j_1 < \cdots < j_{a-1}; k_1 < \cdots < k_b) \\
&\qquad\qquad \times \mathcal{A}(f_{j_1},\cdots,f_{j_{a-1}},\mathcal{B}(f_{k_1},\cdots,f_{k_b})) \\
&\qquad - \sum{}' \operatorname{sgn}(j_1 < \cdots < j_a; k_1 < \cdots < k_{b-1}) \\
&\qquad\qquad \times \mathcal{B}(\mathcal{A}(f_{j_1},\cdots,f_{j_a}),f_{k_1},\cdots,f_{k_{b-1}}) .
\end{aligned}$$

3. Associator of $\lhd$ and super Jacobi identity of $[[\cdot,\cdot]]$

For each skew products $\mathcal{A}$, $\mathcal{B}$ and $\mathcal{C}$ of degree a, b and c respectively, we have

$$\begin{aligned}
&(\mathcal{A} \lhd (\mathcal{B} \lhd \mathcal{C}))(f_1,\cdots,f_{a+b+c-2}) \\
&\quad = \sum{}' \operatorname{sgn}(I;J)\mathcal{A}(f_I,(\mathcal{B} \lhd \mathcal{C})(f_J)) \\
&\quad = \sum{}' \operatorname{sgn}(I;J)\mathcal{A}(f_I,\sum{}' \operatorname{sgn}\binom{J}{K;L}\mathcal{B}(f_K,\mathcal{C}(f_L)))
\end{aligned}$$

$$= \sum{}' \operatorname{sgn}(I;K;L)\mathcal{A}(f_I,\mathcal{B}(f_K,\mathcal{C}(f_L)))$$

and

$$\begin{aligned}
&((\mathcal{A} \lhd \mathcal{B}) \lhd \mathcal{C}))(f_1,\dots,f_{a+b+c-2}) \\
&\quad= \sum{}' \operatorname{sgn}(I;J)(\mathcal{A} \lhd \mathcal{B})(f_I,\mathcal{C}(f_J)) \\
&\quad= \sum{}' \operatorname{sgn}(I;J)(\sum{}' \operatorname{sgn}\begin{pmatrix} I,* \\ K,*;L\end{pmatrix}\mathcal{A}(f_K,\mathcal{C}(f_J),\mathcal{B}(f_L))) \\
&\qquad\quad + \sum{}' \operatorname{sgn}\begin{pmatrix} I,* \\ K;L,*\end{pmatrix}\mathcal{A}(f_K,\mathcal{B}(f_L,\mathcal{C}(f_J)))) \\
&\quad= \sum{}' \operatorname{sgn}(I;J)(\sum{}' (-1)^b \operatorname{sgn}\begin{pmatrix} I \\ K;L\end{pmatrix}\mathcal{A}(f_K,\mathcal{C}(f_J),\mathcal{B}(f_L))) \\
&\qquad\quad + \sum{}' \operatorname{sgn}\begin{pmatrix} I \\ K;L\end{pmatrix}\mathcal{A}(f_K,\mathcal{B}(f_L,\mathcal{C}(f_J)))) \\
&\quad= (-1)^{b-1}\sum{}' \operatorname{sgn}(K;L;J)\mathcal{A}(f_K,\mathcal{B}(f_L),\mathcal{C}(f_J)) \\
&\qquad\quad + \sum{}' \operatorname{sgn}(K;L;J)\mathcal{A}(f_K,\mathcal{B}(f_L,\mathcal{C}(f_J)))) \,.
\end{aligned}$$

Thus we have

$$\begin{aligned}
&((\mathcal{A} \lhd \mathcal{B}) \lhd \mathcal{C})(f_1,\dots,f_{a+b-2}) \\
&\quad= (\mathcal{A} \lhd (\mathcal{B} \lhd \mathcal{C}))(f_1,\dots,f_{a+b-2}) \\
&\qquad + (-1)^{b-1}\sum{}' \operatorname{sgn}(K;L;J)\mathcal{A}(f_K,\mathcal{B}(f_L),\mathcal{C}(f_J)) \qquad (1)
\end{aligned}$$

and the associative law does not hold for $\lhd$. (If $a = \deg \mathcal{A} = 1$, then $(\mathcal{A} \lhd \mathcal{B}) \lhd \mathcal{C} = \mathcal{A} \lhd (\mathcal{B} \lhd \mathcal{C})$ holds good.) The associator $\Delta(\mathcal{A},\mathcal{B},\mathcal{C}) := (\mathcal{A} \lhd \mathcal{B}) \lhd \mathcal{C} - \mathcal{A} \lhd (\mathcal{B} \lhd \mathcal{C})$ satisfies $\Delta(\mathcal{A},\mathcal{B},\mathcal{C}) = (-1)^{(b-1)(c-1)}\Delta(\mathcal{A},\mathcal{C},\mathcal{B})$.

From the definition, using the notation $a' = a-1$ and so on, we have

$$\begin{aligned}
&[[[[\mathcal{A},\mathcal{B}]],\mathcal{C}]] \\
&\quad= [[\mathcal{A},\mathcal{B}]] \lhd \mathcal{C} - (-1)^{(a+b-2)(c-1)}\mathcal{C} \lhd [[\mathcal{A},\mathcal{B}]] \\
&\quad= (\mathcal{A} \lhd \mathcal{B} - (-1)^{a'b'}\mathcal{B} \lhd \mathcal{A}) \lhd \mathcal{C} \\
&\qquad - (-1)^{(a'+b')c'}\mathcal{C} \lhd (\mathcal{A} \lhd \mathcal{B} - (-1)^{a'b'}\mathcal{B} \lhd \mathcal{A}) \\
&\quad= (\mathcal{A} \lhd \mathcal{B}) \lhd \mathcal{C} - (-1)^{a'b'}(\mathcal{B} \lhd \mathcal{A}) \lhd \mathcal{C} \\
&\qquad - (-1)^{(a'+b')c'}\mathcal{C} \lhd (\mathcal{A} \lhd \mathcal{B}) + (-1)^{(a'+b')c'+a'b'}\mathcal{C} \lhd (\mathcal{B} \lhd \mathcal{A})
\end{aligned}$$

and so

$$\begin{aligned}
&(-1)^{a'c'}[[[[\mathcal{A},\mathcal{B}]],\mathcal{C}]] \\
&\quad= (-1)^{a'c'}(\mathcal{A} \lhd \mathcal{B}) \lhd \mathcal{C} - (-1)^{b'c'}\mathcal{C} \lhd (\mathcal{A} \lhd \mathcal{B})
\end{aligned}$$

$$- (-1)^{a'(b'+c')}(\mathcal{B} \lhd \mathcal{A}) \lhd \mathcal{C} + (-1)^{b'(c'+a')}\mathcal{C} \lhd (\mathcal{B} \lhd \mathcal{A})$$

using (1)

$$\begin{aligned}
&= (-1)^{a'c'}(\mathcal{A} \lhd (\mathcal{B} \lhd \mathcal{C}) \\
&\quad + (-1)^{b'} \sum{}' \operatorname{sgn}(J;K;L)\mathcal{A}(f_J, \mathcal{B}(f_K), \mathcal{C}(f_L))) \\
&\quad - (-1)^{b'c'}\mathcal{C} \lhd (\mathcal{A} \lhd \mathcal{B}) - (-1)^{a'(b'+c')}(\mathcal{B} \lhd (\mathcal{A} \lhd \mathcal{C}) \\
&\quad + (-1)^{a'} \sum{}' \operatorname{sgn}(J;K;L)\mathcal{B}(f_J, \mathcal{A}(f_K), \mathcal{C}(f_L))) \\
&\quad + (-1)^{b'(c'+a')}\mathcal{C} \lhd (\mathcal{B} \lhd \mathcal{A}) \\
&= (-1)^{a'c'}\mathcal{A} \lhd (\mathcal{B} \lhd \mathcal{C}) - (-1)^{b'c'}\mathcal{C} \lhd (\mathcal{A} \lhd \mathcal{B}) \\
&\quad - (-1)^{a'(b'+c')}\mathcal{B} \lhd (\mathcal{A} \lhd \mathcal{C}) + (-1)^{b'(c'+a')}\mathcal{C} \lhd (\mathcal{B} \lhd \mathcal{A}) \\
&\quad + (-1)^{a'c'+b'} \sum{}' \operatorname{sgn}(J;K;L)\mathcal{A}(f_J, \mathcal{B}(f_K), \mathcal{C}(f_L)) \\
&\quad - (-1)^{a'(b'+c'+1)} \sum{}' \operatorname{sgn}(J;K;L)\mathcal{B}(f_J, \mathcal{A}(f_K), \mathcal{C}(f_L)) \\
&= (-1)^{a'c'}\mathcal{A} \lhd (\mathcal{B} \lhd \mathcal{C}) - (-1)^{b'c'}\mathcal{C} \lhd (\mathcal{A} \lhd \mathcal{B}) \\
&\quad - (-1)^{a'(b'+c')}\mathcal{B} \lhd (\mathcal{A} \lhd \mathcal{C}) + (-1)^{b'(c'+a')}\mathcal{C} \lhd (\mathcal{B} \lhd \mathcal{A}) \\
&\quad + (-1)^{a'c'+b'} \sum{}' \operatorname{sgn}(J;K;L)\mathcal{A}(f_J, \mathcal{B}(f_K), \mathcal{C}(f_L)) \\
&\quad - (-1)^{b'a'+c'} \sum{}' \operatorname{sgn}(J;L;K)\mathcal{B}(f_J, \mathcal{C}(f_L), \mathcal{A}(f_K)) .
\end{aligned}$$

We then have

Theorem 3

$$\underset{\mathcal{A},\mathcal{B},\mathcal{C}}{\mathfrak{S}} (-1)^{(a-1)(c-1)}[[[[\mathcal{A}, \mathcal{B}]], \mathcal{C}]] = 0 \qquad \text{(super Jacobi identity)}$$

where $\underset{\mathcal{A},\mathcal{B},\mathcal{C}}{\mathfrak{S}}$ is the cyclic sum with respect to $\mathcal{A}, \mathcal{B}, \mathcal{C}$.

4. Derivation rule

Lemma 2 Let $\mathcal{X}$ be a degree 1 skew product satisfying Leibniz' rule, i.e., a vector field. Let $\mathcal{A}$ and $\mathcal{B}$ be skew products satisfying Leibniz' rule and of degree a and b respectively. Then

$$[[\mathcal{X} \wedge \mathcal{A}, \mathcal{B}]] = \mathcal{X} \wedge [[\mathcal{A}, \mathcal{B}]] + (-1)^{a(b-1)}[[\mathcal{X}, \mathcal{B}]] \wedge \mathcal{A} .$$

Using Lemma 2 we obtain the next theorem by mathematical induction on the degree of $\mathcal{P}$.

Theorem 4

$$[[\mathcal{P} \wedge \mathcal{A}, \mathcal{B}]] = \mathcal{P} \wedge [[\mathcal{A}, \mathcal{B}]] + (-1)^{a(b-1)}[[\mathcal{P}, \mathcal{B}]] \wedge \mathcal{A}$$

holds for each skew products $\mathcal{P}$, $\mathcal{A}$ and $\mathcal{B}$ satisfying Leibniz' rule with degree p, a and b respectively.

Proof of Theorem 4: We assume that Lemma 2 holds.

When $\deg \mathcal{P} = 1$ Lemma 2 itself guarantees Theorem 4.

Next, we may assume that the Theorem holds for each $\mathcal{P}$ with $p < q$, and take a $\mathcal{Q} = \mathcal{X} \wedge \mathcal{P}$ where $\mathcal{X}$ is a degree 1 (skew) L-product. Then

$$\begin{aligned}
&[[\mathcal{Q} \wedge \mathcal{A}, \mathcal{B}]] \\
&\quad = [[\mathcal{X} \wedge (\mathcal{P} \wedge \mathcal{A}), \mathcal{B}]] \\
&\quad = \mathcal{X} \wedge [[\mathcal{P} \wedge \mathcal{A}, \mathcal{B}]] + (-1)^{(p+a)(b-1)}[[\mathcal{X}, \mathcal{B}]] \wedge \mathcal{P} \wedge \mathcal{A} \\
&\quad = \mathcal{X} \wedge \left(\mathcal{P} \wedge [[\mathcal{A}, \mathcal{B}]] + (-1)^{a(b-1)}[[\mathcal{P}, \mathcal{B}]] \wedge \mathcal{A}\right) \\
&\qquad\quad + (-1)^{(p+a)(b-1)}[[\mathcal{X}, \mathcal{B}]] \wedge \mathcal{P} \wedge \mathcal{A} \\
&\quad = \mathcal{Q} \wedge [[\mathcal{A}, \mathcal{B}]] \\
&\qquad\quad + (-1)^{a(b-1)}\left(\mathcal{X} \wedge [[\mathcal{P}, \mathcal{B}]] + (-1)^{p(b-1)}[[\mathcal{X}, \mathcal{B}]] \wedge \mathcal{P}\right) \wedge \mathcal{A} \\
&\quad = \mathcal{Q} \wedge [[\mathcal{A}, \mathcal{B}]] + (-1)^{a(b-1)}[[\mathcal{Q}, \mathcal{B}]] \wedge \mathcal{A}\,.
\end{aligned}$$

■

Before going to the proof of Lemma 2, we recall some definitions.

Let $\mathcal{X}$ be a degree 1 (skew) L-product and $\mathcal{B}$ be a degree b skew L-product. Then we have

$$(\mathcal{X} \wedge \mathcal{B})(f_1, \dots, f_{b+1}) = \sum_{j=1}^{b+1}(-1)^{j+1}\mathcal{X}(f_j)\mathcal{B}(f_1, \dots, \widehat{f_j}, \dots, f_{b+1})$$

and

$$\begin{aligned}
[[\mathcal{X}, \mathcal{B}]](f_1, \dots, f_b) &= \mathcal{X}(\mathcal{B}(f_1, \dots, f_b)) \\
&\quad - \sum_{j=1}^{b} \mathcal{B}(f_1, \dots, \mathcal{X}(f_j), \dots, f_b) \\
&= \mathcal{X}(\mathcal{B}(f_1, \dots, f_b)) \\
&\quad + \sum_{j=1}^{b}(-1)^j \mathcal{B}(\mathcal{X}(f_j), f_1, \dots, \widehat{f_j}, \dots, f_b)\,.
\end{aligned}$$

Proof of Lemma 2:

$$\begin{aligned}
&[[\mathcal{X} \wedge \mathcal{A}, \mathcal{B}]](f_1, \dots, f_{a+b}) \\
&\quad = \sum_{j_1 < \cdots < j_a}' \operatorname{sgn}(j_1 < \cdots < j_a; K)(\mathcal{X} \wedge \mathcal{A})(f_{j_1}, \dots, f_{j_a}, \mathcal{B}(f_K))
\end{aligned}$$

$$
\begin{aligned}
&- (-1)^{a(b-1)} \sum_{k_1<\cdots<k_{a+1}}' \operatorname{sgn}(J; k_1 < \cdots < k_{a+1}) \\
&\qquad \times \mathcal{B}(f_J, (\mathcal{X} \wedge \mathcal{A})(f_{k_1}, \dots, f_{k_{a+1}})) \\
= &\sum_{j_1<\cdots<j_a}' \operatorname{sgn}(j_1 < \cdots < j_a; K) \\
&\quad \times \left\{ \sum_{s=1}^{a} (-1)^{s+1} \mathcal{X}(f_{j_s}) \mathcal{A}(\dots, \widehat{f_{j_s}}, \dots, \mathcal{B}(f_K)) \right. \\
&\qquad \left. + (-1)^a \mathcal{X}(\mathcal{B}(f_K)) \mathcal{A}(f_{j_1}, \dots, f_{j_a}) \right\} \\
&- (-1)^{a(b-1)} \sum_{k_1<\cdots<k_{a+1}}' \operatorname{sgn}(J; k_1 < \cdots < k_{a+1}) \\
&\qquad \times \mathcal{B}\left(f_J, \sum_{s=1}^{a+1} (-1)^{s+1} \mathcal{X}(f_{k_s}) \mathcal{A}(\dots, \widehat{f_{k_s}}, \dots) \right) \\
= &\sum_{j_1<\cdots<j_a}' \operatorname{sgn}(j_1 < \cdots < j_a; K) \\
&\quad \times \left\{ \sum_{s=1}^{a} (-1)^{s+1} \mathcal{X}(f_{j_s}) \mathcal{A}(\dots, \widehat{f_{j_s}}, \dots, \mathcal{B}(f_K)) \right. \\
&\qquad \left. + (-1)^a \mathcal{X}(\mathcal{B}(f_K)) \mathcal{A}(f_{j_1}, \dots, f_{j_a}) \right\} \\
&- (-1)^{a(b-1)} \sum_{k_1<\cdots<k_{a+1}}' \operatorname{sgn}(J; k_1 < \cdots < k_{a+1}) \\
&\qquad \times \sum_{s=1}^{a+1} (-1)^{s+1} \Big(\mathcal{B}(f_J, \mathcal{X}(f_{k_s})) \mathcal{A}(\dots, \widehat{f_{k_s}}, \dots) \\
&\qquad\qquad + \mathcal{X}(f_{k_s}) \mathcal{B}(f_J, \mathcal{A}(\dots, \widehat{f_{k_s}}, \dots)) \Big) \\
= &\quad \mathrm{I} + \mathrm{II}
\end{aligned}
$$

where I = the terms without $\mathcal{X}(f_j)$'s and II = the terms of $\mathcal{X}(f_j)$'s. Then

$$
\begin{aligned}
\mathrm{I} = &\sum_{j_1<\cdots<j_a}' \operatorname{sgn}(j_1 < \cdots < j_a; K)(-1)^a \mathcal{X}(\mathcal{B}(f_K)) \mathcal{A}(f_{j_1}, \dots, f_{j_a}) \\
&- (-1)^{a(b-1)} \sum_{k_1<\cdots<k_{a+1}}' \operatorname{sgn}(J; k_1 < \cdots < k_{a+1})
\end{aligned}
$$

$$\times \sum_{s=1}^{a+1} (-1)^{s+1} \mathcal{B}(f_J, \mathcal{X}(f_{k_s})) \mathcal{A}(\dots, \widehat{f_{k_s}}, \dots)$$

$$= (-1)^a \sum_{j_1 < \dots < j_a}{}' \operatorname{sgn}(j_1 < \dots < j_a; K) \mathcal{X}(\mathcal{B}(f_K)) \mathcal{A}(f_{j_1}, \dots, f_{j_a})$$

$$- (-1)^{a(b-1)} \sum_{k_1 < \dots < k_{a+1}}{}' \operatorname{sgn}(J; k_1 < \dots < k_{a+1})$$

$$\times \sum_{s=1}^{a+1} (-1)^{s+1} \mathcal{B}(f_J, \mathcal{X}(f_{k_s})) \mathcal{A}(\dots, \widehat{f_{k_s}}, \dots)$$

$$= (-1)^a \sum_{j_1 < \dots < j_a}{}' (-1)^{ab} \operatorname{sgn}(K; j_1 < \dots < j_a) \mathcal{X}(\mathcal{B}(f_K)) \mathcal{A}(f_{j_1}, \dots, f_{j_a})$$

$$- (-1)^{a(b-1)} \sum_{k_1 < \dots < k_{a+1}}{}' \sum_{s=1}^{a+1} \operatorname{sgn}(J; k_s, k_1 < \dots \widehat{k_s} < \dots < k_{a+1})$$

$$\times \mathcal{B}(f_J, \mathcal{X}(f_{k_s})) \mathcal{A}(\dots, \widehat{f_{k_s}}, \dots)$$

$$= (-1)^{a(b-1)} [[\mathcal{X}, \mathcal{B}]] \wedge \mathcal{A}$$

and

$$\mathrm{II} = \sum_{j_1 < \dots < j_a}{}' \operatorname{sgn}(j_1 < \dots < j_a; K)$$

$$\times \sum_{s=1}^{a} (-1)^{s+1} \mathcal{X}(f_{j_s}) \mathcal{A}(\dots, \widehat{f_{j_s}}, \dots, \mathcal{B}(f_K))$$

$$- (-1)^{a(b-1)} \sum_{k_1 < \dots < k_{a+1}}{}' \operatorname{sgn}(J; k_1 < \dots < k_{a+1})$$

$$\times \sum_{s=1}^{a+1} (-1)^{s+1} \mathcal{X}(f_{k_s}) \mathcal{B}(f_J, \mathcal{A}(\dots, \widehat{f_{k_s}}, \dots)) .$$

On the other hand, we have

$$(\mathcal{X} \wedge [[\mathcal{A}, \mathcal{B}]])(f_1, \dots, f_{a+b})$$

$$= \sum_{s=1}^{a+b} (-1)^{s+1} (\mathcal{X} f_s) [[\mathcal{A}, \mathcal{B}]](\dots, \widehat{f_s}, \dots)$$

$$= \sum_{s=1}^{a+b} (-1)^{s+1} (\mathcal{X} f_s) \left(\sum_I{}' \operatorname{sgn}(I; J) \mathcal{A}(f_I, \mathcal{B}(f_J)) \right.$$

$$\left. - (-1)^{(a-1)(b-1)} \sum_K{}' \operatorname{sgn}(K; L) \mathcal{B}(f_K, \mathcal{A}(f_L)) \right)$$

$$
\begin{aligned}
&= \sum_{s=1}^{a+b} (-1)^{s+1} (\mathcal{X} f_s) \sum_I{}' \operatorname{sgn}(I; J) \mathcal{A}(f_I, \mathcal{B}(f_J)) \\
&\quad - (-1)^{(a-1)(b-1)} \sum_{s=1}^{a+b} (-1)^{s+1} (\mathcal{X} f_s) \sum_K{}' \operatorname{sgn}(K; L) \mathcal{B}(f_K, \mathcal{A}(f_L))
\end{aligned}
$$

We can rewrite II as follows:

$$
\begin{aligned}
\mathrm{II} = &\sum_{j_1<\cdots<j_a}{}' \sum_{s=1}^{a} \operatorname{sgn}(j_s, j_1 < \cdots \widehat{j_s} \ldots < j_a; K) \\
&\quad \times \mathcal{X}(f_{j_s}) \mathcal{A}(\ldots, \widehat{f_{j_s}}, \ldots, \mathcal{B}(f_K)) \\
&- (-1)^{a(b-1)} \sum_{k_1<\cdots<k_{a+1}}{}' \sum_{s=1}^{a+1} \operatorname{sgn}(J; k_1 < \cdots \widehat{k_s} \ldots < k_{a+1}, k_s) \\
&\quad \times (-1)^a \mathcal{X}(f_{k_s}) \mathcal{B}(f_J, \mathcal{A}(\ldots, \widehat{f_{k_s}}, \ldots)) \\
= &\sum_{j_1<\cdots<j_a}{}' \sum_{s=1}^{a} (-1)^{j_s+1} \operatorname{sgn}(j_1 < \cdots \widehat{j_s} \ldots < j_a; K) \\
&\quad \times \mathcal{X}(f_{j_s}) \mathcal{A}(\ldots, \widehat{f_{j_s}}, \ldots, \mathcal{B}(f_K)) \\
&- (-1)^{ab} \sum_{k_1<\cdots<k_{a+1}}{}' \sum_{s=1}^{a+1} (-1)^{a+b-1+k_s+1} \operatorname{sgn}(J; k_1 < \cdots \widehat{k_s} \ldots < k_{a+1}) \\
&\quad \times \mathcal{X}(f_{k_s}) \mathcal{B}(f_J, \mathcal{A}(\ldots, \widehat{f_{k_s}}, \ldots)) \\
= &\sum_{j_1<\cdots<j_a}{}' \sum_{s=1}^{a} (-1)^{j_s+1} \operatorname{sgn}(J'; K) \mathcal{X}(f_{j_s}) \mathcal{A}(f_{J'}, \mathcal{B}(f_K)) \\
&- (-1)^{(a-1)(b-1)} \sum_{k_1<\cdots<k_{a+1}}{}' \sum_{s=1}^{a+1} (-1)^{k_s+1} \operatorname{sgn}(J; K') \\
&\quad \times \mathcal{X}(f_{k_s}) \mathcal{B}(f_J, \mathcal{A}(f_{K'})) \\
= &\ (\mathcal{X} \wedge [[\mathcal{A}, \mathcal{B}]])(f_1, \ldots, f_{a+b})
\end{aligned}
$$

Thus we have obtained

$$
[[\mathcal{X} \wedge \mathcal{A}, \mathcal{B}]] = \mathcal{X} \wedge [[\mathcal{A}, \mathcal{B}]] + (-1)^{a(b-1)} [[\mathcal{X}, \mathcal{B}]] \wedge \mathcal{A}\,.
$$

■

Super skewness of $[[\cdot, \cdot]]$ and Theorem 4 imply:

Corollary 1

$$
[[\mathcal{A}, \mathcal{B} \wedge \mathcal{C}]] = [[\mathcal{A}, \mathcal{B}]] \wedge \mathcal{C} + (-1)^{(a-1)b} \mathcal{B} \wedge [[\mathcal{A}, \mathcal{C}]]
$$

holds for each skew product $\mathcal{A}$, $\mathcal{B}$ and $\mathcal{C}$ satisfying Leibniz' rule of degree a, b and c.

Final Remarks

The Schouten–Nijenhuis bracket $[\mathbf{A}, \mathbf{B}]_{\text{SN}}$ of multi-vector fields $\mathbf{A}$ and $\mathbf{B}$ of degree a and b is characterized as follows:

1. $(\mathbf{A}, \mathbf{B}) \mapsto [\mathbf{A}, \mathbf{B}]_{\text{SN}}$ is a graded Lie algebra bracket with **reduced** grading:so $\deg[\mathbf{A}, \mathbf{B}]_{\text{SN}} = a + b - 1$ and
   $[\mathbf{A}, \mathbf{B}]_{\text{SN}} = -(-1)^{(a-1)(b-1)}[\mathbf{B}, \mathbf{A}]_{\text{SN}}$.
2. For each $\mathbf{C}$, $[\cdot, \mathbf{C}]_{\text{SN}}$ is a degree $(c-1)$ graded derivation from the right of the super-commutative algebra with the natural wedge product, i.e.,
   $$[\mathbf{A} \wedge \mathbf{B}, \mathbf{C}]_{\text{SN}} = \mathbf{A} \wedge [\mathbf{B}, \mathbf{C}]_{\text{SN}} + (-1)^{b(c-1)}[\mathbf{A}, \mathbf{C}]_{\text{SN}} \wedge \mathbf{B}\,.$$

Since we have already proved the same properties for the bracket $[[\mathcal{A}, \mathcal{B}]]$ and have observed that they coincide when $\deg \mathcal{A} = \deg \mathcal{B} = 1$, we can deduce the next theorem:

Theorem 5 The multi-vector field of the bracket $[[\mathcal{A}, \mathcal{B}]]$ is $[\mathbf{A}, \mathbf{B}]_{\text{SN}}$ for each skew products $\mathcal{A}$ and $\mathcal{B}$ satisfying Leibniz' rule.

Remark 1 The explicit formula of the Schouten–Nijenhuis bracket is given by

$$\begin{aligned}&[X_1 \wedge \cdots \wedge X_p, Y_1 \wedge \cdots \wedge Y_q]_{\text{SN}} \\ &\qquad = \sum (-1)^{i+j}[X_i, Y_j] \wedge X_1 \wedge \cdots \wedge \widehat{X_i} \wedge \cdots \wedge \widehat{Y_j} \wedge \cdots \wedge Y_q\end{aligned}$$

where $X_1, \ldots, Y_q$ are vector fields.

There is another definition of Schouten–Nijenhuis bracket: For each $\mathbf{C}$, $[\cdot, \mathbf{C}]_{\text{SN}}$ is a degree $(c-1)$ graded derivation from the **left** of the super-commutative algebra with the natural wedge product. Then the explicit formula is

$$\begin{aligned}&[X_1 \wedge \cdots \wedge X_p, Y_1 \wedge \cdots \wedge Y_q]'_{\text{SN}} \\ &\quad = (-1)^{(p-1)(q-1)} \sum (-1)^{i+j}[X_i, Y_j] \wedge X_1 \wedge \cdots \wedge \widehat{X_i} \wedge \cdots \wedge \widehat{Y_j} \wedge \cdots \wedge Y_q\end{aligned}$$

where $X_1, \ldots, Y_q$ are vector fields (see [1]).

References

[1] M. Gerstenhaber and S. D. Schack, Algebraic cohomology and deformation theory, in Michiel Hazewinkel and Murray Gerstenhaber, editors,

Deformation Theory of Algebras and Structures and Applications, volume 247 of *NATO ASI Series C: Mathematics and Physical Sciences*, pages 11–264, Kluwer Academic Publishers, Dordrecht, Boston, London, 1988.

[2] J.-L. Koszul, Crochet de Schouten–Nijenhuis et cohomologie, in *Élie Cartan et les Mathématiques d'aujourd'hui*, pages 257–271, Société Math. de France, Astérisque, hors série, 1985.

[3] G.Marmo, G.Vilasi and A.M.Vinogradov, The local structure of n-Poisson and n-Jacobi manifolds, *J. Geom. Phys.* **25** (1998), 141–182.

[4] A. Nijenhuis, Jacobi-type identities for bilinear differential concomitants of certain tensor fields, *Indag. Mathematics*, **t.17** (1955) 390–403.

[5] J. A. Schouten. On the differential operators of first order in tensor calculus, *Convegno Internazionale di Geometria Differenziale*, pages 1–7, 1954, Italia, 20–26 September 1953.

REMARKS ON THE CHARACTERISTIC CLASSES ASSOCIATED WITH THE GROUP OF FOURIER INTEGRAL OPERATORS

Dedicated to Professor Hideki Omori on his 60th birthday

Naoya Miyazaki
Department of Mathematical Sciences, Faculty of Science,
Yokohama City University,
22-2 Seto, Kanazawa-ku, Yokohama, 236-0027 JAPAN
Naoya.Miyazaki@math.yokohama-cu.ac.jp

Abstract We introduce secondary characteristic classes associated with symplectic diffeomorphism groups which recapture Maslov classes associated with the linear symplectic group. Analogously we also introduce secondary characteristic classes associated with Lie group FIO(N) of invertible Fourier integral operators on a manifold N ([2, 3, 4] and [18, 19]). As an application we find a nontrivial cycle in FIO(S^n) as a lift of the geodesic flow of a sphere.

Mathematics Subject Classification (2000): 58D05, 22E65, 57S05, 58B25, 17B65

Keywords: Maslov form, characteristic classes, Infinite dimensional Fréchet Lie group.

1. Introduction

In this article we define secondary characteristic classes associated with the group of Fourier integral operators, which may be regarded as a slight extension of the usual Maslov classes associated with the linear symplectic groups.

Y. Maeda, et al. (eds.), Noncommutative Differential Geometry and Its Applications to Physics, 145–154.

In the theory of deformation quantization (cf. [5], [6], [8], [16], [17], [21]), it is known that each element of the automorphism group of a noncommutative associative algebra obtained by deformation of the Poisson algebra of a symplectic manifold induces a symplectic diffeomorphism on the base symplectic manifold. It is viewed as a quantization of the symplectic diffeomrphism group. Conversely, it is a fundamental question whether for any symplectic diffeomorphism there exists a lift (an automorphism which induces the symplectic diffeomorphism). Furthermore, it is natural to ask whether for a periodic Hamilton flow there exists a periodic lift. Since FIO(N) can be regarded as a quantization of the homogeneous symplectic diffeomorphism group on the cotangent bundle, it is also interesting to ask whether there exists a nontrivial periodic lift of the periodic Hamilton (geodesic) flow in FIO(N). As an application of secondary characteristic classes we show the existence of a nontrivial periodic lift in FIO(S^n).

We briefly review the Maslov index. The theory of the Maslov index is initiated by the construction of asymptotic solutions for the Schrödinger equation (cf. [11], [10]). First we recall the definition of the Maslov index of a loop Φ in the linear symplectic group $Sp(2n, \mathbf{R})$. Let $\eta : Sp(2n,\mathbf{R}) \times [0,1] \to Sp(2n,\mathbf{R}) : (X,s) \mapsto (X^tX)^{-\frac{s}{2}}X$ be a retraction of $Sp(2n,\mathbf{R})$ onto $Sp(2n,\mathbf{R}) \cap O(2n) \stackrel{i}{=} U(n)$, where i is the canonical identification. Then we define ρ by $\rho : Sp(2n,\mathbf{R}) \to U(1); X \mapsto \det \circ i((X^tX)^{-\frac{1}{2}}X)$. Using the map ρ, the Maslov index of $[\Phi] \in \pi_1(Sp(2n,\mathbf{R}))$ is defined by

$$\nu([\Phi]) := \deg(\rho \circ \Phi); \pi_1(Sp(2n,\mathbf{R})) \to \mathbf{Z}. \tag{1}$$

Since the retraction η plays a crucial role for this definition, we cannot use the same procedure in order to define the Maslov index associated with the group FIO(N) of Fourier integral operators. In this article we shall extend the notion of Maslov index to the group FIO(N) and give some applications. Moreover, using transgression formulae just as Godbillon–Vey classes, we introduce secondary characteristic classes associated with FIO(N).

2. Secondary characteristic classes associated with a Lagrangian fiber bundle

In this section we give the definition of secondary characteristic classes associated with Lagrangian fiber bundle. First we define the notion of Lagrangian fiber bundle.

Definition 1 *Let S be a symplectic manifold. A Lagrangian fiber bundle $(\mathcal{M} \stackrel{\pi}{\to} G, S)$ is a subbundle of the symplectic fiber bundle $G \times S$ on G satisfying the following conditions:*

(1) *For any $g \in G$, there exists a neighborhood U of g and a Lagrangian submanifold L_0 of the symplectic manifold S such that $\pi^{-1}(U) \cong U \times L_0$.*

(2) *The projection* $pr : \mathcal{M} \to S$ *is smooth.*
The fiber through g *is denoted* * *by* $\mathcal{M}_g$.

Example Let (M, ω) be a symplectic manifold and G the symplectic diffeomorphism group $\mathrm{Diff}_\omega(M)$. Then

$$\mathrm{Diff}_\omega(M) \times M \ni (\varphi, p) \leftrightarrow (\varphi, p, \varphi(p)) \in \bigsqcup_{\varphi \in \mathrm{Diff}_\omega(M)} \mathrm{Graph}(\varphi) =: \mathcal{M} \quad (2)$$

is an example of a Lagrangian fiber bundle, where we regard a canonical graph $\mathrm{Graph}(\varphi)$ as a Lagrangian fiber $\mathcal{M}_\varphi$ and

$$pr : \bigsqcup_{\varphi \in \mathrm{Diff}_\omega(M)} \mathrm{Graph}(\varphi) \ni (\varphi, p, \varphi(p)) \to (p, \varphi(p)) \in M \times M = S. \quad (3)$$

Before introducing secondary characteristic classes associated with Lagrangian fiber bundle, we recall the *Souriau map* (cf. [1], [10], [23], [25]). It is well known that the unitary group $U(n)$ acts on $\Lambda(n)$ transitively, and $\Lambda(n) = U(n)/O(n)$. Let $\lambda_{im} = \{\sqrt{-1}x | x \in \mathbf{R}^n\} \in \Lambda(n)$. Then for any $\lambda \in \Lambda(n)$ there exists $U_\lambda \in U(n)$ satisfying $\lambda = U_\lambda \lambda_{im}$. Using this U_λ, we can define a mapping W of $\Lambda(n)$ into $U(n)$, $W(\lambda) = U_\lambda {}^t U_\lambda$, where W is called the Souriau map.

Next we recall the construction of the *generating function* of a Lagrangian submanifold L of $\mathbf{C}^n$. For example, let p_0 be a point of L such that $T_{p_0}L$ transversely intersects $\lambda_{re} = \{\xi | \xi \in \mathbf{R}^n\}$. Then there is a neighborhood V of p_0 in L parameterized the variable $x \in U \subset \lambda_{im}$, i.e. $L|_V = \{(x, \xi(x)) | x \in U \subset \lambda_{im}\}$.On the other hand, the restriction of standard canonical 1-form θ to L is a closed 1-form. Thus, by Poincaré's lemma, we have a local potential function S of $\theta|_V$ as follows:

$$S(x) = \int_{p_0}^{p} \theta, \quad \text{where} \quad p_0 = (0, \xi(0)), \; p = (x(p), \xi(x(p))). \quad (4)$$

Hence we have$L|_V = \{ (x, \partial S(x)/ \partial x) | \, x \in U\}$. We shall refer to the function S as the generating function of L around p_0. Furthermore, it is well-known in [1] that

$$W(\tau(p)) = \frac{1_n - \sqrt{-1}\partial_x \partial_x S(x)|_{x(p)}}{1_n + \sqrt{-1}\partial_x \partial_x S(x)|_{x(p)}}, \quad (5)$$

*Note that the fiber of $\mathcal{M}$ is not necessarily a Lagrangian vector space. On the other hand, the fiber of Lagrangian subbundle is a Lagrangian vector space.

where $\tau(p)$ indicates a tangent space at p of the Lagrangian submanifold and 1_n is the $n \times n$-identity matrix. That is, the image of the Souriau map is equal to the image of the Hessian matrix of the generating function of the Cayley transformation.

We now introduce the secondary characteristic classes associated with Lagrangian fiber bundle. We can induce a symplectic vector bundle $\mathcal{E} = pr^*(TS)$ on $\mathcal{M}$. Let J be an almost complex structure and h an Hermitian structure on $\mathcal{E}$ induced by pr. Thus we can consider unitary frame bundle $\mathcal{U}$ associated to $\mathcal{E}$. This bundle is principal $U(m)$-bundle where $2m = \dim S$. Moreover, we can consider Lagrangian subbundles as follows: By assumption $\mathcal{M}_g$ is a Lagrangian submanifold in S, then the tangent space $T_{(g,m_g)}\mathcal{M}_g$ is a Lagrangian subspace. Hence we can define a Lagrangian subbundle $\mathcal{L}_1$. On the other hand, if S has a polarization we can define another Lagrangian subbundle $\mathcal{L}_0$. For each Lagrangian subbundle $\mathcal{L}_i$ $(i = 0, 1)$, we consider '$\mathcal{L}$-related' unitary frame bundles $\mathcal{O}_i$ $(i = 0, 1)$ (cf. [23]). Note that these are principal $O(m)$-bundles. Then for each $O(m)$-bundles we fix $o(m)$-valued connection forms ω_i $(i = 0, 1)$. Set $\phi = (1-t)\omega_0 + t\omega_1$, and $m_k := \int_{[0,1]} c_{2k-1}(d\phi + \phi \wedge \phi)$, where c_i is the i-th Chern polynomial.

Lemma 1 *m_k is a closed $(4k-3)$-form.*

Proof Combining Stoke's formula, Bianchi's identity, and skew symmetry of the connection form we can demonstrate the theorem.

From now on we consider the case in which G is an infinite-dimensional Lie group whose elements relate to canonical graphs.

Example 1 If $G = Sp(2n, \mathbf{R})$ and $S = (\mathbf{R}^{2n}, \omega) \ominus (\mathbf{R}^{2n}, \omega)$, then we have secondary characteristic classes associated with linear symplectic group. Using m_1, set $\mu(\Phi) = -\frac{1}{2}\langle m_1|_{Sp(2n,\mathbf{R})}, \Phi\rangle$, then we have the following:

(**homotopy**) Two loops in $Sp(2n, \mathbf{R})$ are homotopic if and only if they have the same index with respect to μ.

(**product**) For any two loops Ψ_1, Ψ_2 in $Sp(2n, \mathbf{R})$ we have $\mu(\Psi_1\Psi_2) = \mu(\Psi_1) + \mu(\Psi_2)$.

(**direct sum**) When $n = n' + n"$, identify $Sp(2n', \mathbf{R}) \times Sp(2n", \mathbf{R})$ in the obvious way with a subgroup of $Sp(2n, \mathbf{R})$. Then $\mu(\Psi' \oplus \Psi") = \mu(\Psi') + \mu(\Psi")$.

(**normalization**) The loop $\Psi : S^1 \to U(1) \subset Sp(2n, \mathbf{R})$ defined by $\Psi(t) = e^{2\pi i t}$ has the index 1 with respect to μ.

Note that since the above axioms characterize Maslov index, $\mu = -\frac{1}{2}\langle m_1|_{Sp(2n,\mathbf{R})}, \ \rangle$ coincides with the Maslov index.

Example 2 Let $G = \mathrm{Diff}_\omega(\mathbf{R}^{2n})$, $S = \mathbf{R}^{2n} \ominus \mathbf{R}^{2n}$, then we can have secondary characteristic classes associated with $\mathrm{Diff}_\omega(\mathbf{R}^{2n}) \times \mathbf{R}^{2n}$. Using $m_1|_{\mathrm{Diff}_\omega(\mathbf{R}^{2n})}$, it is easy to see that the flow of harmonic oscillator gives a nontrivial cycle.

Example 3 Let $G = \mathrm{Diff}_\omega(T^*_\star N)$, $S = (T^*_\star N, d\theta) \ominus (T^*_\star N, d\theta)$, then we can construct secondary characteristic classes associated with $\mathrm{Diff}_\omega(T^*_\star N) \times (T^*_\star N, d\theta)$. Suppose $N = S^n$ and φ_g is a geodesic flow on the n-dimensional sphere S^n, then φ_g is a closed curve of $\mathrm{Diff}_\omega(T^*_\star N)$ and $\int_{\varphi_g} m_1|_{\mathrm{Diff}_\omega(T^*_\star N)} \circ W \circ \tau)^*(d\theta) = n$.

Example 4 Let $G = FIO(N)$, $S = (T^*N, d\theta) \ominus (T^*N, d\theta)$, $\gamma = WF$ (cf. [2, 3, 4], [15], [18, 19]), then we can construct secondary characteristic classes associated with FIO(N)$\times T^*N$. We give an explanation of this example in the next section.

3. Application of secondary characteristic classes associated with the Lie group of invertible Fourier integral operators

As mentioned in Introduction, since FIO(N) can be regarded as a quantization of homogeneous symplectic diffeomorphism group on the cotangent bundle, it is natural to ask whether there exists a nontrivial periodic lift of a periodic Hamilton flow in FIO(N). In this section, by virture of the Maslov form on FIO(N), we show the existence of such a periodic lift. First we recall the property for the wave front set (cf. [7], [9], [20]) from [18, 19]. The following is known:

Proposition 1 *Set*

$$\chi(\phi) = \{(\bar{x},\ d_{\bar{x}}\phi(\bar{x},\theta,x),\ x,\ -d_x\phi(\bar{x},\theta,x)) : d_\theta\phi(\bar{x},\theta,x) = 0\},$$

where ϕ is a phase function of Fourier integral operator. Then we have that $\mathrm{WF}(F(a,\phi)) = \chi(\phi)$ *and the wave front set gives a Lie group homomorphism* WF *of* FIO(N) *into* $\mathrm{Diff}_\omega(T^*_\star N)$*;* $F(a,\phi) \mapsto \chi(\phi)^{-1}$.

Note that in [2, 3, 4] $\chi(\phi)$ is referred as the canonical relation associated with Fourier integral operator.

Suppose that N is an orientable compact Riemannian manifold. Let $\{e_1, \cdots, e_n\}$ be an orthonormal frame on an open neighborhood U of the tangent bundle TN, and $(X, \Xi) = \{e_1, \cdots, e_n, e^1, \cdots, e^n\}$ a symplectic frame on $\pi^{-1}(U)$ of $T(T^*N)$. Then, by the definition, $\omega(e_i, e^j) = \delta_i^j$. Next, we set

$$\varphi_*(X(p), \Xi(p))$$

$$= (\tilde{X}(\varphi(p)), \tilde{\Xi}(\varphi(p))) \left(\begin{array}{c|c} A^i_j(\varphi; e, p) & B^{ij}(\varphi; e, p) \\ \hline C_{ij}(\varphi; e, p) & D^j_i(\varphi; e, p) \end{array} \right), \quad (6)$$

where $(\tilde{X}(\varphi(p)), \tilde{\Xi}(\varphi(p))) = \{\tilde{e_1}, \cdots, \tilde{e_n}, \tilde{e^1}, \cdots, \tilde{e^n}\}$ is a symplectic frame around $\varphi(p)$. With these notations we set

$$\Phi(\varphi; p) \\ = \det\{-D(\varphi; e, p) - A(\varphi; e, p) - \sqrt{-1}(C(\varphi; e, p) - B(\varphi; e, p))\}. \quad (7)$$

Then we obtain the following:

Proposition 2 *(7) is well defined as a function on* $\mathrm{Diff}_\omega(T^*_\star N) \times T^*N$, *i.e.*, Φ *is independent of the choice of symplectic frames.*

Therefore we use the following notation:

$$\Phi(\varphi; p) = \det\{-D(\varphi; p) - A(\varphi; p) - \sqrt{-1}(C(\varphi; p) - B(\varphi; p))\} \quad (8)$$

instead of (7).

Using the wave front set we have the following diagram:

$$\mathrm{FIO}(N) \times T^*N \overset{\mathrm{WF}}{\to} \mathrm{Diff}_\omega(T^*_\star N) \times T^*_\star N \\ \overset{\tilde{\tau}}{\to} \Lambda(2n) = U(2n)/O(2n) \overset{W}{\to} U(2n) \overset{\det}{\to} U(1), \quad (9)$$

where the image of WF is the wave front set, the image of $\tilde{\tau}(\varphi, p)$ is the tangent space of graph of φ at the reference point $(p, \varphi(p))$, and the mapping W is Souriau map. Note that, in general, the canonical graph of symplectic diffeomorphism on symplectic manifold (M, ω) is a Lagrangian submanifold in $(M \times M, \omega \ominus \omega)$. Therefore $\tau(\varphi, p)$ is a Lagrangian subspace of $(\mathbf{R}^{2n} \ominus \mathbf{R}^{2n}, \sigma \ominus \sigma)$. Since τ depends on the choice of orthonormal frame $\{e_1, \cdots, e_n\}$, it does not define a genuine map. However, the composition map obtained via (8) is well defined. We then have the following:

Proposition 3 $m_1 = (1/2\pi)(\det \circ W \circ \tau \circ \mathrm{WF})^*(d\theta)$.

Using the above function (7), we obtain the following:

Proposition 4 $m_1 = \frac{1}{\pi} d \arg \Phi(\mathrm{WF}(F(a, \phi)), p)$,

Proof. We compute right hand side of formula of Proposition 3. Let φ be an element of $\mathrm{Diff}_\omega(T^*_\star N)$ determined by the wave front set of the Fourier integral operator $F(a, \phi)$. Let $\mathbf{H}(x, \bar{\xi}, \xi, \bar{x})$ be a Jacobi involutive system (cf. [25]) of functions whose level surface $\mathbf{H} = \mathbf{0}$ coincides with canonical graph of the symplectic diffeomorphism φ. If $S(x, \bar{\xi})$ is a generating function of the canonical graph φ then

$$\mathbf{H}(x, \bar{\xi}, \partial_{(x,\bar{\xi})} S(x, \bar{\xi})) = 0.$$

Combining this with (5) gives the formula above.

Using these 1-forms we have the following (cf. [12]):

Proposition 5 *Suppose that $N = S^n$. If Φ_P is a solution of the Schrödinger equation with respect to 1st order pseudo-differential operator*

$$P = \sqrt{-\Delta + \left(\frac{n-1}{2}\right)^2},$$

that is,

$$\frac{d}{dt}\Phi_P = -\sqrt{-1}P\Phi_P, \qquad \Phi_P(0) = id,$$

then Φ_P is a closed curve of FIO(N) *and*

$$\int_{\Phi_P} m_1 = \begin{cases} 2n & (n : \text{even}), \\ n & (n : \text{odd}). \end{cases} \tag{10}$$

Proof It is already known that Φ_P is a curve of FIO(N) (see [18, 19]). On the other hand, we know that P is a block-wise diagonal matrix such that k-th block is the diagonal matrix

$$\text{diag}\left(k + \frac{n-1}{2}, \cdots, k + \frac{n-1}{2}\right)$$

of size

$$\frac{k + (n-1)/2}{(n-1)/2} \frac{(k+n-2)!}{k!(n-2)!}$$

(cf. [22]). Therefore,

$$\Phi_P(t) = \exp\left(-\sqrt{-1}\sqrt{-\Delta + \left(\frac{n-1}{2}\right)^2}\,t\right)$$

is a closed curve. If n is an even (odd) number, $\Phi_P(t)$ is a 2- (resp. 1-) fold covering of φ_g in Example 3. Thus we have the conclusion. See also Example 3.

Remark Note that for the Lie group of oscillatory integral transformations, the product integral for the harmonic oscillator $H(x, \xi) = \frac{1}{2}(|x|^2 + |\xi|^2)$ gives the following formula:

$$
\begin{aligned}
\Phi_H &= \exp(-\sqrt{-1}\widehat{H}t) \\
&= \lim_{L\to\infty} \int\cdots\int \prod_{i=1}^{L-1} \bar{d}\mathbf{x}_i \int\cdots\int \prod_{j=1}^{L} \bar{d}\xi_j \\
&\qquad \exp\left[\sqrt{-1}\tau \sum_{k=1}^{L} \xi_k \cdot \frac{\mathbf{x}_k - \mathbf{x}_{k-1}}{\tau} - \left(\frac{|\xi_k|^2}{2} + \frac{|\mathbf{x}_k|^2}{2}\right)\right] \\
&= \lim_{L\to\infty} \int\cdots\int \prod_{i=1}^{L-1} \bar{d}\mathbf{x}_i \\
&\qquad \exp\left[\sqrt{-1}\tau \sum_{k=1}^{L} \left\{\frac{1}{2}\left|\frac{\mathbf{x}_k - \mathbf{x}_{k-1}}{\tau}\right|^2 - \frac{|\mathbf{x}_k|^2}{2}\right\}\right] \\
&= \left(\frac{1}{2\pi\sqrt{-1}|\sin t|}\right)^{\frac{n}{2}} \exp\left\{n\left(-\frac{\sqrt{-1}\pi}{2}k\right)\right\} \\
&\qquad \times \exp\left\{n\left(\frac{1}{2\sqrt{-1}\sin t}\{(x^2+y^2)\cdot\cos t - 2xy\}\right)\right\}, \quad (11)
\end{aligned}
$$

where $k\pi < t < (k+1)\pi, \tau = T/L$, and we have successively used Fresnel's integral formula

$$\int_{-\infty}^{\infty} dp \quad e^{-\sqrt{-1}\alpha p^2} = \sqrt{\frac{\pi}{\sqrt{-1}\alpha}}.$$

The above curve gives a non-trivial cycle (cf. [13]).

Acknowledgments

I would like to express sincere gratitude to Professor H.Omori for his encouragement and helpful advice. I also thank Professors K.Fujii, Y.Maeda, K.Mikami, H.Moriyoshi, D.Sternheimer, A. Weinstein and A.Yoshioka for their valuable comments.

It is a pleasure to acknowledge with the gratitude the support of Grant-in-Aid for Science Research, Ministry of Education, Science and Culture, Japan and of the Grant in Support of Promotion of Research at Yokohama City University.

References

[1] V. I. Arnol'd, On a characteristic class entering in quantization conditions, *Func.Anal.Appl.* **1** (1967), 1–13.

[2] M. Adams, T. Ratiu and R. Schmid, The Lie group structure of diffeomorphism groups and invertible Fourier integral operators with applications, in *Infinite dimensional Groups with applications,* eds. V.Kac, Springer (1985), pp.1–69 .

[3] M. Adams, T. Ratiu and R. Schmid, A Lie group structure for Pseudodifferential Operators, *Math. Ann.* **273** (1986), 529–551.

[4] M. Adams, T. Ratiu and R. Schmid, A Lie group structure for Fourier integral Operators, *Math. Ann.* **276** (1986), 19–41.

[5] F. Bayen, M. Flato, C. Fronsdal, A. Lichnerowicz and D. Sternheimer, Deformation theory and quantization I, II, *Ann.Phys.* **111** (1978), 61–151.

[6] V. Fedosov, A simple geometric construction of deformation quantization, *J.Diff. Geom.* **40-2** (1994), 213–248.

[7] L. Hörmander, Fourier integral operators I, *Acta Math.* **121** (1971), 79–183.

[8] M. Kontsevich, Deformation quantization of Poisson manifolds, I, q-alg/9709040

[9] H. Kumanogo, *Pseudo differential operators*, MIT Press, 1982.

[10] J. Leray, *Lagrangian analysis and mechanics ; a mathematical structure related to asymptotics expansion and Maslov index*, MIT Press, Cambridge, 1981.

[11] V. P. Maslov, *Theory of Perturbations and Asymptotic Methods*, izd. MGU, 1965.

[12] N. Miyazaki, A remark on the Maslov form on the group generated by invertible Fourier integral operators, *Lett. Math. Phys.* **42** (1997), 35–42.

[13] N. Miyazaki, Secondary characteristic classes and Feynman path integrals, submitted.

[14] H. Omori, *Theory of infinite dimensional Lie groups*, Amer. Math. Soc. Trans.158, 1996.

[15] H. Omori, Y. Maeda, and A. Yoshioka, On regular Fréchet-Lie groups I, II, *Tokyo J. of Math.* **3** (1980), 353–390, **4** (1981), 231–253.

[16] H. Omori, Y. Maeda, and A. Yoshioka, Weyl manifolds and deformation quantization, *Adv. in Math.* **85** (1991), 224–255.

[17] H. Omori, Y. Maeda, N. Miyazaki and A. Yoshioka, Poincaré-Cartan class and deformation quantization, *Commun. Math. Phys.* **194** (1998), 207–230.

[18] H. Omori, Y. Maeda, A. Yoshioka, and O. Kobayashi, On Regular Fréchet-Lie Groups III, IV, V, VI, VII, VIII, *Tokyo J. of Math.* **4** (1981), 255–277, **5** (1982), 365–398, **6** (1983), 39–64, **6** (1983), 217–246, **7** (1984), 315–336, **8** (1985), 1–47.

[19] H. Omori, Y. Maeda, A. Yoshioka, and O. Kobayashi, The theory of infinite dimensional Lie groups and its applications, *Acta Appl. Math.* **3** (1985), 71–105.

[20] C. D. Sogge, *Fourier integrals in classical analysis,* Cambridge University Press, 1993.

[21] D. Sternheimer, Deformation Quantizatin Twenty Years After, q-alg/9809056

[22] M. Takeuchi, *Modern spherical functions*, Translations of Mathematical Monographs, Vol.135, American Mathematical Society.

[23] I.Vaismann, *Symplectic geometry and secondary characteristic classes*, Progress in Math. 72, (1987), Birkhäuser,

[24] A. Weinstein, Some questions about the index of quantized contact transformations, *RIMS Kôkyûroku* **1014** (1997), 1–14.

[25] A. Yoshioka, Maslov Quantization Conditions for the Bounded States of the Hydrogen Atom, *Tokyo Journal of Math.* **9** (1986), 415–437.

C^*-ALGEBRAIC DEFORMATION AND INDEX THEORY

Toshikazu Natsume
School of Mathematics, Nagoya Institute of Technology
Showa-ku, Nagoya 466-8555, JAPAN

Abstract In this note we explain how C^*-algebraic deformation of symplectic manifolds yields a down to earth proof of Atiyah–Singer type of index formulae.

Mathematics Subject Classification (2000): 46L65, 58J20

Keywords: quantization, C*-algebra, index theory

1. Introduction

In [6] we studied a family of pseudo-differential operators on $\mathbb{R}^n$ and showed an Atiyah–Singer type of index theorem. The key idea is the use of continuous fields of C^*-algebras and continuous fields of cyclic cocycles. The continuous fields used there come from C^*-algebraic deformation quantization of the Poisson manifolds $T^*\mathbb{R}^n$.

Definition 1.1. By a *C^*-algebraic deformation quantization* of a Poisson manifold M, we mean a continuous field of C^*-algebras $(A_\hbar)$ over $[0,\epsilon)$ for some $\epsilon > 0$ equipped with linear maps $\pi_\hbar : C_c^\infty(M) \longrightarrow A_\hbar$, $\hbar \in [0,\epsilon)$ satisfying the conditions:

(1) $A_0 = C_0(M)$, and π_0 is the canonical inclusion of $C_c^\infty(M)$ into A_0;

(2) for every $f \in C_c^\infty(M)$, the vector field $(\pi_\hbar(f))$ is continuous;

(3) $\|(1/i\hbar)(\pi_\hbar(f)\pi_\hbar(g) - \pi_\hbar(g)\pi_\hbar(f)) - \pi_\hbar(\{f,g\})\| \longrightarrow 0$ as $\hbar \longrightarrow 0$ for any fixed f and g in $C_c^\infty(M)$; and

Y. Maeda, et al. (eds.), Noncommutative Differential Geometry and Its Applications to Physics, 155–167.

(4) the C^*-algebra generated by the linear subspace $\pi_\hbar(C_c^\infty(M))$ is dense in $A_\hbar$ for every $\hbar$.

In the present note we discuss two applications of the idea mentioned above. One is Connes's index theorem for longitudinal elliptic operators for foliations. The other is the index theorem for families of operators.

2. Continuous field of C^*-algebras

We briefly review basic definitions and pertinent properties of continuous fields of C^*-algebras. As for detail see [3], [9].

Suppose that $\mathcal{B} = (B_t)$ is a family of Banach spaces parametrized by the points of a topological space T. Let Θ be a linear subspace of the space $\prod_t B_t$ of vector fields (sections).

Definition 2.1. A structure of a *continuous field of Banach spaces* is given by a pair $(\mathcal{B}, \Theta)$ which satisfies the conditions:

1. For every $t \in T$, the set $\{\xi(t); \xi \in \Theta\}$ is dense in B_t;
2. For every $\xi \in \Theta$, the function $t \mapsto \|\xi(t)\|$ is continuous;
3. Let ξ be a vector field. If, for every $t \in T$ and any $\epsilon > 0$, there exists a $\xi' \in \Theta$ such that $\|\xi'(t') - \xi(t')\| \leq \epsilon$ on some neighbourhood of t, then ξ belongs to Θ.

The condition 3 means that Θ is closed under local uniform convergence. Any $\xi \in \Theta$ is referred to as *continuous vector field.*

In Definition 2.1, if all B_t's are C^*-algebras, and if Θ is a $*$-subalgebra of $\prod B_t$, then the field $(\mathcal{B}, \Theta)$ is called a *continuous field of C^*-algebras.* If B_t's are Hilbert spaces, then the field is a *continuous field of Hilbert spaces.*

If the set $\coprod B_t$ has a structure of topological space, and if the canonical map $\coprod B_t \longrightarrow T$ is continuous, one may define Θ as the space of all continuous sections $T \longrightarrow \coprod B_t$. When this is not possible, we define 'continuity' by specifying the space of continuous vector fields.

Let $\mathcal{A} = (A_t)_{t \in T}$ be a parametrized family of C^*-algebras. Suppose that a $*$-subalgebra $\Delta \subset \prod A_t$ satisfies the conditions 1 and 2 of Definition 2.1. Then the space Θ of sections, locally uniformly approximated by continuous sections, satisfies all the conditions 1–3. Thus it is enough to obtain a $*$-subalgebra with 1 and 2, in order to define a continuous field of C^*-algebras.

Example 2.2. Let X be a topological space, and let A be a C^*-algebra. Set $A_x = A, x \in X$. Let Θ be the space of all continuous maps from X into A. Then $((A_x), \Theta)$ is a continuous field of C^*-algebras, called a *constant field.*

Example 2.3. Let $p : E \longrightarrow X$ be a Hilbert space bundle. For every $x \in X$ set $H_x = p^{-1}(x)$. If Θ is the space of all continuous sections of p, then $((H_x), \Theta)$ is a continuous field of Hilbert spaces.

Example 2.4. Suppose that $((H_x), \Theta)$ is a continuous field of Hilbert spaces. Then the collection $(\mathcal{K}(H_x))$ has a structure of continuous field of C^*-algebras (10.7 of [3]).

Continuous fields which will be used in the index theory are the following.

Example 2.5 (*Kohn–Nirenberg quantization*). Let $A_0 = C_0(T^*\mathbb{R}^n)$, and let $A_\hbar = \mathcal{K}(L^2(\mathbb{R}^n)), 0 < \hbar \leq 1$. For $a \in C_c^\infty(T^*\mathbb{R}^n)$ set, as before, $a_\hbar(x,\xi) = a(x, \hbar\xi),\ \ \hbar > 0$, and set

$$\rho_\hbar(a) = \begin{cases} \mathrm{Op}(a_\hbar), & \hbar > 0, \\ M_a, & \hbar = 0. \end{cases}$$

Then $\Theta_0 = \{(\rho_\hbar(a))_\hbar\ ; a \in C_c^\infty(T^*\mathbb{R}^n)\}$ satisfies the conditions 1 and 2 of Definition 2.1. Thus we find a $*$-subalgebra $\Theta \subset \prod A_\hbar$ so that $(A = (A_\hbar), \Theta)$ is a continuous field. Notice that this field is constant away from 0. The Kohn–Nirenberg quantization is a C^*-algebraic deformation quantization of $T^*\mathbb{R}^n$.

Example 2.6. Suppose that α is an $\mathbb{R}^n$-action on $A = C_0(X)$. Denote by $\alpha^\hbar (0 \leq \hbar \leq 1)$ the rescaled action :

$$\alpha_t^\hbar(a) = \alpha_{\hbar t}(a)\ ,\ t \in \mathbb{R}^n.$$

Define an $\mathbb{R}^n$-action $\tilde{\alpha}$ on $C([0,1]) \otimes A$ by

$$\tilde{\alpha}_t(a)(\hbar) = \alpha_t^\hbar(a)\ ,\ a \in C([0,1]) \otimes A, \hbar \in [0,1].$$

Set $A_\hbar = C_0(X) \rtimes_{\alpha^\hbar} \mathbb{R}^n$. Then $((A_\hbar), \Theta)$ is a C^*-algebraic deformation of $X \times \mathbb{R}^n$, where $\Theta = (C_0([0,1]) \otimes A) \rtimes_{\tilde{\alpha}} \mathbb{R}^n \times \mathbb{R}^n$.

Example 2.7. Let (A, Θ) be a continuous field of C^*-algebras over T . Then the family $\tilde{A} = (\tilde{A_\hbar})$, where $\tilde{A_\hbar}$ is $A_\hbar$ with unit adjoined, has a structure of continuous field of C^*-algebras. The family $(\mathrm{M}_N(A_\hbar))_\hbar$ also has a structure of continuous field of C^*-algebras.

3. Index theorem on $\mathbb{R}^n$

We recall the main result of [6]. The Kohn–Nirenberg quantization assigns an operator P_a on $L^2(\mathbb{R}^n)$ to a smooth function a on $T^*\mathbb{R}^n$.

Definition 3.1. A *symbol of order m* is a C^∞-function a on $T^*\mathbb{R}^n$ such that for arbitrary multi-indices α, β, there exists a constant C with

$$|(\partial_x^\alpha \partial_\xi^\beta a)(x,\xi)| \leq C(1 + |x| + |\xi|)^{m-|\alpha|-|\beta|},$$

where $|x| = \sqrt{x_1^2 + \cdots + x_n^2}$, $|\xi| = \sqrt{\xi_1^2 + \cdots + \xi_n^2}$, and $|\alpha| = \alpha_1 + \cdots + \alpha_n$.

A symbol a defines an operator P_a by the formula :

$$(P_a u)(x) = (2\pi)^{-n} \int_{T^*\mathbb{R}^n} \mathrm{e}^{i\langle x-y,\xi\rangle} a(x,\xi) u(y) \mathrm{d}y \mathrm{d}\xi, \quad u \in C_c^\infty(\mathbb{R}^n).$$

It is known [10] that if $m \leq 0$, then P_a extends to a bounded operator on $L^2(\mathbb{R}^n)$, in particular, if $m < 0$, then P_a is compact. The operator P_a is called a *pseudo-differential operator* (abbreviated ΨDO) *with symbol a.*

Roughly speaking, a symbol of order m is a polynomial function in $x_1, \ldots, x_n, \xi_1, \ldots, \xi_n$ of order m. An elliptic symbol of order m is like a polynomial of order m in every direction.

Definition 3.2. Let a be a symbol of order m. The symbol is *elliptic* if there exist $C, R > 0$ such that

$$|a(x,\xi)| \geq C(|x| + |\xi|)^m, \text{ for } |x| + |\xi| \geq R.$$

The ΨDO's which we are interested in have some of the good properties enjoyed by ΨDO's on compact manifolds, for instance, the existence of a parametrix (pseudo-inverse) for any elliptic operator. Consequently the regularity theorem holds. In particular, if a is elliptic then $\mathrm{Ker}P_a$, $\mathrm{Coker}P_a$ are finite-dimensional, and the index is defined by

$$\mathrm{ind}P_a = \dim \mathrm{Ker}P_a - \dim \mathrm{Coker}P_a.$$

So far we discussed ΨDO's acting on scalar-valued functions. Matrix-valued symbols can be defined. For a symbol $a \in \mathrm{M}_N(C^\infty(T^*\mathbb{R}^n))$, the operator P_a acts on $\mathbb{C}^N$-valued functions. Set $V = \mathbb{C}^N$. If a is elliptic (Definition 2.7 of [6]), then P_a is Fredholm as (possibly unbounded) operator on $L^2(\mathbb{R}^n; V)$. Consequently, $\mathrm{ind}P_a$ is defined.

The main result of [6] is a formula which describes $\mathrm{ind}P_a$ in terms of the symbol a.

Set

$$e_a = \begin{pmatrix} (1+a^*a)^{-1} & (1+a^*a)^{-1}a^* \\ a(1+a^*a)^{-1} & a(1+a^*a)^{-1}a^* \end{pmatrix},$$

and

$$\widehat{e}_a = e_a - \left(\begin{smallmatrix} 0 & 0 \\ 0 & 1 \end{smallmatrix}\right).$$

Now assume that $m > 0$. Then the ellipticity implies that the differential $2n$-form

$$\mathrm{trace}(\widehat{e}_a(\mathrm{d}\widehat{e}_a)^{2n})$$

is integrable over $T^*\mathbb{R}^n$.

THEOREM 3.3. *If P_a is elliptic of positive order, then*

$$\mathrm{ind}P_a = \frac{1}{(2\pi i)^n\, n!}\int_{T^*\mathbb{R}^n} \mathrm{trace}(\widehat{e}_a(\mathrm{d}\widehat{e}_a)^{2n}).$$

Outline of proof We use the continuous field of Example 2.5.

For $0 < \hbar \leq 1$, set $a_\hbar(x,\xi) = a(x,\hbar\xi)$, and denote by $P_\hbar$ the ΨDO with symbol $a_\hbar$. The operator $P_\hbar$, defined initially on $C_c^\infty(\mathbb{R}^n)\otimes\mathbb{C}^N$, extends to a closed operator on $L^2(\mathbb{R}^n\ ;\ \mathbb{C}^N)$. Denote by $e_\hbar$ the orthogonal projection of $L^2(\mathbb{R}^n\ ;\ \mathbb{C}^N)\oplus L^2(\mathbb{R}^n\ ;\ \mathbb{C}^N)$ onto the graph of $P_\hbar$.

Look at the equality in Theorem 3.3. Denote byP_+, P_- the orthogonal projection onto $\ker P_a$, $\mathrm{Coker}P_a$, respectively. The projectionsP_+, P_- defines a K_0-class $[P_+]-[P_-]\in K_0(\mathcal{K}(L^2(\mathbb{R}^n,\mathbb{C}^N))$, and

$$\text{analytic index} = \text{left hand side} = \langle \mathrm{Tr}, [P_+]-[P_-]\rangle.$$

On $C_c^\infty(T^*\mathbb{R}^n, M_{2N}(\mathbb{C}))$ define a cyclic $2n$-cocycle ε by

$$\varepsilon(f^0,\dots,f^{2n}) = \int_{T^*\mathbb{R}^n} \mathrm{trace}(f^0\mathrm{d}f^1\wedge\cdots\wedge\mathrm{d}f^{2n}).$$

The cocycle ε extends to a holomorphically closed $*$-subalgebra.

The C^∞-function e_a takes values in projections of $M_{2N}(\mathbb{C})$, and

$$e_a - \left(\begin{smallmatrix}0&0\\0&1\end{smallmatrix}\right)\in C_0(T^*\mathbb{R}^n, M_{2N}(\mathbb{C})).$$

Therefore

$$[e_a]-[\left(\begin{smallmatrix}0&0\\0&1\end{smallmatrix}\right)]\in K_0(C_0(T^*\mathbb{R}^n, M_{2N}(\mathbb{C}))),$$

and

$$\text{topological index} = \text{right-hand side} = \langle\varepsilon, [e_a]-[\left(\begin{smallmatrix}0&0\\0&1\end{smallmatrix}\right)]\,\rangle.$$

As one can see at first sight, the nature of the analytic and the topological indices are quite different. Firstly, in the analytic index the cyclic cocycle used is Tr, *i.e.* 0-cocycle, whereas a cyclic $2n$-cocycle is used in the topological index. Secondly, the nature of K_0-class is different. As for the class $[P_+]-[P_-]$, the projections P_+, P_- belong to the non-unital C^*-algebra $\mathcal{K}(L^2(\mathbb{R}^n))$, while as for the class $[e_a]-[\left(\begin{smallmatrix}0&0\\0&1\end{smallmatrix}\right)]$, the projection e_a, $\left(\begin{smallmatrix}0&0\\0&1\end{smallmatrix}\right)$ belong to the C^*-algebra $C_0(T^*\mathbb{R}^n, M_{2N}(\mathbb{C}))$ with unit adjoined. These observations suggest that one needs to reformulate either the analytic index or the topological index in a significant way in order to relate two indices to each other.

Consider on $\mathbb{R}^n$ unbounded operators $M_j = x_j, D_j = \partial/\partial x_j, j = 1,\dots,n$. Denote by $\mathcal{K}^\infty$ the space of all integral operators on $L^2(\mathbb{R}^n)$ with integral kernel being a Schwartz function $\in \mathcal{S}(\mathbb{R}^n \times \mathbb{R}^n)$. The space $\mathcal{K}^\infty$ is a $*$-subalgebra of $\mathcal{L}^1(L^2(\mathbb{R}^n))$. Derivations $\delta_1,\dots,\delta_{2n}$ on $\mathcal{K}^\infty$ are defined by

$$\delta_{2j}(a) = [M_j, a],\ \delta_{2j-1}(a) = [D_j, a],\ j = 1,\dots,n,\ a \in \mathcal{K}^\infty,$$

and $\delta_j\delta_k = \delta_k\delta_j$. Moreover, for $\mathrm{Tr}(\delta_j(a)) = 0, a \in \mathcal{K}^\infty$. For $a^0,\cdots,a^{2n} \in \mathcal{K}^\infty$, if we set

$$\omega(a^0,\cdots,a^{2n}) = \frac{(-1)^n}{n!}\sum_{\sigma\in S_{2n}} \mathrm{sgn}(\sigma)\mathrm{Tr}(a^0\delta_{\sigma(1)}(a^1)\cdots\delta_{\sigma(2n)}(a^{2n})),$$

then ω is a cyclic 2n-cocycle.

We use the same letter ω to denote the cup product of ω and the trace on $M_{2N}(\mathbb{C})$. Then ω is a $2n$-trace on $\mathcal{K}(L^2(\mathbb{R}^n, \mathbb{C}^N \oplus \mathbb{C}^N))$ with holomorphically closed domain. For any $\hbar > 0$, set $\omega_\hbar = \omega$, and $\omega_0 = \varepsilon$. Then $(\omega_\hbar)$ is a (strongly) continuous field of $2n$-cocycles. The restriction of the field $(\omega_\hbar)$ onto $(0,1]$ is a trivial field.

Since $\mathrm{S}^n\mathrm{Tr}$ is cohomologous to ω, we have that

$$\langle \mathrm{Tr}, [P_+] - [P_-]\rangle = \langle \omega, [P_+] - [P_-]\rangle.$$

What remains to be done is to replace the K_0-class $[P_+] - [P_-]$ by a class which looks like $[e_a] - [\left(\begin{smallmatrix}0&0\\0&1\end{smallmatrix}\right)]$.

Lemma 3.4. *In $K_0(\mathcal{K}(L^2(\mathbb{R}^n, \mathbb{C}^N))$, for any $\hbar > 0$ we have that*

$$[P_+] - [P_-] = [e_\hbar] - [\left(\begin{smallmatrix}0&0\\0&1\end{smallmatrix}\right)].$$

Set $e_0 = e_a$. The key to the proof of Theorem 3.3is the following.

Lemma 3.5 *The vector field $(e_\hbar)$ is continuous.*

The continuity at $\hbar > 0$ follows from the observation that if symbols are close, then operators are close. The crucial point is to show the continuity at $\hbar = 0$, and one needs to make full use of techniques in the theory of ΨDO's such as asymptotic expansions (see [6]).

Now we can complete the proof of Theorem 3.3.

$$\begin{aligned}
\text{analytic index} &= \langle \mathrm{Tr}, [P_+] - [P_-] \rangle \\
&= \langle \omega_0, [e_1] - [(\begin{smallmatrix} 0 & 0 \\ 0 & 1 \end{smallmatrix})] \rangle \\
&= \langle \omega_\hbar, [e_\hbar] - [(\begin{smallmatrix} 0 & 0 \\ 0 & 1 \end{smallmatrix})] \rangle \\
&= \langle \omega_0, [e_0] - [(\begin{smallmatrix} 0 & 0 \\ 0 & 1 \end{smallmatrix})] \rangle \\
&= \langle \varepsilon, [e_a] - [(\begin{smallmatrix} 0 & 0 \\ 0 & 1 \end{smallmatrix})] \rangle \\
&= \text{topological index.}
\end{aligned}$$

To conclude this section let us restate the crucial fact behind the proof of the index formula.

THEOREM 3.6. *Let $\mathcal{B} = (B_\hbar)$ be a continuous field of C^*-algebras over $[0, \epsilon)$, which is trivial away from 0. Suppose that $\psi = (\psi_\hbar)$ is a strongly continuous field of cyclic $2k$-cocycles with dense holomorphically closed domains. Assume that ψ is trivial away from 0, that is, $(\psi_\hbar)_{\hbar>0}$ is a constant field with respect to the trivialization of $\mathcal{B}$ over $(0, \epsilon)$. Then for any continuous field $(e_\hbar)$ of projections we have the equality $\langle \psi_0, [e_0] \rangle = \langle \psi_1, [e_1] \rangle$.*

4. Foliation index theorem

The idea used in the preceding section can be applied to give an elementary proof of Connes's index theorem [2] for structurally simple foliations.

Let $(X, \mathcal{F})$ be a foliated manifold, and let $P : \Gamma(E) \longrightarrow \Gamma(F)$ be an elliptic longitudinal ΨDO. The principal symbol σ of P defines a class $[\sigma] \in K^*(T^*\mathcal{F})$. Applying Chern character and Thom isomorphism, we get $\mathrm{ch}(\sigma) \in H^*(X, \mathbb{Q})$. The analytic index $\mathrm{ind}P$ of P is defined to be an element of $K_0(C^*(X, \mathcal{F}))$. When the foliation possesses a transverse invariant measure ν, using the trace τ on $C^*(X, \mathcal{F})$ induced from ν we obtain a numerical invariant

$$\mathrm{ind}_\nu P = \tau_*(\mathrm{ind}P).$$

Connes's index theorem says that

$$\mathrm{ind}_\nu P = \varepsilon \langle \mathrm{ch}(\sigma) \cup \mathrm{Td}(T\mathcal{F}_\mathbb{C}), [C] \rangle,$$

where $\mathrm{Td}(\mathcal{F}_\mathbb{C})$ is the Todd class of the complexification of the tangent bundle of the foliation, and C is the Ruelle–Sullivan current (foliation cycle) associated with the invariant measure ν.

We consider a foliation $\mathcal{F}$ on a closed manifold X defined by a locally free action of the Lie group $\mathbb{R}^n$. Assume that the holonomy groupoid of $\mathcal{F}$ is isomorphic to the transformation groupoid $X \times \mathbb{R}^n$. In this case the foliation C^*-algebra is the crossed product $C(X) \rtimes \mathbb{R}^n$.

A normalized trace τ on the C^*-algebra $C(X)$ is nothing but a probability measure μ on X. Suppose that μ is invariant under the $\mathbb{R}^n$-action. Then τ gives rise to a trace $\hat{\tau}$ on $C(X) \rtimes \mathbb{R}^n$ by

$$\hat{\tau}(f) = \int_X f(x,0)\mathrm{d}\mu(x) \ , \ f \in C_c(X \times \mathbb{R}^n).$$

The trace $\hat{\tau}$ is called a *dual trace*.

Under our assumption the cotangent bundle $T^*\mathcal{F}$ is a trivial bundle $T^*\mathcal{F} \cong X \times \mathbb{R}^n$. The tangent bundle of $\mathcal{F}$ is also trivial. In particular, $\mathrm{Td}(T\mathcal{F}_{\mathbb{C}}) = 1$.

We borrow the notation from [7]. Each $t \in \mathbb{R}^n$ acts on X as a diffeomorphism. The displacement of $x \in X$ under this diffeomorphism is denoted $x + t$.

A C^∞-function σ on $T^*\mathcal{F}$ is a *symbol of order* m if for arbitrary multi-indices α, β, there exists a constant C such that

$$|\partial_t^\alpha \partial_s^\beta \sigma(x+t,s)| \leq C(1+|s|)^{m-|\beta|} \ , \ t,s \in \mathbb{R}^n.$$

Set $\sigma_x(t,s) = \sigma(x+t,s)$. For each $x \in X$, define a ΨDO P_x acting on $L^2(\mathbb{R}^n)$ by

$$(P_x\phi)(t) = (2\pi)^{-n} \int_{\mathbb{R}^n \times \mathbb{R}^n} \mathrm{e}^{i\langle t-s,\xi\rangle} \sigma_x(t,\xi)\phi(s)\mathrm{d}s\mathrm{d}\xi.$$

Then $P = (P_x)$ is a longitudinal operator.

As usual we need to define ellipticity. A symbol σ of order m is *elliptic* [7] if there exist $C, R > 0$ such that

$$|\sigma(x,s)| \geq C|s|^m \ , \ |s| \geq R.$$

When σ is elliptic, we say that the longitudinal operator P constructed above is elliptic. Notice that the definition of ellipticity here is different from that for operators on $\mathbb{R}^n$. This is because we do not have to impose any boundary condition at infinity, for X is compact.

THEOREM 4.1. *Let $(X, \mathcal{F})$ be as above. If P is a longitudinal elliptic operator of positive order, then*

$$\mathrm{ind}_\mu(P) = \langle \mathrm{ch}(\sigma), [C] \rangle.$$

Of course, the theorem above is a special case of Connes's index theorem. The point we would like to make here is that Theorem 2.8 can be applied to give a simpler proof.

Outline of proof As in the preceding section, consider the element

$$e_\sigma = \begin{pmatrix} (1+\sigma^*\sigma)^{-1} & (1+\sigma^*\sigma)^{-1}\sigma^* \\ \sigma(1+\sigma^*\sigma)^{-1} & \sigma(1+\sigma^*\sigma)^{-1}\sigma^* \end{pmatrix}.$$

Then

$$[\sigma] = [e_\sigma] - \left(\begin{smallmatrix} 0 & 0 \\ 0 & 1 \end{smallmatrix}\right) \in K_0(C_0(T^*\mathcal{F})) = K^0(T^*\mathcal{F}).$$

As before, we represent the analytic index by the graph projection e of the operator. At this point we must know that the projection belongs to an appropriate C^*-algebra. Applying Theorem 2.8 of [7] to the self-adjoint operator $\left(\begin{smallmatrix} 0 & P^* \\ P & 0 \end{smallmatrix}\right)$, we see that $(1+P^*P)^{-1}, (1+PP^*)^{-1}, P(1+P^*P)^{-1}$ belong to the crossed product $C(X) \rtimes \mathbb{R}^n$. Consequently,

$$e = \begin{pmatrix} (1+P^*P)^{-1} & (1+P^*P)^{-1}P^* \\ P(1+P^*P)^{-1} & P(1+P^*P)^{-1}P^* \end{pmatrix} \in M_2((C(X) \rtimes \mathbb{R}^n)\,),$$

and

$$\mathrm{ind}P = [e] - \left(\begin{smallmatrix} 0 & 0 \\ 0 & 1 \end{smallmatrix}\right) \in K_0(C(X) \rtimes \mathbb{R}^n).$$

The $\mathbb{R}^n$-action on X produces n everywhere linearly independent vector fields $\xi_1, \cdots, \xi_n$. The Ruelle–Sullivan current C is given by the formula

$$\langle C, \eta \rangle = \int_X \eta(\xi_1, \cdots, \xi_n) \mathrm{d}\mu$$

for any n-form η. The current C defines a cyclic n-cocycle ϕ_C in the usual way:

$$\phi_C(f^0, \cdots, f^n) = \int_X (f^0 \mathrm{d}f^1 \wedge \cdots \wedge \mathrm{d}f^n)(\xi_1, \cdots, \xi_n) \mathrm{d}\mu.$$

The machinery invented in [4] manufactures a $2n$-cocycle ω densely defined on $C(X) \rtimes \mathbb{R}^n$ out of ϕ_C. One of the main result in [4] is that the $2n$-cocycle ω and $\tilde{\tau}$ determine the same class in the stabilized cyclic cohomology. Thus they define the same map $K_0(C(X) \rtimes \mathbb{R}^n) \longrightarrow \mathbb{C}$.

As usual, rescale the action α of $\mathbb{R}^n$ on $C(X)$ to get a family of actions $\alpha^\hbar, 0 \leq \hbar \leq 1$. Applying the machine just mentioned to each $\alpha^\hbar$, we obtain a family of $2n$-cocycles $(\omega_\hbar), 0 \leq \hbar \leq 1$.

The family $(\omega_\hbar)$ is a strongly continuous field of $2n$-cocycles with dense holomorphically closed domains. Its restriction onto $(0, 1]$ is constant with respect to the trivialization of the field $(C(X) \rtimes_{\alpha^\hbar} \mathbb{R}^n)_{0<\hbar\leq 1}$.

The cocycle ω_0 is the $2n$-cocycle $[T^*\mathcal{F}]_\mu$:

$$[T^*\mathcal{F}]_\mu(f^0, \cdots, f^{2n}) = \langle C, \Pi_*(f^0 \mathrm{d}f^1 \wedge \cdots \wedge \mathrm{d}f^{2n}) \rangle,$$

where Π_* is the integration along the fibre of $T^*\mathcal{F} \longrightarrow X$.

What remains to be checked is that the field $\hbar \mapsto e_\hbar$ is continuous. As in Example 2.6, define an action $\alpha^\sharp$ on $X \times [0,1]$ by

$$\alpha_t^\sharp(x,\hbar) = (x + \hbar t, \hbar).$$

Define also an elliptic symbol $\sigma^\sharp$ of order m by $\sigma^\sharp(x,\hbar,\xi) = \sigma(x,\xi)$. A ΨDO $P^\sharp$ on $L^2(X \times [0,1] \times \mathbb{R}^n)$ with symbol $\sigma^\sharp$ is given by

$$(P^\sharp f)(x,\hbar,t) = (2\pi)^{-n} \int \mathrm{e}^{i\langle t-s,\xi\rangle} \sigma^\sharp(x,\hbar,\xi) f(x + \hbar(s-t), s) \mathrm{d}s\mathrm{d}\xi.$$

Applying (a modification of) Theorem 2.8 of [7] to the operator

$$\begin{pmatrix} 0 & (P^\sharp)^* \\ P^\sharp & 0 \end{pmatrix},$$

we get an element $e^\sharp$ in $\mathrm{M}_2(C(X \times [0,1]) \rtimes_{\alpha^\sharp} \mathbb{R}^n)\ )$ such that $e_\hbar = \pi_\hbar(e^\sharp)$, where $\pi_\hbar$ is given by the evaluation map

$$C(X \times [0,1]) \rtimes_{\alpha^\sharp} \mathbb{R}^n \longrightarrow C(X) \rtimes_{\alpha^\hbar} \mathbb{R}^n,$$

and, of course, when $\hbar = 0$, we identify $C(X) \rtimes_{\alpha^0} \mathbb{R}^n$ with $C(X) \otimes C_0(\mathbb{R}^n) = C_0(X \times \mathbb{R}^n)$. Now, by the continuity structure of the field we know that the field $\hbar \mapsto e_\hbar$ is continuous.

Finally, putting everything together, we can complete the proof:

$$\begin{aligned} \langle ch(\sigma), [C] \rangle &= \langle [T^*\mathcal{F}]_\mu, [e_\sigma] - [(\begin{smallmatrix} 0 & 0 \\ 0 & 1 \end{smallmatrix})] \rangle \\ &= \langle \omega, [e] - [(\begin{smallmatrix} 0 & 0 \\ 0 & 1 \end{smallmatrix})] \rangle \\ &= \langle \hat{\tau}, \mathrm{ind} P \rangle \\ &= \mathrm{ind}_\mu P. \end{aligned}$$

REMARK 4.2 The main result of the preceding section can be reformulated in terms of crossed products as the result in this section. The Lie group $\mathbb{R}^n$ acts on itself by translations, and $C_0(\mathbb{R}^n) \rtimes \mathbb{R}^n \cong \mathcal{K}(L^2(\mathbb{R}^n))$. By rescaling the action we obtain the continuous field of Example 2.6. Since $\mathbb{R}^n$ is not compact, we cannot apply Theorem 2.8 of [7] to this case in order to show that the field of graph projections is continuous.

5. Families of operators

Let us recall the Atiyah–Singer index theorem for families [1]. Let $\pi : E \longrightarrow X$ be a smooth fibre bundle with compact fibre M over a compact X, and let $P = (P_x)$ be a family of elliptic operators, where P_x is an operator on $E_x = \pi^{-1}(x), x \in X$, of fixed order. Such a family determines a class $\mathrm{ind}P \in K^0(X)$.

Set $T_{E/X} = \ker \mathrm{d}\pi$, where $\mathrm{d}\pi : TE \longrightarrow TX$. The index theorem is the following:

THEOREM 5.1. *Let P be as above. Then*

$$\mathrm{ch}(\mathrm{ind}P) = (-1)^{\frac{n(n+1)}{2}} \int_{E/X} \mathrm{ch}(\sigma(P))\mathrm{Td}(T_{E/X} \otimes \mathbb{C}) \ \in H^{2*}(X;\mathbb{Q}),$$

where $n = \dim M$, $\int_{E/X}$ is the integration along the fibre and $\mathrm{Td}(T_{E/X} \otimes \mathbb{C})$ is the Todd class of the complexification of $T_{E/X}$.

The aim of this section is to prove the index theorem above in a different setting, that is, the fibre of the bundle is non-compact. More precisely, we consider an oriented real Riemannian vector bundle $E \longrightarrow X$. Let P be the associated principal $\mathrm{SO}(n)$-bundle. The group $\mathrm{SO}(n)$ acts on $T^*\mathbb{R}^n$. Denote by $T^*_{E/X}$ the associated $T^*\mathbb{R}^n$-bundle : $T^*_{E/X} = P \times_{\mathrm{SO}(n)} T^*\mathbb{R}^n$. Let π denote the projection $T^*_{E/X} \longrightarrow X$.

We consider the space $\mathcal{P}^{(m)}$ of symbols of order m. The space $\mathcal{P}^{(m)}$ is a Fréchet space, where the Fréchet norm is given by

$$\|a\|_{\alpha,\beta} = \sup \left\{ \frac{|\partial_x^\alpha \partial_\xi^\beta a(x,\xi)|}{(1+|x|+|\xi|)^{m-|\alpha|-|\beta|}} \ ; \ (x,\xi) \in T^*\mathbb{R}^n \right\}.$$

The canonical SO(n)-action on $\mathbb{R}^n$ induces an action on $\mathcal{P}^{(m)}$.

Let $\mathcal{P}^{(m)}(X) \longrightarrow X$ be the $\mathcal{P}^{(m)}$-bundle associated with the vector bundle $E \longrightarrow X$. Denote by $C^\infty(\mathcal{P}^{(m)})$ the space of smooth sections. A section a is called a *smooth family of symbols* of order m. The value of a at $x \in X$ is denoted by a_x. When every a_x is elliptic, the family a is an *elliptic family* of symbols. Every $a \in C^\infty(\mathcal{P}^{(m)})$ determines a smooth family Op(a) of ΨDO's of order m: Op(a) = (Op(a_x)).

Denote by $\mathcal{H} \longrightarrow X$ the $L^2(\mathbb{R}^n)$-bundle associated with the given Riemannian vector bundle $E \longrightarrow X$. The family Op(a) acts on the Hilbert bundle $\mathcal{H}$.

If a is elliptic, then $\mathrm{Ker\ Op}(a_x)$, $\mathrm{Coker\ Op}(a_x)$ are finite-dimensional for all $x \in X$. As in [1], if $\dim \mathrm{Op}(a_x)$ is independent of $x \in X$, then $\mathrm{KerOp}(a)$ is a vector bundle. The same holds for $\mathrm{CokerOp}(a)$, and we obtain

$$\mathrm{ind}(\mathrm{Op}(a) = [\mathrm{KerOp}(a)] - [\mathrm{CokerOp}(a)] \in K^0(X).$$

In general this is not the case. Nevertheless, as in [1] a modification of the definition above works to define $\mathrm{ind}(\mathrm{Op}(a))$.

Let a be an elliptic family of symbols of positive order. As usual, set

$$e_a = \begin{pmatrix} (1+a^*a)^{-1} & (1+a^*a)^{-1}a^* \\ a(1+a^*a)^{-1} & a(1+a^*a)^{-1}a^* \end{pmatrix},$$

and

$$\widehat{e}_a = e_a - \left(\begin{smallmatrix} 0 & 0 \\ 0 & 1 \end{smallmatrix}\right).$$

Since a is of positive order, the $2n$-form $\mathrm{trace}(\widehat{e}(\mathrm{d}\widehat{e})^{2n})$ on $T^*_{E/X}$ is integrable along the fibre of $T^*_{E/X} \longrightarrow X$. Now we are ready to formulate the main result of this section. For simplicity, suppose that X is orientable and is oriented. Let ω_X be the volume form on X.

THEOREM 5.2. *Let a be as above. Then*

$$\mathrm{ch}(\mathrm{ind}\mathrm{Op}(a))[X] = \frac{1}{(2\pi i)^n n!}\int_{T^*_{E/X}} \pi^*\omega_X trace(\widehat{e}(\mathrm{d}\widehat{e})^{2n}),$$

where the fibre of $T^*_{E/X} \longrightarrow X$, when identified with $T^*\mathbb{R}^n$, is oriented so that $\mathrm{d}x_1 \wedge \mathrm{d}\xi_1 \wedge \cdots \wedge \mathrm{d}x_n \wedge \mathrm{d}\xi_n > 0$.

Outline of proof. Our strategy is obvious. We represent $\mathrm{ind}\mathrm{Op}(a)$ as a graph projection, as usual. Let e_x be the graph projection of the closed operator $\mathrm{Op}(a_x)$ on $L^2(\mathbb{R}^n)$ (see Section 3).

The Hilbert bundle $\mathcal{H} \longrightarrow X$ is, in particular, a continuous field of Hibert spaces. Let $\mathcal{K}$ be the associated field (in fact a bundle) of elementary C^*-algebras, and let $\mathcal{K}^\sim$ be the field $\mathcal{K}$ with unit added. Denote by $C(\mathcal{K})$ the C^*-algebra of continuous sections of $\mathcal{K}$.

The family $e = (e_x)$ is a continuous section of $\mathcal{K}^\sim$. The field e of projections defines an element

$$[e] - [\left(\begin{smallmatrix} 0 & 0 \\ 0 & 1 \end{smallmatrix}\right)] \in K_0(C(\mathcal{K})).$$

Through the isomorphism $K_0(C(\mathcal{K})) \cong K_0(C(X)) = K^0(X)$, we have

$$[e] - [\left(\begin{smallmatrix} 0 & 0 \\ 0 & 1 \end{smallmatrix}\right)] = \mathrm{ind}\mathrm{Op}(a).$$

The relation between two K_0-classes is that by adding infinite-dimensional subspaces to both $\mathrm{Ker}\ \mathrm{Op}(a)$ and $\mathrm{Coker}\ \mathrm{Op}(a)$ projections e and $\left(\begin{smallmatrix} 0 & 0 \\ 0 & 1 \end{smallmatrix}\right)$ are obtained.

In order to complete the proof we need three things: continuous fields of C^*-algebras, of projections, and of cyclic cocycles.

Making use of Kohn–Nirenberg quantization fibre-wise, we obtain a continuous field $\mathcal{A} = (A_\hbar)$ on $[0,1]$ such that $A_0 = C_0(T^*_{E/X})$ and $A_\hbar = C(\mathcal{K})$, $\hbar > 0$. By rescaling the symbol a, as in the preceding sections, we obtain a continuous field of projections.

The $2n$-trace ω on $\mathcal{K}(L^2(\mathbb{R}^n))$, used in Section 3, is invariant under the canonical $\mathrm{SO}(n)$-action. Consequently, we can define a $2n$-trace $\widehat{\omega}$ on $C(\mathcal{K})$ by

$$\widehat{\omega}(a^0, \cdots, a^{2n}) = \int_X \omega(a^0, \cdots, a^{2n})\omega_X \,.$$

Set $\widehat{\omega}_\hbar$, $\hbar > 0$. Define a $2n$-trace $\widehat{\omega}_0$ on $C_0(T^*_{E/X})$ by

$$\widehat{\omega}_0(f^0, \cdots, f^{2n}) = \int_{T^*_{E/X}} f^0 \mathrm{d}f^1 \wedge \cdots \wedge \mathrm{d}f^{2n} \wedge \pi^*\omega_X \,.$$

The family $(\widehat{\omega}_\hbar)$ forms a strongly continuous field of $2n$-traces.

Once we obtain the three necessary ingredients, the proof will follow as in the preceding sections.

So far we have studied rather a special case. In [8] we investigate the index formula for families in full generality.

References

[1] M.F. Atiyah, I.M. Singer, The index of elliptic operators: IV, *Ann. Math.* **93** (1971), 119–138.

[2] A. Connes, A survey of foliations and operator algebras, *Operator algebras and Applications, Proc. Symposia Pure math.* **38** (1982) Part I, 521–628.

[3] J. Dixmier, *C^*-algebras,* Northholland, Amsterdam, 1977.

[4] G.A. Elliott, T. Natsume, R. Nest, Cyclic cohomology for one-parameter smooth crossed product, *Acta Math.*.

[5] G.A. Elliott, T. Natsume, R. Nest, The Heisenberg group and K-theory, *K-Theory* **7** (1993), 409–428.

[6] G.A. Elliott, T. Natsume, R. Nest, The Atiyah-Singer index theorem as passage to the classical limit in quantum mechanics, *Comm. Math. Phys.* **182** (1996), 505–533.

[7] L.Kaminker, J. Xia, The spectrum of operators elliptic along the orbits of $\mathbb{R}^n$ actions, *Comm. Math. Phys.* **110** (1987), 427–438.

[8] T. Natsume, R. Nest, C^*-algebraic deformation quantization and the index theorem for families, *in preparation.*

[9] M.A. Rieffel, Continuous fields of C^*-algebras coming from group cocycles and actions, *Math. Ann.* **283** (1989), 631–643.

[10] M.A. Shubin, *Pseudodifferential operators and spectral theory,* Springer-Verlag, Tokyo, 1987.

SINGULAR SYSTEMS OF EXPONENTIAL FUNCTIONS

Hideki Omori
Department of Mathematics, Faculty of Science and Technology
Science University of Tokyo, Noda, Chiba, 278, JAPAN
omori@ma.noda.sut.ac.jp

Yoshiaki Maeda
Department of Mathematics, Faculty of Science and Technology
Keio University, Hiyoshi, Yokohama, 223, JAPAN
maeda@math.keio.ac.jp

Naoya Miyazaki
Department of Mathematics, Faculty of Science
Yokohama City University, Kanazawa-ku, Yokohama, 236, JAPAN
Naoya.Miyazaki@math.yokohama-cu.ac.jp

Akira Yoshioka
Department of Mathematics, Faculty of Engineering
Science University of Tokyo
Kagurazaka, Shinjyuku-ku, Tokyo 162-8601, JAPAN
yoshioka@rs.kagu.sut.ac.jp

Abstract We attempt to establish a calculus of the Moyal product treating the deformation parameter as a parameter moving in positive reals. We show strange phenomenas different from formal deformation quantization by studying the convergence of

Y. Maeda, et al. (eds.), Noncommutative Differential Geometry and Its Applications to Physics, 169–186.

the parameter in computing the product of the exponential functions of the quadratic form. In the case of deformation quantization with the positive real parameters, the associativity for the Moyal product fails for a wider class of functions.

Mathematics Subject Classification (2000): 53D55, 53D17, 46F20

Keywords: deformation quantization, Moyal product.

1. Introduction

In the traditional theory of deformation quantization initiated by [1] the main target has been to consider noncommutative deformations of the usual commutative product by treating the deformation parameter $\hbar$ as a formal parameter (cf. [2], [8], [4]).

However, if $\hbar$ is viewed not as a formal parameter, there appear various strange phenomena which break the associativity (see §4). This is, in fact, a pathological phenomenon, but an amenable fact which will be to open another application.

In this paper we give an example of fields containing all rational functions and the exponential functions of x. We showed in [9] that the exponential functions of quadratic forms have crucial characters for associativity breaking. This example is obtained by deforming the topology of the space of all polynomials, and is naturally obtained from the $*$-exponential functions of quadratic forms in Weyl algebra, where the deformation parameter is not treated as a formal parameter but a positive real parameter.

Though the fields of rational functions are main subjects in number theory, they hardly appear in analysis, but there is some similarity between the Galois group [5] and the gauge group.

2. Singular products

Consider a system of usual exponential functions $\{ae^{itx}; t \in \mathbb{R}, a \in \mathbb{C}\}$, where x is the variable moving in $\mathbb{R}$. These are considered in the space $C^\infty(\mathbb{R})$ with the C^∞-topology. Note that $C^\infty(\mathbb{R})$ is a Fréchet space defined by a countable family of semi-norms.

On this system, we consider a new rule of product with singularities as follows:

$$ae^{isx} * be^{itx} = \frac{4ab}{4+\hbar^2 st}\, e^{\frac{4i(s+t)}{4+\hbar^2 st}x}, \tag{2.1}$$

where $a, b \in \mathbb{C}_* = \mathbb{C}-\{0\}$ and $\hbar$ is a fixed positive parameter. It is easy to check the associativity where products are defined. The inverse of ae^{isx} is given as $\frac{1}{4}a^{-1}(4-\hbar^2s^2)e^{-isx}$ for $4-\hbar^2s^2\neq 0$. If $s=\pm 2/\hbar$, then $e^{\pm\frac{2i}{\hbar}x}$ are not invertible but have the idempotent property:

$$2e^{\pm\frac{2i}{\hbar}x} * 2e^{\pm\frac{2i}{\hbar}x} = 2e^{\pm\frac{2i}{\hbar}x}, \tag{2.2}$$

however, the product $e^{\frac{2i}{\hbar}x} * e^{-\frac{2i}{\hbar}x}$ is not defined. We refer to the elements $2e^{\pm\frac{2i}{\hbar}x}$ as *vacuums*. $2e^{\frac{2i}{\hbar}x}$ and $2e^{-\frac{2i}{\hbar}x}$ will be denoted by $\varpi_{0,0}$, $\bar{\varpi}_{0,0}$, respectively.

In §7 we show that such systems appear naturally when we consider the $*$-exponential functions of quadratic forms in the usual Weyl algebra. This shows the roots of our system and the reason why we are interested in such a system.

In this paper we show that this system contains various strange phenomena (cf. Theorem 3 and §5.1), however interesting properties, by which one can expect several applications (cf. Theorem 4).

Replacing s, t by $-is$, $-it$ in the defining equality (2.1) we rewrite (2.1) as

$$ae^{sx} * be^{tx} = \frac{4ab}{4-\hbar^2st}e^{\frac{4(s+t)}{4-\hbar^2st}x}. \tag{2.3}$$

Notice here that we can not replace x by ix in (2.3); (2.1) should be written strictly in the form

$$ae^{s(ix)} * be^{t(ix)} = \frac{4ab}{4+\hbar^2st}e^{\frac{4(s+t)}{4+\hbar^2st}(ix)}.$$

Keeping (2.3) as the product formula for $\{ae^{tx}\}$, we may view the above as the product formula for the system $\{ae^{t(ix)}\}$.

The product formula of $ae^{s(ix)} * be^{tx}$, $ae^{sx} * be^{t(ix)}$ are also given by setting $e^{s(ix)} = e^{(is)x}$:

$$ae^{s(ix)} * be^{tx} = \frac{4ab}{4-\hbar^2(is)t}e^{\frac{4(is+t)}{4-\hbar^2(is)t}x} = \frac{4ab}{4+\hbar^2s(-it)}e^{\frac{4(s-(it))}{4+\hbar^2s(-it)}(ix)}. \tag{2.4}$$

3. $\mathbb{C}[x]$-module

In this section we consider the product formula defined by (2.3). First of all, setting $a=b=1$ we regard (2.3) as a generating function to obtain the product formula $x^k * x^l$. The power series expansion

$$e^{isx} * e^{itx} = \frac{4}{4+\hbar^2st}e^{\frac{4}{4+\hbar^2st}i(s+t)x} = \sum_{k,l,m,n} c_{k,l,m,n}\hbar^{2k}x^l(is)^m(it)^n \tag{3.1}$$

gives the product form:

$$x^m * x^n = \sum_{k,l} c_{k,l,m,n}\hbar^{2k}x^l.$$

It is easy to see that $c_{k,l,m,n} = c_{k,l,n,m}$ and $c_{k,l,m,n} = 0$, for k, l, m, n except for

$$|m-n| \leq l \leq \min(m,n).$$

To obtain the general product formula, we restrict x in the real axis for a while and we use the Fourier transform. Set

$$f(x) * g(x) = \iint \hat{f}(s)\hat{g}(t)e^{isx} * e^{itx} d\!\!\!{}^{-}s d\!\!\!{}^{-}t.$$

Rewriting the expansion (3.1) in the shape

$$e^{isx} * e^{itx} = \sum \tilde{c}_{k,l,m,n}\hbar^{2k}x^l(-is)^m(-it)^n e^{i(s+t)x},$$

we have the product formula:

$$f(x) * g(x) = \sum \tilde{c}_{k,l,m,n}\hbar^{2k}x^l f^{(m)}(x)g^{(n)}(x). \tag{3.2}$$

Though we have used the Fourier transform, the product formula (3.2) is valid if one of f, g is a polynomial. The Moyal product formula in § 7 gives a more precise product formula.

Proposition 1. (2.3) *gives uniquely a commutative associative product* $*$ *on the space* $\mathbb{C}[x]$ *of all polynomials of* x *of complex coefficients. The product formula is given in the form:*

$$\begin{aligned} &f(x) * g(x) \\ &\quad = f(x)g(x) + \frac{\hbar^2}{4}\left(f''(x)xg'(x) + f'(x)xg''(x) + f'(x)g'(x)\right) + \cdots \\ &\qquad + \tilde{c}_{k,l,m,n}\hbar^{2k}x^l f^{(m)}(x)g^{(n)}(x) + \cdots . \end{aligned}$$

Since this product formula is written as a power series of $\hbar$, whose leading term is the usual product, this product is viewed as a deformation of the ordinary product. If one treats every element as a formal power series in $\hbar$, then this gives a commutative and associative algebra structure on the space $C^\infty(\mathbb{R})[[\hbar]]$ of all formal power series of $\hbar$ with coefficients of C^∞ functions on $\mathbb{R}$. In the case where $\hbar$ is a positive real parameter, indeed this is our main target, the product $*$ does not necessarily defined for all C^∞ functions, as is seen in (2.3).

Recalling that the deformed product is also commutative, we think that what is deformed is not an algebraic structure but the *topology* by which we define converging series and transcendental functions.

By the product formula in Proposition1 we see that if f is a polynomial, then the products $f * g$ and $g * f$ are defined for every C^∞-function g, and $f*$, $*f$ are differential operators, and hence continuous mappings of $C^\infty(\mathbb{R})$ into itself.

However, even if the products $f * g, (f * g) * h, g * h, f * (g * h)$ are defined, the associativity $(f * g) * h = f * (g * h)$ may not hold in general (see Theorem 3).

Differentiating (2.1) by s we see that

$$\left(x + \frac{\hbar i}{2}\right) * \varpi_{0,0} = 0, \quad \left(x - \frac{\hbar i}{2}\right) * \bar{\varpi}_{0,0} = 0. \tag{3.3}$$

By Proposition1, we consider a differential equation

$$\frac{d}{dt} f_t(x) = ip(x) * f_t(x), \quad f_0(x) = f \tag{3.4}$$

for every polynomial $p(x)$.

Though the existence of the solution is not ensured in general, the real analytic solution in t of (3.4) is unique with respect to the initial function f. We now give a sufficient condition for the existence of the solution for the initial condition 1 (cf. Theorem 4). Note that if there exists such a solution, then one may call this solution as $*$-*exponential* of $p(x)$ and denote this by $e_*^{itp(x)}$.

4. Singular $*$-exponentials and inverses

Through the correspondence between the pair (a, s) and the function ae^{sx}, the product (2.3) gives a local complex Lie group structure on a neighborhood U of $(1, 0)$ of $\mathbb{C}_* \times \mathbb{C}$. Namely, the multiplication is given by

$$(a, s) * (b, t) = \left(\frac{4ab}{4 - \hbar^2 st}, \frac{4(s + t)}{4 - \hbar^2 st}\right). \tag{4.1}$$

Let $\mathfrak{g}$ be the complex Lie algebra of the local group $(U, *)$. $\mathfrak{g}$ can be naturally identified with $\mathbb{C}^2$ and $\{(a, 0); a \in \mathbb{C}_*\}$ is a normal local Lie subgroup. The factor space is also a local Lie group with the simplest nontrivial example of singular product:

$$s * t = \frac{s + t}{1 - \frac{\hbar^2}{4} st}.$$

The differential dR_t of the right translation R_t is given as $1 + \frac{\hbar^2}{4}t^2$, hence the differential equation which gives the integral curve of right-invariant vector field is $f'(t) = 1 + \frac{1}{4}\hbar^2 f(t)^2$. Thus, we have $\exp_* t\mathbf{1} = (2/\hbar)\tan(\frac{1}{2}\hbar t)$.

In our system, the exponential mapping is given by

$$e_*^{itx} = \frac{1}{\cosh \frac{\hbar t}{2}} e^{i(\frac{2}{\hbar}\tanh\frac{\hbar t}{2})x}, \quad t \in \mathbb{C} - \{\tfrac{1}{\hbar}(m + \tfrac{1}{2})i;\ m \in \mathbb{Z}\}. \tag{4.2}$$

A direct computation shows that $e_*^{i0x} = 1$ and that the exponential law

$$e_*^{isx} * e_*^{itx} = e_*^{i(s+t)x}$$

holds where both sides are defined. If $s, t \in \mathbb{R}$, then e_*^{itx} forms a one parameter group. Note that by (4.2) e_*^{itx} is rapidly decreasing in $t \in \mathbb{R}$. If $\hbar \to 0$ in the formula (4.2) we then have ordinary exponential functions. Hence (4.2) gives a deformation of exponential functions.

The formula (4.2) is in fact the origin of all strange phenomena. Some of them will be discussed in §7. Here we remark only that e_*^{itx} is rapidly decreasing in t and that $e_*^{\frac{2\pi}{\hbar}x} = -1$ and $e_*^{\frac{\pi}{\hbar}x}$ is singular in the expression (4.2).

Differentiating (2.1) gives that e_*^{itx} is a solution of the differential equation

$$\frac{d}{dt}f_t(x) = ix * f_t(x) = ixf_t(x) + i\frac{\hbar^2}{4}(f_t'(x) + xf_t''(x)). \tag{4.3}$$

Note that the right hand side is the operator of Bessel functions.

The function (4.2) is the unique analytic solution in t of (4.3) with the initial function 1, and hence e_*^{itx} in (4.2) is the $*$-exponential of x.

As is seen in the product formula (2.1), we have the following:

Corollary 2. *By the product formula* (2.1), *the solution of* (4.3) *with the initial function* e^{-iax}, $a\hbar > 2$, *has a singularity at t such that* $\tanh\frac{\hbar t}{2} = \frac{2}{\hbar a}$.

In § 5, we show:

Theorem 3. *For every* $\alpha \in \mathbb{C}$, $x + \alpha$ *has an inverse element. Moreover, if* $\alpha \notin \{i\rho; |\rho| > \frac{\hbar}{2}\}$, *then there are two inverses* $(x+\alpha)_{*+}^{-1}$, $(x+\alpha)_{*-}^{-1}$ *of* $x + \alpha$ *satisfying the following:*

(i) $(x+\alpha)_{*+}^{-1}$ *(resp.* $(x+\alpha)_{*-}^{-1}$*) is holomorphic with respect to* α *on the domain* $\mathbb{C} - \{ai; a > \frac{\hbar}{2}\}$ *(resp.* $\mathbb{C} - \{ai; a < -\frac{\hbar}{2}\}$*);*

(ii) If $\alpha = i\rho$ *and* $\rho > \frac{\hbar}{2}$ *(resp.* $< -\frac{\hbar}{2}$*), then only* $(x+\alpha)_{*+}^{-1}$ *(resp.* $(x+\alpha)_{*-}^{-1}$*) is defined.*

Theorem3 shows that almost every element $x + \alpha$ has two different inverses. Noting

$$\big((x+\alpha)_{*+}^{-1} * (x+\alpha)\big) * (x+\alpha)_{*-}^{-1} \neq (x+\alpha)_{*+}^{-1} * \big((x+\alpha) * (x+\alpha)_{*-}^{-1}\big),$$

we see that the associativity fails in the extended $*$-product. Notice also that $(x+\alpha)_{*+}^{-1} * (x+\alpha)_{*-}^{-1}$ is not defined in general.

On the other hand, since $x + \frac{\hbar i}{2}$ has an inverse, (3.3) gives another example where the associativity breaks down :

$$\big((x + \tfrac{\hbar i}{2})_{*+}^{-1} * (x + \tfrac{\hbar i}{2})\big) * \varpi_{0,0} \neq (x + \tfrac{\hbar i}{2})_{*+}^{-1} * \big((x + \tfrac{\hbar i}{2}) * \varpi_{0,0}\big).$$

Since $x+\alpha$ is invertible for every $\alpha \in \mathbb{C}$ and several product formulas can be established among inverses, our system involves rational function field of x (cf. Theorem9). The product formulas of inverses are given in §6.

Let

$$te^{i\theta} + p(x) = a(\omega_1(t) + x) * (\omega_2(t) + x) * \cdots * (\omega_n(t) + x), \quad a,\ \omega_j(t) \in \mathbb{C}$$

be the factor decomposition for every polynomial $p(x)$ and a fixed $\theta \in \mathbb{R}$. We may assume that $\omega_j(t)$ is continuous in $t \in \mathbb{R}$ for a fixed θ.

$p(x)$ is called a *sectorial polynomial*, if there is θ such that each $\omega_j(t)$ does not meet the pure imaginary axis except the open interval $(-\frac{1}{2}, \frac{1}{2})i\hbar$.

By the similar manner as Hille–Yoshida theorem[11] or the Bromwich integral of inverse Laplace transformation, we show the following theorem.

Theorem 4. *For every sectorial polynomial $p(x)$, there exists a solution of (3.4) for all $t \in \mathbb{R}$ with the initial condition 1, that is, the star exponential $e_*^{itp(x)}$ is defined.*

The solution does not extend to holomorphic one parameter group in general. Though the application of this theorem would be in the forthcoming paper, Theorem 4 would give the possibility of solving the equation of higher order evolution equations.

5. Proof of Theorem 3

We first remark that (4.3) gives the following:

Proposition 5. *The integral $\int_{-\infty}^{\infty} e_*^{itx} dt$ exists in $C^\infty(\mathbb{R})$.*

This crucially gives strange properties of our system. As is seen later, this may be viewed that the delta function $\delta_*(x)$ in the $*$-product is a smooth function. We also show that $\int_0^\infty e_*^{itx} dt$ and $\int_{-\infty}^0 e_*^{itx} dt$ exist and

$$\lim_{t\to\pm\infty} e_*^{itx} = 0.$$

We now show Theorem 3. To obtain the inverse, we consider the integral

$$-ie^{i\theta}\int_0^{\infty} e_*^{i(te^{i\theta})(x+\alpha)}dt \tag{5.1}$$

for a fixed θ such that $-\pi \leq \theta < \pi$, but $\theta \neq \pm\frac{\pi}{2}$. For every $\alpha = re^{i\eta}$ we choose θ such that $\mathrm{Re}(ie^{\theta+\eta}) \leq 0$. Then, the integral (5.1) converges to give the inverse of $x+\alpha$.

The next lemma shows that the inverse given by (5.1) does not depend on the choice of θ.

Lemma 6. *The integral* (5.1) *does not depend on θ, whenever θ moves in the domain* $\mathrm{Re}\, ie^{\theta+\eta} \leq 0$.

Proof. By differentiating (5.1) by θ formally, it suffices to show

$$e^{i\theta}\int_0^{\infty} e_*^{i(te^{i\theta})(x+\alpha)}dt + e^{i\theta}\int_0^{\infty} te^{i\theta}(x+\alpha) * e_*^{i(te^{i\theta})(x+\alpha)}dt = 0.$$

Noting that $e_*^{i(te^{i\theta})(x+\alpha)}$ is rapidly decreasing in t and then $\lim_{t\to\infty} te_*^{i(te^{i\theta})(x+\alpha)} = 0$, we have the above equality by integration by parts. □

If $\eta \neq \pm\frac{\pi}{2}$, then there are θ, θ' such that $\theta+\eta = 0$, $\theta'+\eta = -\pi$. Since

$$e^{ite^{i\theta}\alpha} = e^{itr}, \qquad e^{ite^{i\theta'}\alpha} = e^{-itr},$$

the integrals (5.1) converge in both cases.

Theorem 3 is an easy consequence of above results. This completes the proof of Theorem 3.

Next lemma shows that two integrals $\int_0^{\infty} e_*^{itx}dt$ and $\int_{-\infty}^{0} e_*^{itx}dt$ are not equal.

Lemma 7. *Since $\theta' = \theta - \pi$, $\alpha = re^{-i\theta}$, the difference*

$$-ie^{-i\theta}\int_0^{\infty} e^{irt}e_*^{i(te^{i\theta})x}dt + ie^{-i\theta'}\int_0^{\infty} e^{-irt}e_*^{i(te^{i\theta'})x}dt$$

is given by the integral

$$-i\frac{\alpha}{r}\int_{-\infty}^{\infty} e^{irt}e_*^{i(te^{i\theta})x}dt = -i\frac{\alpha}{r}\int_{-\infty}^{\infty} e^{irt}\frac{1}{\cosh(\frac{\hbar}{2}te^{i\theta})}e^{i\frac{2}{\hbar}\tanh(\frac{\hbar}{2}te^{i\theta})x}dt. \tag{5.2}$$

We show that (5.2) does not vanish in general.

Set $F(\theta) = \int_{-\infty}^{\infty} e^{irt} e_*^{i(te^{i\theta})x} dt$. Then, the integration by parts gives

$$\frac{d}{d\theta}F(\theta) = i\int_{-\infty}^{\infty} e^{irt} tie^{i\theta}x * e_*^{i(te^{i\theta})x} dt = -i\int_{-\infty}^{\infty} \frac{d}{dt}(te^{irt})e_*^{i(te^{i\theta})x} dt.$$

This gives

$$i\int_{-\infty}^{\infty} e^{irt} tie^{i\theta}x * e_*^{i(te^{i\theta})x} dt = -i\int_{-\infty}^{\infty} (1+irt)e^{irt} e_*^{i(te^{i\theta})x} dt.$$

Hence we have

$$\frac{d}{d\theta}F(\theta) = i\int_{-\infty}^{\infty} e^{irt} tie^{i\theta}x * e_*^{i(te^{i\theta})x} dt = -\frac{i}{1+ir}F(\theta).$$

It follows $F(\theta) = F(0)e^{-i\theta(1+ir)^{-1}}$.

$$F(\theta) = \int_{-\infty}^{\infty} e^{irt} e_*^{i(te^{i\theta})x} dt = e^{-i\theta(1+ir)^{-1}} \int_{-\infty}^{\infty} e_*^{it(x+r)} dt. \tag{5.3}$$

We show that the right hand side of (5.3) does not vanish in the next subsection.

5.1. Relation to hyper-functions (cf. [6])

By Lemma 7 the difference of two inverses is given by the integral

$$x_{*+}^{-1} - x_{*-}^{-1} = -i\int_{-\infty}^{\infty} e_*^{itx} dt.$$

This may be written as $-i\delta_*(x)$, for $\int_{-\infty}^{\infty} e_*^{itx} dt$ means the Fourier transform of 1 in the $*$-product. Though the delta function $\delta(x)$ is not a genuine function of x, $\int_{-\infty}^{\infty} e_*^{itx} dt$ is a genuine function of x in the $\cdot$-product. Indeed this is given as $\frac{\pi}{2}J_0(\frac{2}{\hbar}x)$, by using Hansen-Bessel identity of Bessel function (cf. [3]):

$$\delta_*(x) = \int_{-\infty}^{\infty} e_*^{itx} dt = \frac{\pi}{2}J_0\left(\frac{2}{\hbar}x\right).$$

Since the $*$-exponential e_*^{itx} is given as a rapidly decreasing function in t, we see that several distributions in the $*$-expression can be represented as ordinary functions in the expression using the $\cdot$-product.

Using standard notations of distributions, we may write

$$\frac{1}{x+i0*} = -2\pi i\int_0^{\infty} e_*^{itx} dt, \qquad \frac{1}{x-i0*} = 2\pi i\int_0^{\infty} e_*^{-itx} dt.$$

Our nations can stand for

$$\frac{1}{x+i0*}=(2\pi)x_{*+}^{-1}, \qquad \frac{1}{x-i0*}=(2\pi)x_{*-}^{-1}.$$

Using generating functions of Bessel functions we have

$$\int_0^\infty e_*^{itx}dt=\frac{\pi}{2}J_0\left(\frac{2}{\hbar}x\right)+i\sum\frac{1}{2n+1}J_{2n+1}\left(\frac{2}{\hbar}x\right).$$

With the identities of hyper-functions

$$\delta_*(x)=\frac{-1}{2\pi i}\left(\frac{1}{x+i0*}-\frac{1}{x-i0*}\right), \quad \text{p.v.}\frac{1}{x_*}=\frac{1}{2}\left(\frac{1}{x+i0*}+\frac{1}{x-i0*}\right).$$

and our notations, we may write as follows:

$$\delta_*(x)=i(x_{*+}^{-1}-x_{*-}^{-1}), \quad \text{p.v.}\frac{1}{x_*}=\pi(x_{*+}^{-1}+x_{*-}^{-1}).$$

Thus the identity

$$e^{-\frac{t}{2}x^2}=\frac{1}{\sqrt{2\pi t}}\int_{-\infty}^{\infty}e^{-\frac{1}{2t}s^2}e^{isx}ds$$

for $t>0$, defines an element $e_*^{-\frac{t}{2}x^2}$ explicitly by

$$e_*^{-\frac{t}{2}x^2}=\frac{1}{\sqrt{2\pi t}}\int_{-\infty}^{\infty}e^{-\frac{1}{2t}s^2}e_*^{isx}dt$$

for $t>0$.

6. Proof of Theorem 4

In this section, we show Theorem 4. Recall the inverse Laplace transform, called the Bromwich integral:

$$e_*^{-itp(x)}=\frac{1}{2\pi}\int_{-\infty}^{\infty}e^{ist}(s+p(x))_{*+}^{-1}ds.$$

Though $(e^{i\theta}s+p(x))_{*+}^{-1}$ is given in our situation, we first remark that the integral

$$\frac{e^{i\theta}}{2\pi}\int_{-\infty}^{\infty}e^{ise^{i\theta}t}(e^{i\theta}s+p(x))_{*+}^{-1}ds \tag{6.1}$$

gives $e_*^{-itp(x)}$.

We compute $(e^{i\theta}s+p(x))_{*+}^{-1}$ in terms of $\cdot$-product by minding that the topology is given by using the expression of $\cdot$-product.

6.1. Product formula of inverses: Resolvent equations

To compute $(e^{i\theta}s{+}p(x))^{-1}_{*+}$, we first establish the product formulas involving $(x+\alpha)^{-1}_{*\pm}$. The main part of the proof of Theorem 4 is based on the following resolvent identities:

$$\begin{aligned}\frac{1}{(\alpha+x)(\beta+x)} &= \frac{1}{\beta-\alpha}\left(\frac{1}{\alpha+x}-\frac{1}{\beta+x}\right).\\ \frac{1}{(\alpha+x)^{n+1}} &= (-1)^n\frac{d^n}{d\alpha^n}\frac{1}{\alpha+x}.\end{aligned}$$

It follows that

$$(\alpha+x)^{-k}(\beta+x)^{-l} = (-1)^{k+l}\frac{d^{k+l}}{d\alpha^k d\beta^l}\frac{1}{\beta-\alpha}\big((\alpha+x)^{-1}-(\beta+x)^{-1}\big). \tag{6.2}$$

Proposition 8. *For every $\alpha\in\mathbb{C}$, $((x-\alpha)^{-1}_{*+})^m_*$, $((x-\alpha)^{-1}_{*-})^m_*$ are well-defined. However, $(x-\alpha)^{-1}_{*+}*(x-\alpha)^{-1}_{*-}$ diverges.*

We define the $*$-product of inverses by the same formula (6.2). Hence, we define for instance

$$(x-\alpha)^{-1}_{*+}*(x-\beta)^{-1}_{*-} = \frac{1}{\beta-\alpha}\big((x-\alpha)^{-1}_{*+}-(x-\beta)^{-1}_{*-}\big).$$

Remark that this is a critical definition. For $\alpha,\beta\in\mathbb{R}$ we see that

$$\int_0^\infty e_*^{it(x-\alpha)}dt * \int_0^\infty e_*^{-is(x-\beta)}ds = \int_0^\infty e^{is(\beta-\alpha)}\int_{-s}^\infty e_*^{i\tau(x+\alpha)}d\tau.$$

This integral does not converge in the ordinary sense. In spite of this, it is obvious that

$$\frac{1}{\beta-\alpha}\big((x-\alpha)^{-1}_{*+}-(x-\beta)^{-1}_{*-}\big)$$

is an inverse of $(x-\alpha)*(x-\beta)$.

The next theorem will play a fundamental role in the future application.

Theorem 9. *Let S_+ be any subset of $\mathbb{C}-\{i\rho;\rho<-\frac{\hbar}{2}\}$ such that $S_+\supset\{i\rho;\rho>\frac{\hbar}{2}\}$. The system of $\mathbb{C}[x]$ and*

$$\{(\alpha+x)^{-1}_{*+};\alpha\in S_+\}\cup\{(\alpha+x)^{-1}_{*-};\alpha\in\mathbb{C}{-}S_+\}$$

generate a commutative associative field of all rational functions.

Denote by $\mathcal{F}(S_+)$ the field given by Theorem 9. Since the inverse of $\alpha+x$ is discontinuous in α at the boundary of S_+, we cannot make $\mathcal{F}(S_+)$ a complete topological field.

In the proof of Theorem 9, we see that the inverse $(\alpha+x)_{*\pm}^{-1}$ is holomorphic in α. Hence we have also

Corollary 10. *For a sectorial polynomial $p(x)$, there is θ such that $(e^{i\theta}s+p(x))_*^{-1}$ is holomorphic with respect to s in a neighborhood of real axis.*

Estimating $(e^{i\theta}s+p(x))_*^{-1}$ by semi-norms, we see that the integral (6.1) converges. This completes the proof of Theorem4.

7. $*$-exponentials of quadratic functions in Weyl algebra

In this section we show that the singular system of exponential functions appears naturally in the system of $*$-exponentials of quadratic functions in Weyl algebra.

We refer general properties for Moyal products in [7], [8]. The Moyal product is given as follows:

$$f(u,v)*g(u,v)=fe^{\frac{\hbar i}{2}\overleftarrow{\partial_v}\wedge\overrightarrow{\partial_u}}g. \tag{7.1}$$

This defines an associative product on the space $C^\infty(\mathbb{R}^2)[[\hbar]]$. Being $\hbar$ a parameter moving in positive reals, the product $*$ has the following properties:

- $f*g$ is defined if either of f,g is a polynomial.
- For any polynomial $p(u,v)$, the right-multiplication $*p$, and the left multiplication $p*$ give continuous linear mapping on the space $C^\infty(\mathbb{R})$ into itself with respect to the C^∞-topology.

For every quadratic form au^2+bv^2+2cuv, we consider the following differential equation:

$$\frac{\partial}{\partial t}F(t,u,v)=(au^2+bv^2+2cuv)*F(t,u,v),\quad F(0,u,v)=1 \tag{7.2}$$

The right hand side of (7.2) is written as

$$\begin{aligned}(au^2+bv^2&+2cuv)*F(t,u,v)\\ &=(au^2+bv^2+2cuv)F+\hbar i\{(bv+cu)\partial_uF-(au+cv)\partial_vF\}\\ &\quad-\frac{\hbar^2}{4}\{b\partial_u^2F-2c\partial_v\partial_uF+a\partial_v^2F\}\end{aligned}$$

If we assume that $F(t,u,v)=f_t(au^2+bv^2+2cuv)$, then

$$\begin{aligned}(au^2+bv^2+2cuv)&*f_t(au^2+bv^2+2cuv)\\ &=(au^2+bv^2+2cuv)f_t(au^2+bv^2+2cuv)\\ &\quad-\hbar^2(ab-c^2)(f_t'(au^2+bv^2+2cuv)\\ &\quad+f_t''(au^2+bv^2+2cuv)(au^2+bv^2+2cuv)).\end{aligned}$$

It is remarkable that the equation depends only on the discriminant. By setting $x = au^2 + bv^2 + 2cuv$ the equation is written as

$$\frac{d}{dt}f_t(x) = xf_t(x) - \hbar^2(ab - c^2)(f_t'(x) + xf_t''(x)), \tag{7.3}$$

and the solution of (7.2) is given explicitly as

$$f_t(x) = \frac{1}{\cosh(\hbar\sqrt{ab-c^2}t)}\exp\{\frac{x}{\hbar\sqrt{ab-c^2}}\tanh(\hbar\sqrt{ab-c^2}\,t)\}.$$

If $au^2 + bv^2 + 2cuv = (\alpha u + \beta v)(\gamma u + \delta v)$, then $[\alpha u + \beta v, \gamma u + \delta v] = -\hbar i(\alpha\delta - \beta\gamma)$ and $ab - c^2 = -\frac{1}{2}(\alpha\delta - \beta\gamma)^2$. If $\alpha\delta - \beta\gamma = 1$, a pair (u', v') such that

$$u' = \alpha u + \beta v, \quad v' = \gamma u + \delta v$$

is canonically conjugate. The replacing (u', v') by (u, v) as above gives an isomorphism of the algebra. Hence e_*^{ituv} is a generic normal form of the $*$-exponential function.

In the case of generic normal form, the differential equation is given as

$$\frac{\partial}{\partial t}F_t(x) = ixF_t(x) + i\frac{\hbar^2}{4}F_t'(x) + i\frac{\hbar^2}{4}F_t''(x)x, \qquad F_0(x) = 1, \tag{7.4}$$

whose solution is given as

$$e_*^{ituv} = \frac{1}{\cosh\frac{\hbar t}{2}}e^{(\frac{2i}{\hbar}\tanh\frac{\hbar t}{2})uv}. \tag{7.5}$$

Remark that (7.5) is valid even if the variable u, v are restricted in some open subset of $\mathbb{R}^2$. Moreover, these properties nothing to do with the properties of Laplacian but the generic character of the quadratic exponentials.

Note that $uv = u{*}v + \frac{\hbar i}{2} = \hbar i(u\partial_u + \frac{1}{2})$, and $uv = v{*}u - \frac{\hbar i}{2} = -\hbar i(v\partial_v + \frac{1}{2})$. By Theorem 3, we have $(u{*}v)_{*-}^{-1}$ and $(v{*}u)_{*+}^{-1}$. This causes another example where the associativity breaks down. Let

$$u^\bullet = (v{*}u)_{*+}^{-1} * v, \quad u^\circ = v * (u{*}v)_{*-}^{-1}.$$

Then these are a left and a right inverse respectively such that

$$u * u^\bullet = 1 - \varpi_{0,0}, \quad u^\circ * u = 1 - \bar{\varpi}_{0,0}.$$

We then have

$$(u^\bullet * u) * u^\circ \neq u^\bullet * (u * u^\circ).$$

Remark the following:

Lemma 11. *For every* $\alpha \notin \mathbb{Z}_{\geq 0}$, $u\partial_u - \alpha : Hol(\mathbb{C}) \to Hol(\mathbb{C})$ *is invertible, and the inverse is holomorphic in* α. *Similarly, for every* $\alpha \notin \mathbb{Z}_{\leq 0}$, $-v\partial_v - \alpha : Hol(\mathbb{C}) \to Hol(\mathbb{C})$ *is invertible.*

Proof. Entire functions are written as converging power series $\sum a_k u^k$. Obviously,

$$(u\partial_u - \alpha)\sum a_k u^k = \sum (k-\alpha) a_k u^k.$$

If α is not a non-negative integer, then $u\partial_u - \alpha$ is invertible. The inverse is holomorphic in α.

□

Recalling (2.2) we set $\varpi_{0,0} = 2e^{\frac{2i}{\hbar}uv}$, and $\bar{\varpi}_{0,0} = 2e^{-\frac{2i}{\hbar}uv}$. By the product formula (7.1) we easily see that

$$v * \varpi_{0,0} = 0 = \varpi_{0,0} * u, \qquad u * \bar{\varpi}_{0,0} = 0 = \bar{\varpi}_{0,0} * v$$

Consider regular representations:

$$f(u)*\varpi_{0,0} \mapsto (uv)*f(u)*\varpi_{0,0} = (\hbar i(u\partial_u + 1/2)f(u)) * \varpi_{0,0},$$
$$g(v)*\varpi_{0,0} \mapsto (uv)*g(v)*\bar{\varpi}_{0,0} = (-\hbar i(v\partial_v + 1/2)f(u)) * \bar{\varpi}_{0,0}$$

on the space of all entire functions. Lemma 11 shows that if $\alpha \notin (\mathbb{Z}_{\geq 0} + \frac{1}{2})\hbar i$, $uv - \alpha$ is invertible as an operator on $Hol(\mathbb{C}) * \varpi_{0,0}$, and that if $\alpha \notin -(\mathbb{Z}_{\geq 0} - \frac{1}{2})\hbar i$, $uv - \alpha$ is invertible as an operator on $Hol(\mathbb{C}) * \bar{\varpi}_{0,0}$.

Setting $2i\eta = \frac{2i}{\hbar}\tanh\frac{\hbar t}{2}$, we have

$$(e^{2i\eta uv} * f(u)) = e^{2i\eta uv} f((1-\hbar\eta)u), \quad (f(u)*e^{2i\eta uv}) = e^{2i\eta uv} f((1+\hbar\eta)u).$$

Thus we obtain

$$e_*^{ituv} * f(u) = \frac{1}{\cosh\frac{\hbar t}{2}} e^{(\frac{2i}{\hbar}\tanh\frac{\hbar t}{2})uv} f((1-\tanh\tfrac{1}{2}\hbar t)u),$$

which gives the exchanging rule

$$e^{2i\eta uv} * f(u) = f\left(\frac{1-\hbar\eta}{1+\hbar\eta}u\right) * e^{2i\eta uv}.$$

Noting $(uv) * \varpi_{0,0} = \frac{\hbar i}{2}\varpi_{0,0}$, we have

Lemma 12. $e_*^{ituv} * \varpi_{0,0} = e^{-\frac{\hbar t}{2}}\varpi_{0,0}$, *and* $e_*^{ituv} * \bar{\varpi}_{0,0} = e^{\frac{\hbar t}{2}}\bar{\varpi}_{0,0}$.

By Lemma 12, we define vacuum representations as follows:

$$e_*^{ituv} * f(u) * \varpi_{0,0} = e^{-\frac{\hbar t}{2}} f\left(\left(\frac{1-\tanh\frac{\hbar t}{2}}{1+\tanh\frac{\hbar t}{2}}\right)u\right) * \varpi_{0,0} \qquad (7.6)$$

$$e_*^{ituv} * g(v) * \bar{\varpi}_{0,0} = e^{\frac{\hbar t}{2}} g\left(\left(\frac{1+\tanh\frac{\hbar t}{2}}{1-\tanh\frac{\hbar t}{2}}\right)v\right) * \bar{\varpi}_{0,0} \qquad (7.7)$$

Remark that (7.6) may be rewritten as

$$\frac{1}{\cosh \frac{\hbar t}{2}} e^{(\frac{2i}{\hbar} \tanh \frac{\hbar t}{2})uv} * f(u) * \varpi_{0,0} = e^{-\frac{\hbar t}{2}} f\left(\left(\frac{1-\tanh \frac{\hbar t}{2}}{1+\tanh \frac{\hbar t}{2}}\right) u\right) * \varpi_{0,0}.$$

If t is restricted in the real numbers, this gives a representation of the one parameter group e_*^{ituv}, which is extendible to the complex numbers, if $f(u)$ is an entire function. If we identify $x = uv$, the identity such as $e_*^{\frac{2\pi}{\hbar}x} = -1$ does make sense as the identity of operators acting on the space of all entire functions. It is easy to see that

$$e_*^{\frac{\pi}{\hbar}x} * f(u) * \varpi_{0,0} = if(u) * \varpi_{0,0}, \quad e_*^{\frac{\pi}{\hbar}x} * g(v) * \bar{\varpi}_{0,0} = -ig(v) * \bar{\varpi}_{0,0}.$$

Thus, $e_*^{\frac{\pi}{\hbar}x}$ depends on the choice of representation.

Moreover, we have

$$\lim_{t\to\infty} e_*^{itx} * f(u) * \varpi_{0,0} = 0, \quad \lim_{t\to-\infty} e_*^{itx} * f(u) * \varpi_{0,0} = \infty f(-\infty) * \varpi_{0,0}$$

as an operator acting on $Hol(\mathbb{C}) * \varpi_{0,0}$. This suggests that $\delta_*(x)$ is not defined as an operator.

8. Harmonic oscillators, and a discrete picture

In this section we give two models analogous to our system that are not exactly isomorphic.

Though $e_*^{\frac{\pi}{\hbar}x}$ is not defined in the expression using the usual $\cdot$-product, we view formally $e_*^{\frac{\pi}{\hbar}x}$ as an element. We set $\mathfrak{I} = e_*^{\frac{\pi}{\hbar}x}$. Since $\mathfrak{I}^2 = -1$, the set $\{a + b\mathfrak{I}; a, b \in \mathbb{R}\}$ is naturally identified with $\mathbb{C}$.

First, we show that our system is similar to harmonic oscillators. Consider two correspondences

$$\psi(ae^{tx}) = \frac{1}{a}(1 - \tfrac{\hbar}{2}t\mathfrak{I}), \quad \bar{\psi}(ae^{tx}) = \frac{1}{a}(1 + \tfrac{\hbar}{2}t\mathfrak{I}), \quad a, t \in \mathbb{C}. \tag{8.1}$$

We easily see that ψ, $\bar{\psi}$ give $*$-homomorphisms

$$\begin{aligned} \psi((ae^{tx}) * (be^{sx})) &= \psi(ae^{tx})\psi(be^{sx}), \\ \bar{\psi}((ae^{tx}) * (be^{sx})) &= \bar{\psi}(ae^{tx})\bar{\psi}(be^{sx}) \end{aligned} \tag{8.2}$$

if the $*$-product in the left hand sides of the above equalities are defined.

Remark also that

$$\begin{aligned}\psi(\varpi_{0,0}) &= \frac{1}{2}(1-i\mathfrak{J}),\ \psi(\bar{\varpi}_{0,0}) = \frac{1}{2}(1+i\mathfrak{J}),\\ \bar{\psi}(\varpi_{0,0}) &= \frac{1}{2}(1+i\mathfrak{J}),\ \bar{\psi}(\bar{\varpi}_{0,0}) = \frac{1}{2}(1-i\mathfrak{J}).\end{aligned} \tag{8.3}$$

Since $e^t = \cosh t + \sinh t = \cosh t(1+\tanh t)$, we have

$$\psi\left(\frac{1}{\cosh\frac{\hbar t}{2}}e^{i\frac{2}{\hbar}(\tanh\frac{\hbar t}{2})x}\right) = \cosh\tfrac{1}{2}\hbar t(1-(\tanh\tfrac{1}{2}\hbar t)i\mathfrak{J}).$$

It follows that

$$\psi(e_*^{tx}) = \psi\left(\frac{1}{\cos\frac{\hbar t}{2}}e^{\frac{2}{\hbar}\tan\frac{\hbar t}{2}x}\right) = \cos\tfrac{1}{2}\hbar t + \sin\tfrac{1}{2}\hbar t\mathfrak{J}.$$

We have also that

$$\psi(re_*^{tx}) = r^{-1}e^{\frac{\hbar t}{2}\mathfrak{J}},$$

which make us an analogy to a harmonic oscillator.

However, if we identify e_*^{tx} with $e^{\frac{\hbar t}{2}\mathfrak{J}}$, then differentiating by t, we must set $x = \frac{1}{2}\hbar\mathfrak{J}$ and hence $x^2 + \frac{1}{4}\hbar^2 = 0$. This contradicts $(x+\frac{1}{2}\hbar i)*(x-\frac{1}{2}\hbar i)$ possession of an inverse

$$\frac{1}{\hbar i}\left(\left(x-\frac{1}{2}\hbar i\right)_{*-}^{-1} - \left(x+\frac{1}{2}\hbar i\right)_{*+}^{-1}\right).$$

Hence we cannot identify e_*^{tx} with $e^{\frac{\hbar t}{2}\mathfrak{J}}$.

8.1. A discrete picture

In this subsection we show that our system is similar to the exponential functions of quaternions. Let $\mathbb{H}$ be the quaternion field written by $q = x_0 + x_1e_1 + x_2e_2 + x_3e_3$. We view x as

$$x = \{(m+\tfrac{1}{2})\hbar(\alpha e_2 + \beta e_3);\ m\in\mathbb{Z},\ \alpha^2+\beta^2=1\}. \tag{8.4}$$

The product of a quaternion is regarded as the $*$-product in the previous sections.

Since $e_1 * e_2 = e_3 = -e_2 * e_1$ in $\mathbb{H}$, we see that

$$e_1 * x = \{(m+\tfrac{1}{2})\hbar(\alpha e_3 - \beta e_2);\ m\in\mathbb{Z},\ \alpha^2+\beta^2=1\} = -x*e_1$$

For any $\eta = \alpha e_2 + \beta e_3$ such that $\alpha^2+\beta^2 = 1$, we have $\eta_*^2 = -1$ and

$$e_*^{t\eta} = \cos t + \eta\sin t,\quad e_*^{t(m+\frac{1}{2})\hbar\eta} = \cos t(m+\tfrac{1}{2})\hbar + \eta\sin t(m+\tfrac{1}{2})\hbar$$

The second identity may be written as

$$e_*^{tx}=\cos\left(\frac{t}{||x||}\right)+\frac{x}{||x||}\sin(t||x||),\quad e^{t(e_1x)}=\cos\left(\frac{t}{||x||}\right)+\frac{e_1x}{||x||}\sin(t||x||)$$

It follows $e_*^{\frac{2\pi}{\hbar}x}=e_*^{\frac{2\pi}{\hbar}e_1x}=-1$ but

$$e_*^{\frac{\pi}{\hbar}x}=(-\eta)^m,\quad e_*^{\frac{\pi}{\hbar}(e_1x)}=(-e_1\eta)^m.$$

Recalling (8.4) we see that $r+se_1+x$ is invertible quaternion for every $r,s\in\mathbb{R}$. The inverse $(r+se_1+x)_*^{-1}$ is unique and depends smoothly on r,s. This is a very similar property to our singular system of exponential functions. However, we see that

$$\int_{-\infty}^{\infty}e_*^{\eta t(m+\frac{1}{2})\hbar}dt=\delta_0((m+\frac{1}{2})\hbar)$$

vanishes because $(m+\frac{1}{2})\hbar\neq 0$.

This gives also that

$$\lim_{t\to\pm\infty}e_*^{\eta t(m+\frac{1}{2})\hbar}=0$$

in the distribution sense, or the WKB sense.

Recall now that $\lim_{t\to\infty}e_*^{\eta t(x-\frac{\hbar i}{2})}=2e^{\frac{2i}{\hbar}x}$ in our singular system, and $2e^{\frac{2i}{\hbar}x}$ may be viewed as a projection operator.

This corresponds in the space of quaternions to the identity:

$$\lim_{t\to\pm\infty}e_*^{it((m+\frac{1}{2})\hbar-\frac{\hbar i}{2})}=e_*^{itm\hbar}=\begin{cases}1, & m=0,\\ 0, & m\neq 0,\end{cases}$$

in the distribution sense.

References

[1] F.Bayen, M.Flato, C.Fronsdal, A.Lichnerowicz, D.Sternheimer Deformation theory and quantization I, *Ann. Phs.* **111** (1978) 61–110.

[2] D.Deligne, *Déformation de l'algèbre des fonctions dúne variété symplectique*, Select Math. New series 1, 1995.

[3] Japanese Math. Soc. *Encyclopedia of mathematics*, Iwanami, 1985.

[4] M.Kontsevich, Deformation quantization of Poisson manifold I, q-alg/9709040.

[5] R.Kolchin, On the Galois theory of differential fields, *Amer. J. Math.*, **77** (1955) 868–894.

[6] M.Morimoto, An introduction to Sato's hyperfunctions, AMS Trans. Mono. 129, 1993.

[7] H.Omori, Infinite dimensional Lie groups, AMS Trans.Mono.,158, 1997.

[8] H.Omori, Y.Maeda, A.Yoshioka, Weyl manifolds and deformation quantization, *Adv. Math.* vol 85, No 2, 224–255, 1991

[9] H.Omori, Y.Maeda, N.Miyazaki A.Yoshioka, Anomalous exponent for convergent star-product on Fréchet–Poisson algebras, preprint.

[10] B. de Witt, *Supermanifolds,* Cambridge Univ. Press, 1984

[11] K.Yoshida, *Functional analysis*, Springer 1966.

DETERMINANTS OF ELLIPTIC BOUNDARY PROBLEMS IN QUANTUM FIELD THEORY

Simon G. Scott
Department of Mathematics, King's College
London WC2R 2LS, U.K.
sscott@mth.kcl.ac.uk

Krzysztof P. Wojciechowski
Department of Mathematics, IUPUI (Indiana/Purdue)
Indianapolis IN 46202–3216, U.S.A.
kwojciechowski@math.iupui.edu

Abstract We review some recent advances in understanding the zeta determinant of an elliptic boundary value problem for the Dirac operator. We discuss a recent adiabatic pasting formula for the determinant with respect to a partition of the underlying manifold, and outline some of the applications to geometric anomalies.

Mathematics Subject Classification (2000): 58J, 35J, 81Q, 35P, 35Q, 53C

Keywords: Dirac operator, ζ determinant, Grassmannian of elliptic boundary conditions, Canonical determinant, Pasting of determinants, Path integral, Topological Quantum Field Theory, Relative geometric anomaly.

Introduction

The purpose of this exposition is to review recent results of the authors and collaborators in understanding the nature of the ζ determinant and some

Y. Maeda, et al. (eds.), Noncommutative Differential Geometry and Its Applications to Physics, 187–215.

applications of these ideas to *Quantum Field Theory*. This report covers some of the results in the works [21, 25, 31, 32, 34, 35, 41]. We refer to those papers for proofs and precise statements. Here we concentrate on ideas behind the results and we also discuss some examples.

The second author wants to express his deepest gratitude to the organizers of the *Workshop on Noncommutative Differential Geometry and its Application to Physics* in Shonan-Kokusaimura for the extreme hospitality and for the creative atmosphere of the meeting, which lead to this paper.

Zeta function regularization and determinant line bundles have come to represent something of a Zeitgeist for the intense interchange of ideas in recent years between geometric analysis, topology and theoretical physics. On the physical side the ζ-regularized determinant of an elliptic differential operator over a closed manifold M defines a putative value for path integral computations in string theory and QFT, while on the mathematical side it is a delicate highly non-local analytic object encoding fundamental spectral and geometric information about the underlying manifold, bundle, and elliptic operator. In either view point ζ determinants are somewhat mysterious, and their computation is of great interest. Non-locality means that such computations are extremely difficult on all but the simplest symmetric spaces, and a general elliptic operator will not even have a canonically defined ζ determinant.

Quantum theory, on the other hand, tells us that anyway the meaningful quantity to think about is rather the democratic 'sum over all histories'. Mathematically one considers a submersion of Riemannian spin manifolds $\pi : Z \to B$ with fibre diffeomorphic to M, defining a family of twisted Dirac operators $\mathbb{D} = \{\mathcal{D}_b : b \in B\} : \mathcal{H}^+ \to \mathcal{H}^-$. Here $\mathcal{H}^\pm$ are the infinite-dimensional Frechet bundles with fibre $\mathcal{H}_b^\pm = C^\infty(M_b; S_b^\pm \otimes E_b)$, with S the vertical spinor bundle and E a vertical Hermitian coefficient bundle with compatible connection. The family $\mathbb{D}$ defines two families of finite-dimensional vector spaces $\mathrm{Ker}\,\mathbb{D} = \cup_b \mathrm{Ker}\,\mathcal{D}_b$ and $\mathrm{Cok}\,\mathbb{D} = \cup_b \mathrm{Cok}\,\mathcal{D}_b$ associated to which one has an index bundle element $\mathrm{Ind}\,\mathbb{D} = \mathrm{Ker}\,\mathbb{D} - \mathrm{Cok}\,\mathbb{D}$ and determinant line bundle $\mathrm{DET}\,\mathbb{D}$ over B with fibre canonically isomorphic to the complex line $\mathrm{DetInd}\mathcal{D}_b := \wedge^{\max}(\mathrm{Ker}\,\mathcal{D}_b)^* \otimes \wedge^{\max}\mathrm{Cok}\,\mathcal{D}_b$.

The prototypical physical problem is for Yang–Mills theory with gauge fields associated to some compact Lie group G coupled to fermions over a compact even-dimensional Riemannian spin manifold M. A gauge field is represented by a connection on a principal G-bundle L over M, and a chiral fermion as a section of a Hermitian bundle $S^\pm \otimes E$, where E is an associated G-bundle to

L. The aim is to evaluate the (heuristic) path integral

$$Z(A,\psi) = \int_{\mathcal{A}} e^{-S(A,\psi)}\, dA d\psi d\overline{\psi}, \tag{0.1}$$

with respect to the action $S(A,\psi) = \int_M \overline{\psi}\mathcal{D}_A\psi + \int_M \|F_A\|^2$, where $\mathcal{D}_A$ is the Dirac operator, and the parameter space $\mathcal{A}$ is the space of connections cross the space of sections of $S^+ \otimes E$. *Anomalies* arise due to the first term in the action coupling fermions to the gauge bosons. Indeed, one is immediately faced with a number of difficulties. First of all, the measure in (0.1) is entirely fictitious. Second the action $S(A,\psi)$ is invariant under the action of the group $\mathcal{G}$ of gauge transformations of P, and so the integral should really be evaluated over the moduli space $\mathcal{M} = B/\mathcal{G}$, which with appropriate constraints may be a respectable finite-dimensional manifold.

The initial method of dealing with the first problem is to ignore it entirely, and pretend we are evaluating a finite-dimensional Gaussian integral. That is, we suppose the ψ are fermionic variables in a finite-dimensional vector space and evaluate according to elementary formulae for finite-dimensional fermionic integrals

$$Z(A,\psi) = \int_{\mathcal{A}} e^{\int_M \|F_A\|^2} \cdot \det\mathcal{D}_A\, dA. \tag{0.2}$$

To proceed we must understand the integrand as a function and that means we must understand $\det\mathcal{D}_A$ as a function. However, *a priori* $\det\mathcal{D}_A$ arises as a section of the complex line bundle DET $\mathbb{D}$ and to realize it as a function over B therefore requires a trivialization of DET $\mathbb{D}$. Using, for example, ζ function regularization one defines a metric and compatible connection on the bundle whose curvature, holonomy, and Chern class represent respectively local, global and topological obstructions to the existence of of a regularized determinant function on B. These quantities may be understood as anomalies for an appropriate QFT, while mathematically they are local, global and topological expressions for the degree 2 part of the family's index [3, 13, 28].

The ζ function metric is defined using the determinant of the Dirac–Laplacian $\Delta_b = \mathcal{D}_b^*\mathcal{D}_b$. Since Δ_b is a self-adjoint elliptic operator it has a discrete spectral resolution $\{\lambda \in \mathbb{R}, \phi_\lambda\}$, and so formally its determinant is the infinite product $\prod \lambda$, which is undefined. To define a regularized product one uses a spectral ζ function. This is based upon a remarkable generalization of the method by which the classical Riemann ζ function, defined initially only in a half-plane, is extended meromorphically to the whole complex plane, and the identity for a finite rank operator $A \in \mathrm{End}(V)$

$$\det A = e^{-\zeta_A'(0)}, \tag{0.3}$$

where s is a complex number

$$\zeta_A(s) := \text{tr } A^{-s} \quad \text{and} \quad \zeta_A^{'}(0) := \frac{d}{ds}\{\text{Tr } A^{-s}\}|_{s=0}.$$

For a general Laplacian $\det_\zeta(\Delta)$ is defined in a similar way. In Seeley's paper [36] it was explained that by standard operator norm estimates Δ^{-s} is trace class for $\text{Re}(s) > \dim(M)/2$, and so in this half-plane $\zeta_\Delta(s) := \text{Tr}\,(\Delta^{-s})$ is a well defined holomorphic function and, moreover, one has, via the Mellin transform, the identity

$$\zeta_\Delta(s) = \frac{1}{\Gamma(s)} \int_0^\infty t^{s-1}\text{Tr } e^{-t\Delta}\, dt, \tag{0.4}$$

where Γ is the gamma function. Since Δ is positive, the heat operator $e^{-t\Delta}$ has a smooth kernel $P_t(x, y)$, and is of trace class. As $t \to 0$ there is an asymptotic expansion

$$\text{Tr}\,[P_t(x,x)] = \sum_{j=-n/2}^{k} a_j(x)t^j + o(t^k, x), \tag{0.5}$$

where $a_j(x)$ is a C^∞ function depending only on the local symbol of D. In particular, we can therefore use the right side of (0.4) to define a meromorphic continuation of $\zeta_\Delta(s)$ to the whole complex plane with only simple poles. In particular $s = 0$ is not a pole. Hence $\zeta'_\Delta(0) = \frac{d}{ds}\{\zeta_\Delta(s)\}|_{s=0}$ is well defined and we may define the ζ determinant by $\det_\zeta(\Delta) := e^{-\zeta_\Delta^{'}(0)}$. The metric on $\text{Det}\,\mathcal{D}$ over the region of B where the D_b are invertible is defined by

$$\|\det\,\mathcal{D}_b\|_\zeta^2 = \det{}_\zeta\Delta. \tag{0.6}$$

Let us end this Section with some explicit examples

Dimension 1: Let us consider the simplest case of interest. Take

$$\mathcal{D} = i\frac{d}{dx} + a : C^\infty(S^1;\mathbb{C}) \to C^\infty(S^1;\mathbb{C}), \quad a \in (0,1). \tag{0.7}$$

Then $\Delta = \mathcal{D}^2$ has spectrum $\{\lambda_n = (n+a)^2 : n = 0, 1, \dots\}$ and so

$$\zeta_\Delta(s) := \text{Tr}\,(\Delta^{-s}) = \sum_{n=0}^{\infty} \frac{1}{(n+a)^{2s}} \tag{0.8}$$

which is precisely the classical Riemann–Hurwitz zeta function $\zeta(2s, a)$. This is well defined for $\text{Re}(s) > \frac{1}{2}$, and has a meromorphic continuation to the entire complex plane with only simple poles. Around $s = 0$ the continuation is holomorphic with

$$\zeta_\Delta'(0) = -\log(4\sin^2(\pi a)). \tag{0.9}$$

Hence we obtain

$$\det_\zeta(\Delta) := e^{-\zeta_\Delta'(0)} = 4\sin^2(\pi a). \tag{0.10}$$

More generally, in dimension one there is a canonical family of Dirac operators. Take $M = S^1$, and consider a rank n Hermitian bundle E over S^1. Let B be the Banach manifold of unitary connections on E, parameterizing the family $\mathbb{D} = \{\mathcal{D} = i\nabla_{d/dx} : \nabla \in B\}$. The determinant line bundle is canonically trivial in this case, and using formulas for the Riemann–Hurwitz ζ function it is easy to prove that

$$\det_\zeta \mathcal{D} = \det(I + h_\nabla), \tag{0.11}$$

where h_∇ is the holonomy of ∇ around the circle. In particular, we see therefore that $\det_\zeta \mathcal{D}$ is a smooth function invariant under the action of the group $\mathcal{G}$ of unitary gauge transformations acting on B, and hence passes to a function on the finite-dimensional moduli space $B/\mathcal{G} = U(n)/conjugation$. So no $\mathcal{G}$ anomaly arises, reflecting that $\nabla^{(\zeta)}$ is $\mathcal{G}$-invariant and the curvature of the ζ-connection is identically zero. In fact,

$$\det_\zeta(\mathcal{D}^*\mathcal{D}) = |\det_\zeta D|^2, \tag{0.12}$$

means that $\|\det\mathcal{D}\|_\zeta^2$ is the norm square of a holomorphic function, and so the metric is flat.

Dimension 2: Consider first the case of an elliptic curve $\Sigma_\tau = \mathbb{R}^2/(\mathbb{Z}\times\tau\mathbb{Z})$, where $\tau = \tau_1 + i\tau_2$ is in the upper-half plane $\tau_2 > 0$, parameterizing the Teichmüller space of complex structures on the torus. Then for $\bar\partial_\tau$ acting on the space $\Omega^{m,0}(\Sigma_\tau)$ of differentials of order m it is well-known that

$$\det_\zeta(\bar\partial_\tau) = q^{\frac{6m^2-6m+1}{12}} \prod_{n=1}^{\infty}(1-q^n)^2, \quad \det_\zeta(\bar\partial_\tau^*\bar\partial_\tau) = \frac{1}{\tau_2}|\det_\zeta(\bar\partial_\tau)|^2, \tag{0.13}$$

where $q = e^{2\pi i\tau}$. This was first computed in [29], as analytic Reidemeister Torsion, but is popular in the String Theory literature because its explicit form allows one to construct the Polyakov integral exactly. Notice that $\det_\zeta(\bar\partial_\tau)$ is a holomorphic function of τ, while, unlike dimension one, the Laplacian fails to be an absolute function squared due to zero mode.

For higher genus surfaces no general closed formula for $\det_\zeta(\bar\partial_\tau)$ is known, though one can say it will be a generalized theta function. However, there is the relative 'conformal anomaly' formula:

$$\frac{\det_\zeta(\bar\partial^*\bar\partial)_{g'}}{\det_\zeta(\bar\partial^*\bar\partial)_g} = \exp\left[\frac{6m^2-6m+1}{6\pi i}L(g,f)\right], \tag{0.14}$$

where $L(g,f) = \int_\Sigma [\bar\partial f \wedge \bar\partial f + fR]$. Here $(\bar\partial^*\bar\partial)_g$ is the $\bar\partial$-Laplacian defined by the complex structure associated to a choice of metric g on Σ, with curvature R, and $g^{'}$ is the conformally equivalent metric $e^{2f}g$.

Higher dimensions: There is a straightforward generalization of (0.13) to operators of the form $\mathcal{D} = d/du + A$ on higher dimensional cylinders $Y \times S^1$, where A is a self-adjoint elliptic operator of order 1, with discrete real spectrum $\{\lambda_k : k \in \mathbb{Z}\}$. One finds, up to a constant, $\det_\zeta \mathcal{D} = \prod_k (1 - e^{-\sigma(\lambda_k)})$, where $\sigma(z) = \mathrm{sgn}(\mathrm{Re}\ z)$.

0.1. Elliptic boundary Value Problems

Topological (or functorial) QFT tells that in fact our theories must include not only histories defined by closed manifolds, but also manifolds with boundary, and that axiomatically this should be realizable as a functor from geometric fibrations to vector bundles (of Hilbert spaces) satisfying pasting laws suggested by heuristic path integral formulae. Whatever the merits of such a theory may be, it has presented the need for a mathematical theory of determinant line bundles for families of elliptic operators over manifolds with boundary and pasting laws for the ζ determinant with respect to a partition of the manifold. We return in Section 4 with a discussion of the functorial quantum field theory (FQFT) description of determinant anomalies, while in Section 3 we describe recent results on an adiabatic pasting law.

First, we need to review briefly some facts about elliptic boundary value problems (EBVPs) for a Dirac type of operator over a compact Riemannian spin manifold M with boundary $\partial M = Y$. Let $\mathcal{D}$ be the corresponding Dirac operator. We assume that the Riemannian metric on M and the Hermitian structure on the Clifford bundle S over M are products in a collar neighborhood $N = [0,1] \times Y$ of the boundary. In N the operator $\mathcal{D}$ has the form

$$\mathcal{D} = G(\partial_u + B)\ , \tag{0.15}$$

where $G : S|Y \to S|Y$ is a unitary bundle isomorphism and $B : C^\infty(Y; S|Y) \to C^\infty(Y; S|Y)$ is the corresponding Dirac operator on Y.

In order to obtain an unbounded Fredholm operator with good elliptic regularity properties we have to impose a boundary condition on the operator $\mathcal{D}$. The space $L^2(Y; S|Y)$ has a preferred orthogonal decomposition

$L^2(Y; S|Y) = H^+ \oplus H^-$, where subspace $H^\pm$ are the subspaces spanned by the eigenvectors corresponding to the $\pm$ eigenvalues of B. Let $\Pi_>$ denote the spectral projection onto the subspace H^+, and $\Pi_< = I - \Pi_>$ the projection onto H^-. It is well known that $\Pi_>$ is an elliptic boundary condition for the operator $\mathcal{D}$ (see [1], [7]). More, generally, any pseudo-differential projection (self-adjoint idempotent) operator P on $L^2(Y; S|Y)$ such that $P - \Pi_>$ is a pseudo-differential operator of order -1 defines a good boundary condition for $\mathcal{D}$. This means that the unbounded operator

$$\mathcal{D}_P = \mathcal{D} : \mathrm{dom}(\mathcal{D}_P) \to L^2(M; S) \tag{0.16}$$

with domain

$$\mathrm{dom}\, \mathcal{D}_P = \{s \in H^1(M; S) \ : \ P(s|Y) = 0\},$$

with H^1 the first Sobolev space, is a Fredholm operator with kernel and cokernel consisting only of smooth sections. The parameter space of all such projections is an infinite-dimensional Grassmannian $Gr_{-1}(\mathcal{D})$, which has the homotopy type $\mathbb{Z} \times BU$ of the space of Fredholm operators; two boundary conditions P_1 and P_2 belong to the same connected component if and only if

$$\mathrm{index}\, \mathcal{D}_{P_1} = \mathrm{index}\, \mathcal{D}_{P_2} .$$

For analytical reasons associated with the existence of the ζ determinant, we shall restrict our comments to the *smooth Grassmannian*, defined by

$$Gr_\infty(\mathcal{D}) = \{P \in Gr(\mathcal{D}) \ : \ P - \Pi_\geq \ \text{has a smooth kernel}\} \ . \tag{0.17}$$

The crucial fact is that, as a consequence of the *Unique Continuation Property* for the Dirac operators, there is a canonical identification between $H(\mathcal{D})$ and the *infinite-dimensional* solution space

$$\{s \in C^\infty(M : S) \ ; \ \mathcal{D}s = 0 \ \text{in} \ M \setminus Y\}.$$

The restriction of smooth sections to the boundary extends to a continuous map:

$$\gamma_0 : H^s(M; S) \to H^{s-\frac{1}{2}}(Y; S|Y) \ ,$$

which is well defined for $s > 1/2$ (see [7]). For any real s the *Poisson operator* of $\mathcal{D}$ is a mapping

$$\mathcal{K} : C^\infty(Y; S|Y) \to C^\infty(M; S)$$

which extends to a continuous map $\mathcal{K} : H^{s-1/2}(Y; S|Y) \to H^s(M; S)$, with range equal to the space $\ker(\mathcal{D}, s) = \{f \in H^s(M; S) \ : \ \mathcal{D}f = 0 \text{ in } M \setminus Y\}$ and

$$\begin{aligned} \mathcal{K} &: \mathcal{H}(\mathcal{D}, s) \\ &= \mathrm{Ran}\{P(\mathcal{D}) : H^{s-1/2}(Y; S|Y) \to H^{s-1/2}(Y; S|Y)\} \to \ker(\mathcal{D}, s) \end{aligned} \tag{0.18}$$

is an isomorphism (see [7]). The Poisson operator defines the Calderon projection:

$$P(\mathcal{D}) = \gamma_0 \mathcal{K} \tag{0.19}$$

We refer to [7] for the details of the construction, which is originally due to Calderon and Seeley. The Poisson operator $\mathcal{K}$ implies that constructing solutions for the boundary operator

$$\mathcal{S}(P) = P \circ P(\mathcal{D}) : H(\mathcal{D}) \longrightarrow \text{range}\,(P),$$

is equivalent to constructing solutions to the elliptic boundary value problem $\mathcal{D}_P$ (and the same for the adjoints). The Fredholm alternative then gives us:

$$\text{index}\,\mathcal{D}_P = \text{index}\,\mathcal{S}(P), \tag{0.20}$$

and the relative index formula

$$\text{index}\,\mathcal{D}_{P_1} - \text{index}\,\mathcal{D}_{P_2} = \text{index}\,(P_1, P_2), \tag{0.21}$$

where $(P_1, P_2) = P_2 \circ P_1 : \text{range}\,(P_1) \longrightarrow \text{range}\,(P_2)$.

In particular $\mathcal{D}_P$ is an invertible operator if and only if the operator $\mathcal{S}(P) = PP(\mathcal{D}) : \mathcal{H}(\mathcal{D}) \to \mathrm{Ran}\,P$ is invertible. The crucial fact proved in [35] is the following exact relative inverse formula giving a a far more delicate analytic realization of (0.21):

Theorem 0.1. *[35] Let $P_1, P_2 \in Gr_\infty(\mathcal{D})$ such that the EBVPs $\mathcal{D}_{P_i}$ are invertible. Then one has the relative inverse formula*

$$\mathcal{D}_{P_1}^{-1} = \mathcal{D}_{P_2}^{-1} - \mathcal{K}\mathcal{S}(P_1)^{-1} P_1 \gamma_0 \mathcal{D}_{P_2}^{-1} \ . \tag{0.22}$$

In particular, $\mathcal{D}_{P_1}^{-1} - \mathcal{D}_{P_2}^{-1}$ is a smoothing operator.

This tells us that if the ζ determinants $\det_\zeta \mathcal{D}_{P_i}$ exist then we can expect the relative determinant $\det_\zeta \mathcal{D}_{P_1} / \det_\zeta \mathcal{D}_{P_2}$ to be a Fredholm determinant associated to the boundary operators $\mathcal{S}(P_1), \mathcal{S}(P_2)$.

Observe furthermore that in the examples outlined earlier for closed manifolds, in each case where one has an explicitly computable zeta determinant the problems are naturally posed as elliptic boundary value problems on the appropriate cylinder (apart possibly from the relative conformal anomaly). Furthermore, the regularized determinants are expressed *purely in terms of boundary Cauchy data*, i.e., there is a canonical basis for the solution space of the Dirac operator on the cylinder constructed from the spectral resolution of the boundary operator. In summary:

- The ζ determinant (and other spectral invariants) for preferred classes of elliptic boundary value problems is a more computable invariant than for the Dirac operator over a closed manifold. This is because of an isomorphism between the (infinite-dimensional) solution space of the Dirac operator and the boundary Cauchy data, defined by the *Poisson operator*.

- In general, one cannot hope to explicitly compute the ζ determinant over topologically non-trivial manifolds. Nevertheless, it may be possible to compute relative determinants, i.e., relative to a basepoint. cf equation (0.14).

These points are made concrete in a recent formula proved by the authors, which explicitly computes the relative ζ determinant for an elliptic boundary value problem for the Dirac operator over an odd-dimensional manifold with boundary in terms of a Fredholm determinant of a canonically defined operator over the *boundary*. The objective is to compute the ζ determinant on a manifold by computing it on 'simpler' codimension 0 submanifolds and then using pasting formulae. This is a process suggested by Topological QFT. The precise formula is explained in Section 1. First, we make some brief remarks on some global properties of determinants associated to families of EBVPs.

0.2. The Determinant Line Bundle

Since the boundary condition is something one must choose from an appropriate parameter space of projections, the sum over histories must therefore include a sum over boundary data (fermionic variables) in addition to the sum over geometric (bosonic) data. The corresponding family parameterized by the manifold B becomes a smooth family of elliptic boundary value problems (EBVPs). Thus in addition to the family of Dirac operators $\mathbb{D}$ over a compact manifold with boundary X, we define a smoothly varying family $\mathbb{P}$ of pseudo-differential projections, called a *Grassmann section* (these include the 'spectral sections' of [19]), assigning to each operator $\mathcal{D}_b$ an elliptic boundary condition P_b. In particular, $\mathbb{D}$ defines canonically the Calderon (Grassmann) section $P(\mathbb{D})$ equal to the Calderon projection $P(\mathcal{D}_b)$ at $b \in B$. We denote the corresponding smooth family of EBVPs by $(\mathbb{D}, \mathbb{P})$. Associated to such a family one has a well defined determinant line bundle $\mathrm{DET}(\mathbb{D}, \mathbb{P})$ and index bundle $\mathrm{Ind}(\mathbb{D}, \mathbb{P})$. The topological realization of the relative inverse formula, and relative index property are the following global identifications for families of EBVPs:

Proposition 0.2. *[31] Let $\mathbb{P}, \mathbb{P}_1, \mathbb{P}_2$ be Grassmann sections for a family of Dirac operators $\mathbb{D}$. There are canonical isomorphisms of determinant line bundles*

$$\mathrm{DET}(\mathbb{D}, \mathbb{P}) \cong \mathrm{DET}(\mathbb{S}(\mathbb{P})), \tag{0.23}$$

and

$$\mathrm{DET}(\mathbb{D}, \mathbb{P}_0) \cong \mathrm{DET}(\mathbb{D}, \mathbb{P}_1) \otimes \mathrm{DET}(\mathbb{P}_1, \mathbb{P}_0) \; . \tag{0.24}$$

If B is compact, as elements of K-theory $K(B)$

$$\mathrm{Ind}(\mathbb{D}, \mathbb{P}_0) = \mathrm{Ind}(\mathbb{D}, \mathbb{P}_1) + \mathrm{Ind}(\mathbb{P}_1, \mathbb{P}_0) \; . \tag{0.25}$$

Here $(\mathbb{P}_1, \mathbb{P}_0)$ is the family of boundary Fredholm operators $P_{1,b} \circ P_{0,b}$: $\mathrm{range}\,(P_{0,b}) \to \mathrm{range}\,(P_{1,b})$ and $\mathbb{S}(\mathbb{P}) := (P(\mathbb{D}), \mathcal{P})$. We refer to [31] for details of these constructions.

Since the anomalies discussed earlier are statements about the local geometric and global topological structure of the determinant bundle, we can see that the presence and determination of anomalies depends explicitly on the choice of boundary conditions, that is, on the fermionic degrees of freedom. In particular, notice that one can always choose $\mathbb{P}$ such that $\mathrm{DET}(\mathbb{D}, \mathbb{P})$ is trivial, so in such a case no topological anomaly is present. In practice, physical considerations in general prohibit such choices.

1. Odd-dimensions

We assume now that M is an odd-dimensional manifold. In this case G and B in (0.15) satisfy the identities

$$G^2 = -Id \quad \text{and} \quad GB = -BG \; . \tag{1.1}$$

Since Y is even-dimensional the boundary spinor bundle $S|Y$ decomposes into its positive and negative chirality components $S|Y = S^+ \bigoplus S^-$ leading to the orthogonal decomposition $F = F^+ \oplus F^-$ of the boundary spinor fields. Equation (0.15) can then be rewritten in the form

$$\begin{pmatrix} i & 0 \\ 0 & -i \end{pmatrix} \left(\partial_u + \begin{pmatrix} 0 & B_- \\ B_+ & 0 \end{pmatrix} \right) \tag{1.2}$$

where B_+ is the chiral Dirac operator over the boundary. To simplify the exposition we assume that B is invertible.

Consider first the case where M is closed. In this case in odd-dimensions, since the Dirac operator D is self-adjoint, the index theorem has nothing to say. However, new more subtle secondary invariants arise. From the viewpoint of determinants and index theory, the most important of these is the η-invariant, which measures the difference between the positive and negative (real) spectrum of D, and may be regarded as the analogue of the index in odd-dimensions. It is a holomorphic function for $\mathrm{Re}(s) > \dim(M)$ defined by

$$\eta_D(s) = \mathrm{Tr}\,[D(D^2)^{(s+1)/2}] = \sum_{\lambda} \frac{\mathrm{sign}(\lambda)}{\lambda^s}. \tag{1.3}$$

From the heat kernel representation

$$\eta_D(s) = \frac{1}{\Gamma(\frac{s+1}{2})} \int_0^\infty t^{\frac{s-1}{2}}\, \mathrm{Tr}\,(De^{-tD^2})dt \tag{1.4}$$

one again sees that $\eta_D(s)$ has a meromorphic continuation to $\mathbb{C}$ with isolated simple poles along the real axis. The point $s = 0$ is not a pole and so the η-invariant $\eta_D(0)$ is defined. If D is a compatible (true) Dirac operator then $\eta_D(s)$ is actually a holomorphic function for $\mathrm{Re}(s) > -2$ [4] and so then

$$\eta_D(0) = \frac{1}{\sqrt{\pi}} \int_0^\infty \frac{1}{\sqrt{t}} \mathrm{Tr}\,(De^{-tD^2})dt. \tag{1.5}$$

This leads to the following ζ function regularization of the determinant of D:

$$\zeta_D(s) = \tfrac{1}{2}(1 + e^{\pm i\pi s})\zeta_{D^2}(\tfrac{s}{2}) + \tfrac{1}{2}(1 - e^{\pm i\pi s})\eta_D(s), \tag{1.6}$$

and hence that

$$\begin{aligned} \det{}_\zeta D := e^{-\zeta_D'(0)} &= e^{\pm\frac{i\pi}{2}(\zeta_{D^2}(0)-\eta_D(0))}.e^{-\frac{1}{2}\zeta_{D^2}'(0)} \\ &= e^{\pm\frac{i\pi}{2}(\zeta_{D^2}(0)-\eta_D(0))}|\det{}_\zeta D|. \end{aligned} \tag{1.7}$$

(see [35, 38] for more details). In particular, one has

$$\det{}_\zeta \Delta = |\det{}_\zeta D|^2, \tag{1.8}$$

and so the ζ metric (see [3, 28] is the norm square of a function, and therefore is the flat metric providing no curvature. This is precisely because D is self-adjoint. Though there is therefore no geometric anomaly present, there is another more subtle parity anomaly due to the ambiguity in the choice of the sign in equation (1.6) and hence in the definition of $\det_\zeta D$. We choose the '+' sign.

Now return to the case where M has boundary. In the odd-dimensional case self-adjoint realizations of the operator $\mathcal{D}$ are of particular interest. The involution $G : S|Y \to S|Y$ equips $L^2(Y; S|Y)$ with a symplectic structure, and by Green's formula (see [7]) we have that that the boundary condition $P \in Gr(\mathcal{D})$ provides a self-adjoint realization $\mathcal{D}_P$ of the operator $\mathcal{D}$ if and only if

$$-GPG = Id - P \ . \tag{1.9}$$

(See [5], [7], [12].) We therefore introduce the submanifold $Gr^*_\infty(\mathcal{D})$ of $Gr_\infty(\mathcal{D})$ parameterizing such projections.

For any $P \in Gr^*_\infty(\mathcal{D})$ the operator $\mathcal{D}_P$ has a discrete spectrum nicely distributed along the real line (see [5], [12]) and the second author proved:

Theorem 1.1. *[41] For any projections $P, P_1, P_2 \in Gr^*_\infty(\mathcal{D})$:*

1. *$\eta_{\mathcal{D}_P}(s)$ and $\zeta_{\mathcal{D}^2_P}(s)$ are holomorphic functions of s in the neighborhood of $s = 0$;*
2. *The value of ζ function at $s = 0$ is constant on $Gr^*_\infty(\mathcal{D})$, i.e.,*

$$\zeta_{\mathcal{D}^2_{P_1}}(0) = \zeta_{\mathcal{D}^2_{P_2}}(0);$$

3. *The function $\eta_{\mathcal{D}_P}(s)$ is a holomorphic function of s in the half-plane $\mathrm{Re}(s) > -1$;*
4. *The function $\Gamma(s)\zeta_{\mathcal{D}_P}(s) = \int_0^\infty t^{s-1}\mathrm{Tr}\ e^{-t\mathcal{D}^2_P} dt$ has a simple pole at $s = 0$. Hence $\zeta_{\mathcal{D}^2_P}(0)$ and, according to formula (1.7), $\ln\ \det_\zeta(\mathcal{D}_P)^2)$ $= -(d/ds)(\zeta_{\mathcal{D}^2_P})|_{s=0}$ are well defined.*

Therefore $\det_\zeta \mathcal{D}_P$ is a well defined, smooth function on $Gr^*_\infty(\mathcal{D})$. The question now is how to compute it. To see this, first observe that the grading $F = F^+ \oplus F^-$ is defined by the self-adjoint involution

$$B^{-1}|B| = \begin{pmatrix} 0 & B_+^{-1}(B_+B_-)^{1/2} \\ (B_+B_-)^{-1/2}B_+ & 0 \end{pmatrix}, \tag{1.10}$$

where $|B| = +\sqrt{B^2} = B\Pi_> - B\Pi_<$, then

$$\Pi_{\geq} = \frac{1}{2}(I + B^{-1}|B|) = \frac{1}{2}\begin{pmatrix} I & w_+^{-1} \\ w_+ & I \end{pmatrix}, \tag{1.11}$$

where

$$w_+ = (B_+B_-)^{-1/2}B_+. \tag{1.12}$$

Equivalently,

$$\text{range}\,(\Pi_>) = \text{graph}(w_+ : F^+ \to F^-). \tag{1.13}$$

In particular, we see that $\Pi_>$ is a pseudo-differential operator of order 0, and inverting the boundary value problem (0.16) is the same thing as inverting the boundary value problem

$$(\partial_u - |B|)f = k, \quad f \in \text{graph}(-w_+ : F^+ \to F^-). \tag{1.14}$$

Anyway, we have that H^+ is the graph of $w_+ : F^+ \to F^-$. Moreover, as we have assumed that ker $B = \{0\}$, then $\Pi_>$ is an element of $Gr^*_\infty(\mathcal{D})$. A generalization [10, 30] of this, extends the graph$\longleftrightarrow$ self-adjoint boundary condition correspondence for H^+ to the whole Grassmannian $Gr^*_\infty(\mathcal{D})$: There is a $1 - to - 1$ correspondence of $Gr^*_\infty(\mathcal{D})$ with unitary maps $g : F^+ \to F^-$, such that the difference $g - g_+$ is an operator with a smooth kernel. The corresponding orthogonal projection P is given by the formula

$$P = \frac{1}{2}\begin{pmatrix} Id_{F^+} & g^{-1} \\ g & I_{F^-} \end{pmatrix}, \quad g = w_+ + S,$$

where $S : F^+ \to F^-$ is a smoothing operator. By choosing a basepoint, the correspondence defined above allows us to establish an isomorphism between $Gr^*_\infty(\mathcal{D})$ and the group $U^\infty(F^-)$ of unitaries acting on $F^- = L^2(Y; S^-)$ which differ from Id_{S^-} by an operator with a smooth kernel. The Calderon projection defines a preferred basepoint, and hence, letting $K : C^\infty(Y; S^+) \to C^\infty(Y; S^-)$ denote the unitary such that $\mathcal{H}(\mathcal{D})$ is equal to the graph(K), we have a natural isomorphism $Gr^*_\infty(\mathcal{D}) \cong U^\infty(F^-)$ defined by the map $P \to TK^{-1}$ (where Ran P = graph(T)). This means that the determinant line bundle associated to the family $\{\mathcal{D}_P : P \in Gr^*_\infty(\mathcal{D})\}$ is canonically trivial. The following theorem asserts that up to a natural constant this is precisely the trivialization defining the ζ determinant:

Theorem 1.2. *[34, 35] Let $P_1, P_2 \in Gr^*_\infty(\mathcal{D})$ such that $\mathcal{D}_{P_1}$,$\mathcal{D}_{P_2}$ are invertible. If* range (P_i) = graph(T_i), *then*

$$\frac{\det_\zeta \mathcal{D}_{P_1}}{\det_\zeta \mathcal{D}_{P_1}} = \frac{\det_{Fr} \frac{1}{2}(Id + KT_1^{-1})}{\det_{Fr} \frac{1}{2}(Id + KT_2^{-1})}. \tag{1.15}$$

Here $\det_{Fr}$ denotes the Fredholm determinant, defined for any operator of the form Id plus trace class. The determinants on the right side are the Fredholm determinants of the boundary operators $\mathcal{S}(P_i)$ computed in the above trivialization.

In particular, it is easy to see that Fredholm determinants are invariant under the action of the gauge group $\mathcal{G}$, and hence this regularization reduces to the moduli space $\mathcal{M}$, as required, corresponding to the fact that for families of self-adjoint EBVPs in odd-dimensions there is no gauge anomaly.

2. Relative Determinant on the Infinite Cylinder

In this Section we consider the simplest non-compact case in which we can define *relative* ζ determinant. We discuss an explicit computation of formula (1.15) in the case of the non-compact half-infinite cylinder $[0, \infty) \times Y$. This is possible because we have in this case explicit knowledge of the Schwartz kernel and heat kernel for $\mathcal{D}_P$. Details will appear elsewhere.

Again, let Y be a compact spin manifold of any dimension. We consider the operator

$$D = G\left(\frac{\partial}{\partial u} + B\right), \tag{2.1}$$

on the half-infinite cylinder $Y \times [0, \infty)$, where the Dirac operator B over Y we take to be invertible, and with the usual identities (1.1). We assume that B is an invertible operator to simplify the exposition. Associated to B we have the Grassmannian parameterizing 'self-adjoint' elliptic boundary conditions for D

$$Gr^*_\infty(D) = \{P : P^2 = P, P^* = P, P - \Pi_> = \text{smoothing}, -\Gamma P \Gamma = I - P\}.$$

Let P_r be a path in $Gr^*_\infty(D)$ connecting $\Pi_>$ and P. Then we can find a corresponding path of unitaries $U_r \in \mathcal{U}_\infty(F) = \{U \in \mathcal{U}(F) : g = I + \text{smoothing}\}$ with

$$P_r = U_r \Pi_> U_r^{-1}. \tag{2.2}$$

We may assume that g_r commutes with G, so that

$$-U_r = G U_r G. \tag{2.3}$$

Because we are working on a non-compact manifold the spectrum of $\mathcal{D}_P$ will in general be continuous. However, we can still define a relative ζ determinant using the relative ζ function, which is determined by the *couple* of boundary conditions

$$\zeta(s;\mathcal{D};P,\Pi_>) = \frac{1}{\Gamma(s)}\int_0^\infty t^{s-1}\mathrm{Tr}\,(e^{-t\mathcal{D}_P} - e^{-t\mathcal{D}_{\Pi_>}})dt. \tag{2.4}$$

The function $\zeta(s;\mathcal{D};P,\Pi_>)$ exists and shares the standard properties of the ζ function on a closed manifold, since $\mathcal{D}_P^{-1}$ and $\mathcal{D}_{\Pi_>}^{-1}$ differ by only a smoothing operator. In the case of the compact manifold M of Section 2 , $\zeta(s;\mathcal{D};P,\Pi_>)$ reduces to the difference of the individual ζ functions. For spectral invariants on manifolds with cylindrical ends we refer to [15] and [24].

The operators D_{P_r} have domains varying with r and we use a 'unitary twist' $U_{rf(u)}$. Fix $\delta_1 > \delta_0 > 0$ and define a smooth non-decreasing function $f(u)$ such that

$$f(u) = 1 \text{ for } u < \delta_0 \text{ and } f(u) = 0 \text{ for } u > \delta_1 .$$

For each r introduce the 2-parameter family

$$U_{r,u} = U_{rf(u)}, \quad u \in [0,\infty) \quad -\epsilon \le r \le \epsilon.$$

The operators D_{P_r} and $(U_{rf(u)}^{-1}DU_{rf(u)})_{\Pi_>}$ are unitary equivalent operators and hence their ζ determinants have the same variation. We write

$$D^r = U_{rf(u)}^{-1}DU_{rf(u)}.$$

Again we have that $U_{rf(u)}$ commutes with G:

$$U_{rf(u)} = GU_{rf(u)}G. \tag{2.5}$$

Then we may define the ζ determinant of D_{P_r} as the relative ζ determinant. We can define both phase and the modulus of such a determinant. The phase is determined by the relative ζ function defined above and by the value at $s = 0$ of the relative η function, which is defined similarly to the equation (2.4), with ζ function replaced by the η function

$$\eta(s;\mathcal{D};P,\Pi_>) = \frac{1}{\Gamma(\frac{s+1}{2})}\int_0^\infty t^{\frac{s-1}{2}}\mathrm{Tr}\,(\mathcal{D}e^{-t\mathcal{D}_P^2} - \mathcal{D}e^{-t\mathcal{D}_{\Pi_>}^2})dt. \tag{2.6}$$

It is not difficult to show that $\eta(s;\mathcal{D};P,\Pi_>)$ is well defined (see [33] for related computations on the cylinder) and in particular that it is regular at $s = 0$.

As usual the modulus of the determinant is more difficult to handle. Its value is given by

$$\det_{\zeta}(D^2_{P_r}; D^2_{\Pi_>}) := e^{-(d/ds)(\zeta(s;D^2_{P_r};D^2_{\Pi_>}))|_{s=0}} . \tag{2.7}$$

When $Y \times [0,\infty)$ is odd-dimensional the tangential operator splits into its positive and negative parts $B_\pm$ according to the chiral splitting $H_Y = F^+ \oplus F^-$ and D has the form (1.2) and $P \in Gr^*_\infty$ has the form

$$P := P_T = \frac{1}{2}\begin{pmatrix} I & T^{-1} \\ T & I \end{pmatrix}, \tag{2.8}$$

where $T = w_+ + S : F^+ \longrightarrow F^-$ is unitary in the L^2-metric, S is a smoothing operator. The unitary twist has the form

$$U_{rf(u)} = \begin{pmatrix} I & 0 \\ 0 & g_{rf(u)} \end{pmatrix} \tag{2.9}$$

with $g_{rf(u)} \in U_\infty(F^-)$, so

$$U_{rf(u)} P_T U^{-1}_{rf(u)} = P_{g_r T}. \tag{2.10}$$

Now a path $\{P_r\}$ between $\Pi_>$ and P_T can be given by choosing a path $\{T_r\}$ between w_+ and T and we can set

$$g_r = T_r w_+^{-1} \tag{2.11}$$

More general, for any pair $\{P_1, P_2\}$ of elements of the self-adjoint smoothing Grassmannian, where Ran P_i = graph T_i we can find a path between those two projections of the form $\{U_r P_1 U_r^{-1}\}$ with

$$U_r = \begin{pmatrix} I & 0 \\ 0 & h_r \end{pmatrix} \tag{2.12}$$

where $h_r \in U_\infty(F^-)$. Now we can follow the method of [35] (see also [34]) and prove the following result, which is an analogue of Theorem 1.2.

Theorem 2.1. *Let $Y \times [0,\infty)$ be odd-dimensional and let $P_i = P_{T_i} \in Gr^*_\infty(D)$. Then*

$$\det_{\zeta}(\mathcal{D}; P_1, P_2) = \det_{Fr}[\tfrac{1}{2}(Id + T_2 T_1^{-1})]. \tag{2.13}$$

One can also prove this Theorem by direct computation following the method presented in [6] using the property that as in the one-dimensional case we know an explicit formula for the kernel of the inverse operators $\mathcal{D}_{P_i}^{-1}$ and the heat kernel of the Atiyah–Patodi–Singer problem.

3. An Adiabatic Pasting Formula

Given that we have a precise formula for a preferred class of self-adjoint elliptic boundary value problems, it is natural to look for a pasting formula for the determinant on a closed manifold endowed with a partition. In this Section we give a brief review of recent formulas obtained by Wojciechowski and Park.

We consider the case of a closed odd-dimensional manifold M. Let $\mathcal{D}$: $C^\infty(M;V) \to C^\infty(M;V)$ be a compatible Dirac operator acting on sections of a bundle of Clifford modules V over M. Assume that we have a partition of M as $M_0 \cup_Y M_1$, where M_0 and M_1 are compact manifolds with boundary such that

$$M_0 \cap M_1 = Y = \partial M_0 = \partial M_1 \ . \tag{3.1}$$

We also assume that the Riemannian metric on M and the Hermitian product on V are products in the bicollar neighborhood $\widetilde{N} = [-1,1] \times Y$ of Y, where $M_0 \cap \widetilde{N} = [-1,0] \times Y$. The operator $\mathcal{D}$ is given by the formula (0.15) in $\widetilde{N}$. Let $\mathcal{D}^i = \mathcal{D}|M^i$ $(i = 0,1)$ and $P_0, P_1 \in Gr_\infty^*(\mathcal{D})$ and let $\eta(P_0, P_1)$ denote the η-invariant of the operator $G(\partial_u + B)$ on the manifold $[-1,1] \times Y$ subject to the boundary condition P_0 at $u = -1$ and the boundary condition $Id - P_1$ at $u = 1$. The second author proved the following additivity formula for the η-invariant (see also [12], [15], [17], [39] for partial results and related topics).

Theorem 3.1. *[40][41] For any $P_0, P_1 \in Gr_\infty^*(\mathcal{D})$ one has the following formula*

$$\eta_{\mathcal{D}} = \eta_{\mathcal{D}^0_{Id-P_0}} + \eta_{\mathcal{D}^1_{P_1}} + \eta(P_0, P_1) \mod \mathbf{Z} \ . \tag{3.2}$$

Results analogous to Theorem 3.1 of [40] were obtained and discussed by other authors. We refer especially to the papers [9], [11], [14], [18], [22], [23]. Theorem 3.1 extends the formula on the variation of the η-invariant under a change of boundary condition from the work [17].

The crucial result from which the Theorem 3.1 follows is:

Theorem 3.2. *[41] Let $P_0, P_1 \in Gr_\infty^*\mathcal{D}$, then*

$$\eta_{\mathcal{D}_{1 P_0}} - \eta_{\mathcal{D}_{1 P_1}} = -\frac{1}{\pi}\int_0^1 dr \int_0^1 du \, \mathrm{Tr}\, G\left(g^{-1}\frac{\dot{\partial g}}{\partial u}|_r\right) \mod \mathbf{Z} \ , \tag{3.3}$$

$$\eta(P_0, P_1) = -\frac{1}{\pi}\int_0^1 dr \int_0^1 du \text{ Tr } G\left(g^{-1}\frac{\partial g}{\partial u}|_r\right) \mod \mathbf{Z} \ , \tag{3.4}$$

where $\{g_{r,u}\}$ is any family connecting P_0 with P_1 in the way described above.

Thus, in this case we get an exact decomposition of the η-invariant $\eta_{\mathcal{D}} = \eta_{\mathcal{D}}(0)$ of the operator $\mathcal{D}$ into contributions coming from the different parts of the manifold. An interesting consequence of Theorem 3.1 and Theorem 1.1(2) is that they imply:

Corollary 3.3. *The phase of the ζ determinant is additive under the pasting of two manifolds with the same boundary.*

That might lead one to expect that the ζ determinant might be multiplicative under pasting of manifolds, but that is not true. The ζ determinant on a closed manifold is independent of the partition, and indeed any choice of boundary conditions. Hence we should expect a formula which 'averages away' the choice of boundary condition. This is suggested formally by path integral formulae (see Section 4), and in fact in the correct analytic formula the boundary condition is scaled away adiabatically. The problem here is with the modulus of the determinant

$$|\det_{\zeta}\mathcal{D}| = e^{-\frac{1}{2}\zeta'_{\mathcal{D}^2}(0)} \ .$$

This is not a local quantity and its variation is not local either. Assuming that dim M is odd it is equal to (see [38, 41]

$$\ln \det_{\zeta}(\mathcal{D}^2) = -\left.\frac{d}{ds}(\zeta_{\mathcal{D}^2})\right|_{s=0} = -\int_0^{\infty}\frac{1}{t}\text{ Tr } e^{-t\mathcal{D}^2}dt \ . \tag{3.5}$$

However, we can discuss the adiabatic splitting formula in this case. We replace the bicollar $\widetilde{N}$ by $\widetilde{N}_R = [-R, R] \times Y$ where $R \to \infty$. In other words we stretch the manifold M to M_R by replacing $\widetilde{N}$ by a cylinder of length $2R$. Due to the construction of the heat kernel the right side of (3.5) now splits into the contribution from the interior, the cylinder contribution, and an error term. If we assume that tangential operator B is invertible then the error term disappears as $R \to \infty$. Therefore we can study the decomposition of the contribution coming from the different part of the manifolds. The first study of this type was made in [16]. The focus of the Klimek–Wojciechowski paper was on the η-invariant of chiral boundary problems and on the analytic torsion, hence there was no direct reference to the ζ determinant. However, the results of [16] allow us to study the adiabatic decomposition of the modulus of the ζ determinant of chiral boundary problems.

This is explained in recent work of Park and Wojciechowski (see [25]). Park and Wojciechowski study also a pasting law, which involves the Atiyah–Patodi–Singer conditions. We describe the result and refer to [25] for more details and proofs. Let $\mathcal{D}_R$ denote a Dirac operator $\mathcal{D}$ on the manifold M_R, which is the manifold M with the bicollar $\widetilde{N}$ replaced by $\tilde{n}_R = [-R, R] \times Y$. Let $\mathcal{D}_{i,R}$ denote the Dirac operator on the manifold $M_{i,R}$ (equal to M_i with a cylinder of length R attached). Now the ζ determinant of the Dirac Laplacian $\mathcal{D}_R^2$ blows up as $R \to \infty$. The same happens with the determinant of the operator $(\mathcal{D}_{i,R})^2_{\Pi_>}$. The corresponding quotient, however, is well defined and we have the following formula

Theorem 3.4 (Park–Wojciechowski).

$$\lim_{R\to\infty} \frac{\det_\zeta \mathcal{D}_R^2}{\det_\zeta((\mathcal{D}_{0,R})^2_{\Pi_>})\cdot\det_\zeta((\mathcal{D}_{0,R})^2_{\Pi_>})} = 2^{-\zeta_{B^2}(0)} . \tag{3.6}$$

Remark 3.5. (1) Of course formula (3.6) works only if the operators $\mathcal{D}_R$ and $(\mathcal{D}_{i,R})_{\Pi_>}$ are invertible (say for large R). We have already assumed that the tangential operator B is invertible, hence the sufficient condition here is that operators $\mathcal{D}_{i,\infty}$ (= the operator $\mathcal{D}$ on the manifold M_i with an infinite cylinder attached) do not have L^2 solutions (see for instance the discussion in [39]).

(2) Besides the techniques introduced and used in [12] (see also [16]) the proof of the Theorem 3.4 employs technical results of the beautiful work [8].

We end this Section with the discussion of the simplest possible pasting situation. Let us recompute the determinant of the $\bar{\partial}$-operator over a mapping cylinder, using the canonical regularization above.

Let $M = M^0 \cup_Y M^1$ be a partitioned manifold. And suppose that that $H(D^0) = \text{graph}(K^0 : H_Y^+ \to H_Y^-)$ and $H(D^1) = \text{graph}(K^1 : H_Y^- \to H_Y^+)$, where the K^i are at least Hilbert–Schmidt operators. From the above identifications we have a canonical isomorphism

$$\text{Det}(D) \cong \text{Det}(P(D^0), I - P(D^1)).$$

Hence, as before, we obtain a canonical regularization of the determinant of D as the regularized determinant of the operator $\mathcal{S}(D) = (I - P(D^1))P(D^0) : H(D^0) \to H(D^1)^\perp$ which we denote $\det_{\mathcal{C}}(D)$. This yields [30]

$$\det_{\mathcal{C}}(D) = \det_{Fr}(I - K^1 K^0). \tag{3.7}$$

We consider the cylinder $M = S^1 \times Y$ with the operator $D = \partial/\partial u + B$, where B is an invertible first-order elliptic self-adjoint operator over Y acting on sections of a bundle E. Now decompose M as the sum of two cylinders $M_R = [0, R] \times Y$ and $M'_R = [R, 2\pi] \times Y$ with restricted operators D^R and $D^{R'}$. Let $\{\lambda_n, \phi_n\}$ be a spectral resolution of B. Then

$$\text{Ker } D^R = \text{span}\{e^{-u\lambda_n}\phi_{\lambda_n}(y) : n \in \mathbb{Z}\},$$

so that

$$H(D) = \text{span}\{(\phi_{\lambda_n}, e^{-R\lambda_n}\phi_{\lambda_n}) : n \in \mathbb{Z}\}. \tag{3.8}$$

The cylinder has boundary $\partial M_R = Y \sqcup \overline{Y}$, where the second component is Y with orientation reversed. Relative to this $H_Y = H_0 \oplus H_R$ and the tangential component of D^R is $B^R = B_0 \oplus B_R$, where $B_0 = B$ and $B_R = -B$. Hence the energy grading of H_Y defined by B^R is

$$H^+ = H_0^+ \oplus H_R^- \quad H^- = H_0^- \oplus H_R^+, \tag{3.9}$$

where $H_i^{\pm}$ are the positive and negative gradings associated to the B_i. From (3.8) and (3.9) we have that $H(D^R)$ is the graph of

$$K : H_0^+ \oplus H_R^- \longrightarrow H_0^- \oplus H_R^+, \quad K(f_0^+, f_R^-) = (e^{-RB}f_0^+, e^{RB}f_R^-).$$

That is,

$$H(D^R) = \text{graph}(K = e^{-R|B|} : H^+ \longrightarrow H^-). \tag{3.10}$$

Applying the same argument to $D^{R'}$, we obtain from (3.7) and (3.10)

$$\det_{\mathcal{C}}(D) = \det_{Fr}(I - e^{-2\pi|B|}) = \prod_{n\in\mathbb{Z}}(1 - e^{-2\pi\lambda_n}). \tag{3.11}$$

If we apply this to the case of the $\bar{\partial}$-operator

$$\bar{\partial}_\tau = \frac{d}{du} + \frac{1}{\tau}\frac{d}{dy}$$

over an elliptic curve, defined by τ in the upper half-plane, we obtain

$$\det_{\mathcal{C}}(\bar{\partial}_\tau) = c \cdot \det_{Fr}(I - e^{-2\pi|B|}) = c \cdot \prod_{n=1}^{\infty}(1 - q^n)^2, \tag{3.12}$$

where $q = e^{2\pi i\tau}$. The operator $\bar{\partial}_\tau$ is not invertible, so we choose an isomorphism between the kernel and cokernel of $\mathcal{S}(\bar{\partial}_\tau)$, the constant c is the determinant of this isomorphism.

Notice that the choice of a decomposition of $M \cong S^1 \times S^1$ defines a homology 1-cycle on M and hence an element γ of $H_1(M, \mathbb{Z})$. Because $\mathrm{Det}(\bar{\partial}_\tau)$ is the dual line to the holomorphic differentials on M, we obtain a canonical element $\xi_\gamma \in \mathrm{Det}(\bar{\partial}_\tau)$. In [37] Segal showed that

$$\det\bar{\partial}_\tau = \prod_{n=1}^{\infty}(1-q^n)^2 \cdot \xi_\gamma \,.$$

Thus we see the canonical trivialization is essentially the trivialization ξ_γ.

From (0.13)

$$\det{}_\zeta(\bar{\partial}_\tau) = \frac{q^{\frac{1}{12}}}{c}\det{}_C(\bar{\partial}_\tau).$$

Similarly one can compute the canonical determinant of $\bar{\partial}_\tau$ acting on holomorphic differentials of degree m, which introduces an additional factor of $q^{(m^2-m)/2}$, and the case where $\bar{\partial}_\tau$ is coupled to a flat connection on a coefficient bundle. Details of these computations will appear elsewhere.

4. Functorial QFT and Gauge Anomalies

Functorial quantum field theory (FQFT) is an attempt to abstract the algebraic framework that a path integral would create if it existed as a rigorous mathematical object. Thus one begins with a myth, and aims to work towards some kind of technology. Roughly speaking, a FQFT is kind of generalized cohomology theory defined by a functor from manifolds with boundary to vector (Hilbert) spaces satisfying axioms formally satisfied by the path integral.

More precisely, a $(d+1)$-dimensional FQFT assigns to a closed manifold Y of dimension d a vector space $Z(Y)$, while to a manifold of dimension $d+1$ it assigns a vector $Z_X \in Z(Y)$, where $Y = \partial X$ is the boundary of X. By *fiat* $Z(\emptyset) = \mathbb{C}$, so that if X is closed then Z_X is a complex number. In practice, the class of manifolds being considered needs to be specified and usually additional data, such as geometric data and boundary conditions.

For $(d+1)$-dimensional manifolds X, X_0, X_1 and d-dimensional manifolds Y, Y_0, Y_1, the 'functor' Z is required to satisfy certain axioms; notably, that

$$Z_{X_0 \sqcup X_1} = Z_{X_0} \otimes Z_{X_1}, \quad Z(Y_0 \sqcup Y_1) = Z(Y_0) \otimes Z(Y_1),$$

while if $\overline{Y}$ denotes Y with reversed orientation then

$$Z(\overline{Y}) = Z(Y)^*.$$

This means that a cobordism X induces a linear transformation $Z_X : Z(Y_0) \to Z(Y_1)$ through the identifications

$$Z_X \in Z(\overline{Y_0} \sqcup Y_1) = Z(\overline{Y_0}) \otimes Z(Y_1) = Z(Y_0)^* \otimes Z(Y_1) = \mathrm{Hom}(Z(Y_0), Z(Y_1)).$$

One further requires that if $M = X_0 \cup_Y X_1$ with $\partial X_0 = \overline{Y_0} \sqcup Y$ and $\partial X_1 = \overline{Y} \sqcup Y_1$, then

$$Z_M = Z_{X_1} \circ Z_{X_0}.$$

This in turn induces a canonical pairing $Z(\overline{Y_0}) \otimes Z(Y) \otimes Z(\overline{Y}) \otimes Z(Y_1) \longrightarrow Z(\overline{Y_0}) \otimes Z(Y_1)$, and in the case when $Y_0 = Y_1 = \emptyset$, so M is a closed manifold, this becomes a pairing

$$(\ ,\ ) : Z(Y) \otimes Z(\overline{Y}) \longrightarrow \mathbb{C}, \tag{4.1}$$

with the *sewing property*

$$Z_M = (Z_{X_0}, Z_{X_1}). \tag{4.2}$$

Thus the number Z_M can be computed by evaluating over the submanifolds and then sewing together the results via the bilinear pairing. These axioms are 'idealized', and for each case modifications are needed.

In the case where X_0 has connected boundary equal to Y_0, so $Y = \emptyset$, then $Z_X \in \mathrm{Hom}(Z(Y_0), \mathbf{C})$. Formally setting $\mathcal{E}_f(X)$ to be a space of fields on X with boundary value $f \in \mathcal{E}(Y_0) := Z(Y_0)$, the vector Z_X corresponds to a partition function given by a path integral

$$Z_X : \mathcal{E}(Y) \longrightarrow \mathbb{C}, \quad Z_X(f) = \int_{\mathcal{E}_f(X)} e^{-S(\psi)}\, d\psi, \tag{4.3}$$

where $d\psi$ is a formal measure and $S : \mathcal{E}_f(X) \to \mathbb{C}$ an action functional. The path integral version of the algebraic sewing formula (4.2) then takes the form

$$\begin{aligned} \int_{\mathcal{E}(M)} e^{-S(\psi)}\, d\psi &= \int_{\mathcal{E}(Y)} df \int_{\mathcal{E}_f(X_0)} e^{-S(\psi_0)}\, d\psi_0 \int_{\mathcal{E}_f(X_1)} e^{-S(\psi_1)}\, d\psi_1 \\ &= \int_{\mathcal{E}(Y)} Z_{X_0}(f) Z_{X_1}(f)\, df. \end{aligned} \tag{4.4}$$

To see how to describe the chiral anomaly in this framework, consider first the closed manifold M which we assume to be even-dimensional, and spin. The chiral Dirac operator acts between positive and negative chirality fields,

$$\mathcal{D} = \mathcal{D}_A^+ : C^\infty(M; E \otimes S^+) \longrightarrow C^\infty(M; E \otimes S^-)$$

coupled to a coefficient bundle E with connection A. Evaluating the Euclidean path integral (0.1), we seek a regularized determinant of $\mathcal{D}_A$ considered as a function on the affine space B of gauge potentials on E. To do that we must choose an operator

$$T_A : C^\infty(M; E \otimes S^-) \longrightarrow C^\infty(M; E \otimes S^+),$$

and then try to regularize $\det T_A \mathcal{D}$. However, there is no reason to expect $\det_r \mathcal{D} := \det_r T_A \mathcal{D}$ to transform equivariantly under gauge transformations, and indeed in general it does not. One finds

$$\det{}_r \mathcal{D}^+_{A^g} = \det{}_r \mathcal{D}^+_A \, \omega(g, A),$$

where $\log \omega$ is an integral over M of a local differential polynomial in g, A and the metric on M. ω defines a one-cocycle on the gauge transformation group $\mathcal{G}$, the transgression of ω to a 2-form on the moduli space $\mathcal{M}$ is the first Chern class of the determinant line bundle pushed-down to $\mathcal{M}$. The non-triviality of this class is the topological obstruction to the existence of $\mathcal{G}$-invariant determinant regularization [2]. It is called the chiral anomaly. From the gauge transformation group view point, the 1-form ω defines a representation of an abelian extension $\widehat{\mathcal{G}}$ of $\mathcal{G}$ for which ω is the extension cocycle [20].

This description of the chiral anomaly corresponds to the left side of equation (4.3). Let us turn then to the right-side of that formula. (For a detailed presentation of the following constructions we refer to [21].) Let $\mathcal{D}^i$, $i = 0, 1$, denote the restrictions of $\mathcal{D}$ to the two halves X_i of M. We assume the geometric set-up of previous sections. For analytic reasons, we must now replace the local boundary condition f for $\mathcal{D}^0$ by a global condition $W = \text{range}\,(P)$ with P is a restricted Grassmannian of the boundary fields, and we shall choose $Gr_\infty(\mathcal{D}^0)$. The path integral (0.1) now becomes, ignoring the pure gauge component,

$$Z_X(P) := \det(\mathcal{D}^0_P) = \int_{\mathcal{E}_P} e^{\int_X \psi^* \mathcal{D}^0 \psi \, dm} \, d\psi d\psi^*, \tag{4.5}$$

where $\mathcal{E}_P = \text{dom}(\mathcal{D}_P)$, while the sewing formula becomes

$$\det(\mathcal{D}) = \int_{Gr_Y} \det\mathcal{D}^0_P \cdot \det\mathcal{D}^1_{I-P} dP \ . \tag{4.6}$$

This is entirely formal, but it asserts that $\det(\mathcal{D})$ is obtained by integrating away the boundary data. Indeed, the determinant over the closed manifold clearly does not depend on boundary conditions, and so this is formally what we would expect. In practice, to tell the ζ function over the closed manifold where the partition lies we have to send it an 'impulse function', the rigorous formulation of this idea is the adiabatic pasting formula of the previous section.

However, (4.6) can be given a precise formulation in terms of a FQFT. Notice first that there is no regularization involved in the formula, it is a relation between sections of appropriate determinant line bundles. Hence the corresponding FQFT pairing (4.2) should be a pairing on spaces of sections of those determinant line bundles.

To see what those spaces are, return for a moment to the action of the gauge group. For a manifold with boundary the essential data needed to detect the chiral anomaly can be seen from the action of the boundary gauge transformation group. To begin with, consider the case of a Riemann surface with connected boundary diffeomorphic to S^1. The boundary gauge group in this case is the loop group LG where G is the (compact) structure group of the (necessarily trivial) complex bundle over the circle. It was explained in the text [27] that a LG has a fundamental projective representation in the Fock space of holomorphic sections of the dual of the determinant line bundle DET_{H^+} over the Grassmannian. Here DET_{H^+} is the bundle based at H^+ with fibre at P equal to the complex line $\mathrm{Det}(H^+, W) := \mathrm{Det}(P\Pi_> : H^+ \to W)$, $W = \mathrm{range}\,(P)$. This suggests the general construction for the Dirac operator $\mathcal{D}^0$ over the manifold X_0 with boundary. First, for each $P \in Gr_\infty(\mathcal{D}^0)$ one has a determinant line bundle DET_W based at $W = \mathrm{range}\,(P)$ with fibre at $W' = \mathrm{range}\,(P')$ equal to $\mathrm{Det}(W, W')$. The Fock space based at W is defined by

$$\mathcal{F}_W = \Gamma_{\mathrm{hol}}(Gr_\infty(\mathcal{D}^0); \mathrm{DET}_W^*),$$

where Γ_{hol} denotes holomorphic sections. (Here we can replace $Gr_\infty(\mathcal{D}^0)$ by the Hilbert–Schmidt Grassmannian, but we ignore this point.) Under a change of base point one has from (0.23) a canonical isomorphism

$$\mathcal{F}_{W_1} \cong \mathcal{F}_{W_2} \otimes \mathrm{DET}(W_1, W_2), \tag{4.7}$$

where the second factor on the right side is the trivial bundle with fibre $\mathrm{Det}(W_1, W_2)$. Moreover, under the Plücker embedding

$$\mathrm{DET}_W \longrightarrow \mathcal{F}_W$$

the determinant section of DET_W is identified with the 'vacuum vector' $\nu_W \in \mathcal{F}_W$. The basic fact is the following Theorem:

Theorem 4.1. *[21] There are functorial bilinear pairings*

$$(\ ,\ ) : \mathcal{F}_{K(D^0)}(H_Y) \times \mathcal{F}_{W^\perp}(\overline{H}_Y) \longrightarrow \mathrm{Det}(D^0_P), \tag{4.8}$$

($W = \text{range}\,(P)$) with

$$(\nu_{K(D^0)}, \nu_{W^\perp}) = \det(D^0_P), \tag{4.9}$$

and

$$(\ ,\ ) : \mathcal{F}_{K(D^0)}(H_Y) \times \mathcal{F}_{K(D^1)}(\overline{H}_Y) \longrightarrow \mathrm{Det}(D), \tag{4.10}$$

with

$$(\nu_{K(D^0)}, \nu_{K(D^1)}) = \det(D). \tag{4.11}$$

These pairings are of course naturally the bilinear pairing needed to define the FQFT we are seeking. This requires that to each closed d-dimensional manifold Y we associate an admissible (W an element of the Grassmannian) polarization $H_Y = W^+ \oplus W^-$ and then to the pair (Y, W) we assign the Fock space $Z(Y, W) := \mathcal{F}_W$. While to a $(d+1)$-dimensional cobordism X with boundary $Y_0 \sqcup Y_1$ with assigned polarizations W_0, W_1 one obtains a canonical vector space homomorphism $Z_X \in \mathrm{Hom}(\mathcal{F}_{W_0}, \mathcal{F}_{W_1})$ via the Plücker embedding. In the case that X is a closed manifold one defines $Z_X = \det \mathcal{D}_X \in \mathrm{Det}(\mathcal{D}_X)$. Comparing (4.8) with (4.2) we see this defines the required FQFT. In particular, in the graph trivialization the pairing (4.11) coincides with the canonical determinant regularization (3.7).

All of these constructions have obvious generalizations to families of Dirac operators. For a smooth family of EBVPs $(\mathbb{D}, \mathbb{P})$ parameterized by a manifold B one has associated *Fock bundles* $\mathcal{F}_{K(\mathbb{D})}$ and $\mathcal{F}_{\mathbb{P}}$ and the pairing (4.8), for example, becomes a pairing from Fock bundles to the determinant line bundle $\mathrm{DET}(\mathbb{D}, \mathbb{P})$. Now return to the action of the gauge group where B is as before the parameter space of gauge potentials. We have the gauge transformation group $\mathcal{G}$ acting on the base B and using the ideas of [20] it is not hard to see how to lift the action to an induced *projective* action of $\mathcal{G}$ in the Fock bundle intertwining with the family of of quantized Dirac Hamiltonians in the fibres. The essential difference here to [20] is that the action is not a true projective representation of $\mathcal{G}$, but rather a linear isometric action between the different fibres of the Fock bundle. This leads to a new description of the chiral anomaly for a family of EBVPs and, when applied to (4.10), for a family of Dirac operators over a closed manifold (see [21]). However, rather than pursue that here we shall finish by mentioning a differential geometric description of gauge anomalies via the ζ function geometry of the determinant bundle.

The essential point is that for a family $(\mathbb{D}, \mathbb{P})$ of EBVPs there are two natural ways to put a metric and connection on the determinant line bundle $\mathrm{DET}(\mathbb{D}, \mathbb{P})$. First, by pullback through the isomorphism (0.23) the determinant bundle $\mathrm{DET}(\mathbb{S}(\mathbb{P}))$ has a canonical metric defined over the open subset U where the operators are invertible by

$$\|\det(\mathcal{D}_P)\|_{\mathcal{C}}^2 = \det_{\mathcal{C}}(\Delta_P) := \det_{Fr}(\mathcal{S}(P)^*\mathcal{S}(P)), \tag{4.12}$$

where $\Delta_P = (\mathcal{D}_P)^*\mathcal{D}_P$ and, recall, $\mathcal{S}(P) := PP(\mathcal{D}) : K(\mathcal{D}) \to \text{range}\,(P)$ with $P(\mathcal{D})$ the Calderon projection of the operator $\mathcal{D}$. On the other hand, $\mathrm{DET}(\mathbb{D}, \mathbb{P})$ has a Quillen metric, defined over U by

$$\|\det(\mathcal{D}_P)\|_{\zeta}^2 = \det_{\zeta}(\Delta_P) := e^{-\zeta'_{\Delta_P}(0)} \tag{4.13}$$

where $\zeta_{\Delta_P}(s) = \mathrm{Tr}\,(\Delta_P^{-s})$ is defined around 0 by analytic continuation.

Furthermore, in each case there are natural connections $\nabla^{\mathcal{C}}, \nabla^{\zeta}$ defined on $\mathrm{DET}(\mathbb{D}, \mathbb{P})$ compatible with these metrics, whose curvatures may represent geometric anomalies, as described in the Introduction. Naturally, we would like to know the relation between the quite simple construction of the metric and connection $\|\,.\,\|_{\mathcal{C}}, \nabla^{\mathcal{C}}$ and their delicate and complicated ζ function partners.

Theorem 4.2. *[32] Let $\mathbb{P}_1, \mathbb{P}_2$ be Grassmann sections for the family of Dirac operators $\mathbb{D}$. Then over the subset of B where the operators $\mathcal{D}_{P_i}$ are invertible one has*

$$\frac{\|\det(\mathcal{D}_{P_1})\|_{\zeta}}{\|\det(\mathcal{D}_{P_2})\|_{\zeta}} = \frac{\|\det(\mathcal{D}_{P_1})\|_{\mathcal{C}}}{\|\det(\mathcal{D}_{P_2})\|_{\mathcal{C}}}. \tag{4.14}$$

Let $\mathbf{R}_{\zeta}^i, \mathbf{R}_{\mathcal{C}}^i \in \Omega^2(B)$ denote the respective curvature 2-forms of the ζ and $\mathcal{C}$-connections on $\mathrm{DET}(\mathbb{D}, \mathbb{P}_i)$. Then:

$$\mathbf{R}_{\zeta}^1 - \mathbf{R}_{\zeta}^2 = \mathbf{R}_{\mathcal{C}}^1 - \mathbf{R}_{\mathcal{C}}^2. \tag{4.15}$$

Thus (4.15) is a *relative geometric anomaly* formula. This immediately, for example, leads to an identification between the $\mathcal{G}$ cocycle and ζ function curvature description of gauge anomalies.

Finally, we point out that a corresponding adiabatic pasting formula will hold for the curvatures with respect to a partition of a closed manifold along the lines of the Park–Wojciechowski formula of Theorem 3.4. That will be explained in a future publication, for the b-calculus analogue of this see [26].

References

[1] Atiyah, M.F., Patodi, V.K., and Singer, I.M.: 1975, 'Spectral asymmetry and Riemannian geometry. I', *Math. Proc. Cambridge Phil. Soc.* **77**, 43–69.

[2] Atiyah, M.F. and Singer, I.M.: Dirac operators coupled to vector potentials. Proc. Nat. Acad. Sci. USA **81**, 2597 (1984).

[3] Bismut, J-M and Freed, D.S.: 1986, D., The analysis of elliptic families I, *Comm. Math. Phys.* **106**, 159–176.

[4] Bismut, J-M and Freed, D.S.: 1986, 'The analysis of elliptic families.II. Dirac operators, eta invariants, and the holonomy theorem', *Comm. Math. Phys.* **107**, 103–163.

[5] Booß–Bavnbek, B., and Wojciechowski, K.P.: 1989, 'Pseudo-differential projections and the topology of certain spaces of elliptic boundary value problems', *Comm. Math. Phys.* **121**, 1–9.

[6] Booß–Bavnbek, B., Scott, S. G., and Wojciechowski, K. P.: 1998, 'The ζ-determinant and $\mathcal{C}$-determinant on the Grassmannian in dimension one', *Letters in Math. Phys.* **45**, 353–362.

[7] Booß–Bavnbek, B., and Wojciechowski, K.P.: 1993, *Elliptic Boundary Problems for Dirac Operators*, Birkhäuser, Boston.

[8] Brünning, J. and Lesch, M.: 1999, On the η-invariant of certain nonlocal boundary value problems, *Duke Math. J.* **96**, 425–468.

[9] Bunke, U.: 1995, 'On the glueing problem for the η-invariant', *J. Diff. Geom.* **41/2**, 397–448.

[10] Daniel, M. and Kirk, P., with an appendix by Wojciechowski, K. P.: 1999, 'A general splitting formula for the spectral flow', *Michigan Math. J.* **46**, 589–617.

[11] Dai, X., and Freed, D.: 1994, 'η-invariants and determinant lines', *J. Math. Phys.* **35**, 5155–5195.

[12] Douglas, R.G., and Wojciechowski, K.P.: 1991, 'Adiabatic limits of the η–invariants. The odd–dimensional Atiyah–Patodi–Singer problem', *Comm. Math. Phys.* **142**, 139–168.

[13] Freed, D.: 1986, 'Determiants, torsion and strings', *Commun. Math. Phys.* **107**, 483–513.

[14] Hassell, A., Mazzeo, M. and Melrose, R.B.: 1997, 'A signature formula for manifolds with corners of codimension 2', *Topology* **36**, 1055–1075.

[15] Klimek, S., and Wojciechowski, K.P.: 1993, 'η–invariants on manifolds with cylindrical ends', *Diff. Geom. and Appl.* **3**, 191–201.

[16] Klimek, S., and Wojciechowski, K.P.: 1996, 'Adiabatic cobordism theorems for analytic torsion and η–invariant', *J. Funct. Anal.* **136**, 269–293.

[17] Lesch, M., and Wojciechowski, K.P.: 1996, 'On the η–invariant of generalized Atiyah–Patodi–Singer problems', *Illinois J. Math.* **40**, 30–46.

[18] Mazzeo, R.R. and Melrose, R.B.: 1995, 'Analytic surgery and the η-invariant', *GAFA* **5**, 14–75.

[19] Melrose, R.B., Piazza, P.: 1997, 'Families of Dirac operators, boundaries and the b-calculus', *J. Diff. Geom.* **46**, 99–167.

[20] Mickelsson, J.: *Current algebras and groups.* London and New york: Plenum Press, 1989.

[21] Michelsson, J., Scott, S.G.: 1999, 'Functorial QFT, gauge anomalies and the Dirac determinant bundle', preprint; hep-th/9908207.

[22] Müller, W.: 'Eta invariants and the manifolds with boundary', *J. Diff. Geom.* **40**, 311–377.

[23] Müller, W.: 'On the index of Dirac operator on manifolds with corners of codimension two.I.', *J. Diff. Geom.* **44**, 97–177.

[24] Müller W.: 1997, 'On the relative determinant', *Comm. Math. Phys.* **192**, 309–347.

[25] Park, J. and Wojciechowski, K. P.: 2000, *Relative ζ-determinant and Adiabatic Pasting Formula for the ζ-determinat of the Dirac Laplacian*, Preprint.

[26] Piazza, P.: 1996, 'Determinant bundles, manifolds with boundary and surgery', I *Comm. Math. Phys.* **178**, 597–626; 1998, II, *Comm. Math. Phys.* **193**, 105-124.

[27] Pressley, A. and Segal, G.B.: *Loop Groups.* Oxford: Clarendon Press, 1986.

[28] Quillen, D. G.: 1985, 'Determinants of Cauchy–Riemann operators over a Riemann surface', *Funkcionalnyi Analiz i ego Prilozhenya* **19**, 37–41.

[29] Ray, D., and Singer, I.M.: 1971, 'R–torsion and the Laplacian on Riemannian manifolds', *Adv. Math.* **7**, 145–210.

[30] Scott, S.G.: 1995, 'Determinants of Dirac boundary value problems over odd–dimensional manifolds', *Comm. Math. Phys.* **173**, 43–76.

[31] Scott, S.G.: 'Splitting the curvature of the determinant line bundle', *Proc. Am. Math. Soc.*, to appear.

[32] Scott, S.G.: 'Relative Zeta determinants and the Quillen metric', preprint; math.AP/9910148.

[33] Scott, S.G., and Wojciechowski, K.P.: 1997, 'Abstract Determinant and ζ–Determinant on the Grassmannian', *Letters in Math. Phys.* **40**, 135–145.

[34] Scott, S.G., and Wojciechowski, K.P.: 1999, 'ζ-determinant and the Quillen determinant on the Grassmannian of elliptic self-adjoint boundary conditions', *C. R. Acad. Sci., Serie I,* **328**, 139–144.

[35] Scott, S.G., and Wojciechowski, K.P.: 'The ζ-determinant and Quillen determinant for a Dirac operator on a manifold with boundary', *GAFA*, to appear.

[36] Seeley, R. T.: 1967 'Complex powers of an elliptic operator'. AMS *Proc. Symp. Pure Math. X*. AMS Providence, 288–307.

[37] Segal, G.B.: 'The definition of conformal field theory', Oxford preprint, 1990.

[38] Singer, I.M.: 1985, 'Families of Dirac operators with applications to physics', *Asterisque, hors série*', 323–340.

[39] Wojciechowski, K.P.: 1994, 'The additivity of the η-invariant: The case of an invertible tangential operator', *Houston J. Math.* **20**, 603–621.

[40] Wojciechowski, K.P.: 1995, 'The additivity of the η-invariant. The case of a singular tangential operator', *Comm. Math. Phys.* **169**, 315–327.

[41] Wojciechowski, K.P.: 1999, 'The ζ-determinant and the additivity of the η-invariant on the smooth, self-adjoint Grassmannian', *Comm. Math. Phys.* **201**, 423–444.

ON GEOMETRY OF NON-ABELIAN DUALITY

Pavol Ševera

I.H.É.S., Le Bois-Marie, 35, Route de Chartres
F-91440 Bures-sur-Yvette, FRANCE

Abstract We consider several pictures connected with non-abelian duality in 2d quantum field theory. A discrete version leads to 3d coloured pictures of quantum groups, using a non-abelian version of Poincaré duality. A continuous classical version (Poisson–Lie T-duality) have some simple mechanical analogs. Finally a connection between Poisson–Lie T-duality and Courant algebroids is presented, which conjecturaly gives a unified picture.

Mathematics Subject Classification (2000): 58F05, 81R50, 81T25

Keywords: Kramers–Wannier duality, quantum groups, contact reduction, Courant algebroids

1. The discrete version: Kramers–Wannier duality and topological field theory

Kramers and Wannier discovered their duality as a way of determining the critical temperature of the 2d Ising model. It is a discrete version of abelian dualities in QFT. It is worthwhile to find a description of KW duality that would suggest non-abelian generalization. After a while one would arrive at the following picture:

Y. Maeda, et al. (eds.), Noncommutative Differential Geometry and Its Applications to Physics, 217–226.

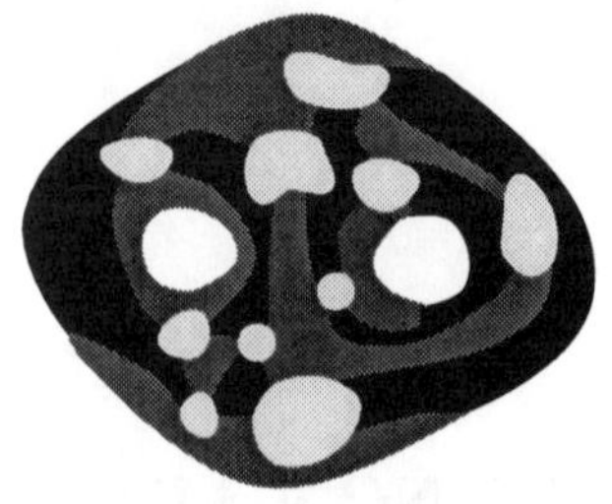

It represents a 3d body with the surface coloured in yellow, red, and black (in a locally nice way). More generally, we choose an integer n and consider an $(n+1)$-dimensional oriented compact manifolds Ω with boundary in these three colours. We also choose integers $k+l=n$ (mostly $1+1=2$) and a finite abelian group G and its dual $\widetilde{G}$. Let y be the yellow part of the boundary; it is an oriented n-dimensional manifold with the boundary coloured in black and red. The relative cohomology groups $H^k(y,r;G)$ and $H^l(y,b;\widetilde{G})$ are mutually dual via Poincaré duality (in expressions like $H^k(X,r;G)$, r denotes the red part of X, and b the black part). Let $\rho : H^k(\Omega,r;G) \to H^k(y,r;G)$ and $\tilde{\rho} : H^l(\Omega,b;\widetilde{G}) \to H^l(y,b;\widetilde{G})$ be the restriction maps. KW duality claims that their images are each other's annihilators. It is an immediate consequence of Poincaré duality and of exactness of

$$H^k(\Omega,r;G) \to H^k(y\cup r,r;G) \to H^{k+1}(\Omega,y\cup r;G).$$

In statistical models it is used in a dual form, i.e., using functions on groups rather than groups themselves. We pick up a function f on $H^k(y,r;G)$ (the Boltzmann weight) and compute the partition sum

$$Z(f) = \sum_{x\in H^k(\Omega,r;G)} f(\rho(x)). \tag{1}$$

Let $\hat{f}$ denote the Fourier transform of f. We can compute

$$\widetilde{Z}(\hat{f}) = \sum_{x\in H^l(\Omega,b;\widetilde{G})} \hat{f}(\tilde{\rho}(x)). \tag{2}$$

KW duality says (via Poisson summation formula) that up to an inessential factor we have $Z(f)=\widetilde{Z}(\hat{f})$.

The point is now that the expression (1) looks like a topological field theory (TFT), with boundary coloured in red and black. For any oriented yellow n-dimensional manifold Σ with $\partial\Sigma$ in black and red we have a state space $\mathcal{H}(\Sigma)$ — the space of functions on $H^k(\Sigma,r;G)$. And any

Ω yields a linear form on $\mathcal{H}(y_\Omega)$ — the one given by (1) (actually, the normalization has to be changed slightly for the gluing property to hold, but this is just a technicality). KW duality now says that exchanging k with l, red with black, and G with $\widetilde{G}$, yields the same TFT.

This TFT point of view readily gives a non-abelian generalization. First, let us have a look if we can recover k, l and G from the TFT. Let $B^{k,l}$ be yellow n-dimensional ball with a tubular neighbourhood of $S^{k-1} \subset S^{n-1} = \partial B^{k,l}$ painted in red and the rest of S^{n-1} (a tubular neighbourhood of S^{l-1}) in black. Then $\mathcal{H}(B^{k',l'})$ is trivial (i.e., 1-dimensional) unless $k' = k$ and $l' = l$; in that case, it is the space of functions on G. The reader will readily recover the Hopf algebra structure on $\mathcal{H}(B^{k,l})$ using TFT; the $1 + 1 = 2$-case is drawn below (it is somewhat difficult to draw 3d objects by mouse: the pictures represent 3d balls and their invisible sides are yellow; the antipode is not drawn — it is simply a half turn):

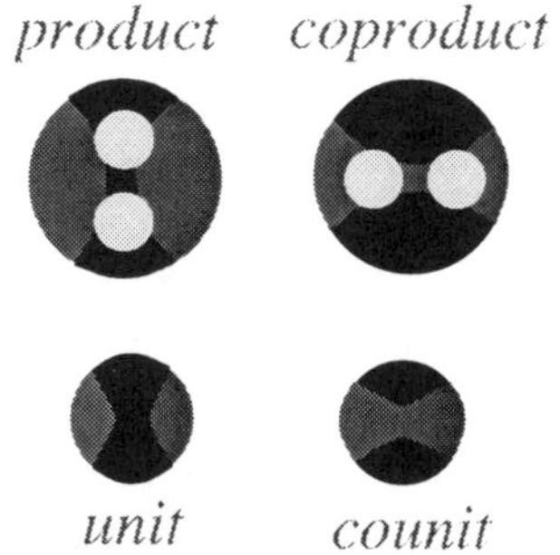

Before stating our non-abelian generalization we need to notice that the excision property of relative cohomology tames our TFT's considerably. It gives rise to the following *squeezing property*: if you put one finger somewhere on the red part of $\partial\Omega$ and another somewhere on the black part, press the fingers against each other and remove the surface (red from one side and black from the other) that has appeared between your fingers, then you don't change anything. (For example, you can change

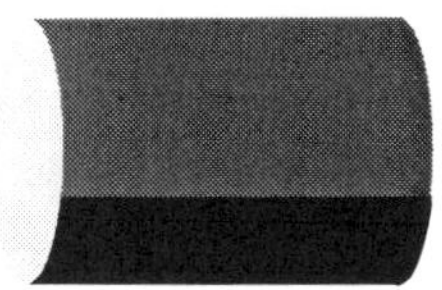

to

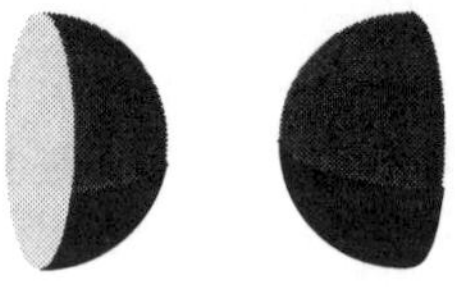

).

Now the conclusion: it seems that TFT's satisfying the squeezing property are fine non-abelian generalizations of KW duality. At least in the $1+1=2$ case: the usual $R \leftrightarrow 1/R$-duality has a generalization (Poisson–Lie T-duality) replacing the two circles (or tori) by a pair of mutually dual Poisson–Lie groups. One therefore expects that the finite abelian groups G and $\widetilde{G}$ could be replaced by a pair of mutually dual (finite dimensional, presumably) Hopf algebras $\mathcal{H}$ and $\tilde{\mathcal{H}}$. And indeed, this is what happens: the pictures above define a Hopf algebra structure on $\mathcal{H} = \mathcal{H}(B^{1,1})$ whenever the TFT satisfies the squeezing property (this 3d way of representing Hopf algebras is due to Maxim Kontsevich, [3]). Moreover, whenever $\mathcal{H}(B^{2,0})$ and $\mathcal{H}(B^{0,2})$ are both trivial, the TFT is specified by its $\mathcal{H}$ uniquely (i.e., the pictures of operations are basic building block of our TFT's). Conjecturally, there is 1-1 correspondence between these TFT's and finite-dimensional involutive (i.e., with $S^2 = 1$, where S is the antipode) Hopf algebras.

One problem that has to be removed in the future is involutiveness: since S^2 is represented by a complete turn, it has to be 1, at least for the straightforward definition of TFT. It is this definition that has to be changed.

One can show that the boundary-free part of our TFT (i.e., without red and black) is the Chern–Simons theory coming from the Drinfeld double of $\mathcal{H}$. Although quantum groups at roots of unity are not Drinfeld doubles themselves, they are quotients of doubles (of Borel quantum groups) by abelian ideals. If we can have $\mathcal{H} =$ Borel, nice interconnections will appear.

Finally, I don't know what happens for higher n. If the picture is conceptually correct then most interesting things should happen in the $2+2=4$ case (the electric-magnetic duality).

2. A mechanical intermezzo

Let us change the topic and look at the question how to draw one-dimensional Lagrangians. The people involved are (among others) Monge,

Hamilton, Jacobi, and Lie, cf [4].* According to them, mechanics studies (bi)characteristics of a field of cones. Suppose P is a manifold with a field of cones in the tangent spaces and consider hypersurfaces in P tangent to these cones. (Bi)characteristics are the curves along which these hypersurfaces touch the cones (or more precisely, they are infinitesimally narrow strips of these hypersurfaces). If we suppose that the field of cones is invariant with respect to action of a 1-parameter group H on P that makes P into a principal bundle $P \to M$, we can formulate a variational problem in M, whose solutions are the projections of characteristics. Namely, we choose (locally) a flat connection α on P; the intersections of the levels $\alpha = 1$ with the cones give us indicatrices in M (indicatrix can be then seen as the set of allowed velocities and we are interested in trajectories with extremal times):

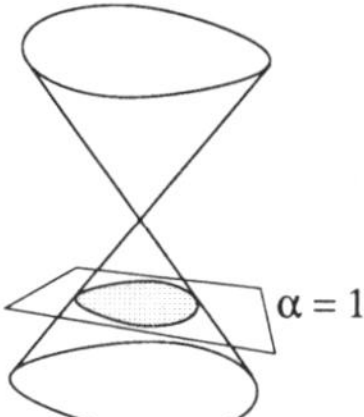

The choice of flat connection is not natural (and it need not exist globally). A different choice differs by a closed 1-form. This is the familiar non-naturalness of Lagrangians; only the field of cones is natural.

Let us now proceed to symmetries and reductions. Symmetry is simply a transformation of P preserving the field of cones. Classically, it is hidden unless it preserves the action of H; nevertheless, we can apply reduction regardless of this property (and thus go somewhat beyond classical reduction). However, let us first consider the most classical case — Maupertuis principle. We suppose that a 2-parameter abelian G preserves the cones, $H \subset G$, and we choose another 1-parameter $\widetilde{H} \subset G$ (this is the choice of energy) We look at P in the direction of the orbits of $\widetilde{H}$: we see $P/\widetilde{H}$ together with a field of cones — the contours of cones on P. $P/\widetilde{H}$ with its field of cones (and action of H) is the reduced problem.

Let us now describe a classically inaccessible example. This example is solved via reduction in contact geometry. It is fairly close to symplectic reduction, but for applications in mechanics it has one virtue — if we

*Unfortunately, Klein died before he could write about Lie in this unique book, but one can look in his 'Elementary mathematics from higher viewpoint'.

can solve reduced problems, the original problem is solved just by computing some derivatives. Our example is as follows: we take the group $SE(2)$ and consider a right-invariant field of cones. It should specify a variational problem in $SE(2)/SO(2)$, i.e., in the plane. The problem looks as follows:

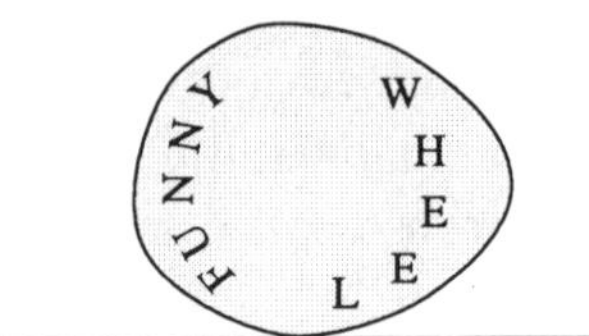

First consider only the closed curve. It specifies a field of indicatrices in the plane in the following weird way: For whatever point x in the plane, rotate the curve around x for 90 degrees; the result is (identifying the tangent space at x with the plane itself) the indicatrix at x. The $SE(2)$-symmetry is well hidden in this picture.

Now look at the picture again. The wheel is really funny: it stays still, and instead, it is the road that rolls around the wheel. Consider the trajectories of the plane of the road (this plane lies on the wheel's plane); they are the extremals we were looking for (among them the evolvents of the wheel — these are the trajectories of the points of the road).

Instead of $SO(2) \subset SE(2)$ we could choose another 1-parameter H; for example, $SE(2)/H$ may be the space of lines in our plane. These two variational problems are in some sense identical. This is despite the fact that they are on different spaces, even their phase spaces are different. Nevertheless, they have the same prequantization (i.e., contact instead of symplectic) space.

More generally, suppose a group G acts on a manifold P, preserving a field of cones. We can choose a 1-parameter subgroup $H \subset G$ and arrive at 1-dimensional variational problem on P/H. For whatever choice of H, these variational problems are equivalent. Moreover, the G-symmetry is obscured in P/H. We shall meet a 2-dimensional analogue in the following section.

3. Poisson–Lie T-duality ...

Poisson–Lie T-duality (introduced in [5]) is a classical picture behind the $1+1=2$ duality of §1, and also a 2-dimensional analogue of the Funny Wheel. Suppose G and $\widetilde{G}$ are mutually dual Poisson–Lie groups and D is their Drinfeld double. We shall describe two mutually equivalent 2d variational problems, one on G and the other on $\widetilde{G}$. They are connected by certain first-order right-invariant differential equation for surfaces in D (this equation will be described below). We identify G with $D/\widetilde{G}$ and $\widetilde{G}$ with D/G (suppose for simplicity that it is possible). Extremal surfaces in $G = D/\widetilde{G}$ are precisely the projections of the surfaces in D that solve our equation, and the same is valid for $\widetilde{G} = D/G$.

The right D-invariance is not visible in G nor in $\widetilde{G}$. Right action of G on itself does not preserve the Lagrangian, so it does not yield any conservation law. Yet, it does yield something: Given an extremal surface in G, a corresponding surface in D is unique up right multiplication by $\widetilde{G}$. In fact, it is obtained by integrating a flat $\widetilde{G}$ connection on the surface in G and this flat connection is the momentum density coming from the right action of G on itself. We may say that this action yields a non-abelian conservation law, and Poisson–Lie T-duality reveals the geometry behind such conservation laws.

Finally, we should say something about the first-order differential equation in D and how to get the Lagrangians on G and $\widetilde{G}$ from it. The question is really what should replace the cones appearing in 1d variational problems. A first order differential equation for surfaces is simply a set of planes in the tangent spaces; since it is right-invariant, we only have to know it at one point.

Here is a way how to draw $2d$ Lagrangians. Let V be a vector space with inner product (examples to be kept in mind are $V = T_xM \oplus T_x^*M$, with inner product given by the pairing between T_xM and T_x^*M (here M is a manifold on which the Lagrangian is supposed to live), and the Lie algebra of D). Let $\bigwedge_{DI}^2 V \subset \bigwedge^2 V$ be the space of bivectors of the form $u \wedge v$, $\langle u, u\rangle = \langle u, v\rangle = \langle v, v\rangle = 0$ (i.e., elements of isotropic planes). $\bigwedge_{DI}^2 V$ is a symplectic manifold (it is a coadjoint orbit of $SO(V)$). A $2d$ Lagrangian is simply a homogeneous Lagrangian submanifold of $\bigwedge_{DI}^2 V$ (here homogeneous means invariant with respect to multiplication by scalars, so that the Lagrangian is really a set of planes — this is what we want).[†] This claim is not clear without explanation, but it is analogous to the representation of $1d$ Lagrangians by cones. If we decompose V as $U_1 \oplus U_2$ with isotropic U's (that's what we have for $V = T_xM \oplus T_x^*M$), the projection $\bigwedge_{DI}^2 V \to \bigwedge_D^2 U_1$ is a Lagrangian fibration ($\bigwedge_D^2$ denotes decomposable bivectors, i.e., plane

[†]Here is an alternative description for people who want to see pictures. Take the quadric $Q \subset \mathbb{P}(V)$ defined by the inner product on V. As a quadric, it has a natural pseudoconformal structure (its lightcones are the intersections of Q with its tangent hyperplanes). In T^*Q take only the isotropic covectors; after reduction (by isotropic geodesics in Q) this space becomes $\bigwedge_{DI}^2 V$. Therefore, up to singularities, a Lagrangian is the same as a light-like hypersurface in Q.

elements); together with $\bigwedge^2_D U_1 \subset \bigwedge^2_{DI} V$ it makes $\bigwedge^2_{DI} V$ to $T^* \bigwedge^2_D U_1$ (there are some problems at 0, but pictures know better than words). A homogeneous Lagrangian submanifold of $\bigwedge^2_{DI} V$ is then the same as a (perhaps partially defined or multivalued) 1-homogeneous function on $\bigwedge^2_D U_1$, which is what you would call a $2d$ Lagrangian.

PL T-duality should be clear now: We choose a Lagrangian in the Lie algebra of D and translate it right invariantly; it gives us the differential equation on D. At any point $x \in G$ the tangent space $T_x D$ is decomposed into $T_x G$ and the fibre of the projection $D \to G = D/\widetilde{G}$. We use this decomposition to define our $2d$ variational problem on G. Finally, we exchange the roles of G and $\widetilde{G}$.

4. ... and Courant algebroids

In this section we shall briefly describe the geometry behind Poisson–Lie T-duality, and more generally, behind arbitrary 2-dimensional variational problems. The question that comes to mind first is: what should replace the principal $U(1)$-bundle of the last section, if we pass to two-dimensional problems? The suggestion of Brylinski [6] is $U(1)$-gerbe, or principal $BU(1)$-bundle. We shall only identify infinitesimal part of this object. It turns out to be a Courant algebroid, as defined in [7].

Let us start with a down to earth approach, using analogy with §2. Infinitesimal part of a principal $U(1)$-bundle $P \to M$ is its Lie algebroid (it is a vector bundle over M, namely $TP/U(1)$). We could arrive at it in this down to earth way, inspired by Noether's theorem: vector fields on M give us conservation laws, provided they are infinitesimal symmetries, i.e., if they preserve the Lagrangian *up to total derivative* (recall that the notion of Lagrangian is not a natural one). A symmetry is therefore a pair (v, f) of a vector field and a function, rather then a vector field alone. It is a section of $TM \oplus \mathbb{R}$, and this is the Lie algebroid. In fact, its splitting is not natural (as a result of non-naturality of Lagrangians); only its Lie algebroid structure is important.

If we do the same for 2-dimensional problems, symmetry becomes a pair (v, θ) of a vector field and a 1-form, i.e., a section of $TM \oplus T^*M$; the flow of θ should change the Lagrangian with velocity $d\theta$. This structure on $TM \oplus T^*M$ is nothing but the Courant algebroid.

All this may serve just as a motivation, if we take into account the discovery of Dmitry Roytenberg of the geometry behind Courant algebroids [8] (actually, the reader is not supposed to know what a Courant algebroid is). Namely, they are non-negatively graded symplectic supermanifolds $(\mathcal{M}, \omega)$, with $\deg \omega = 2$ and with chosen function θ of degree 3 satisfying $\{\theta,\theta\}=0$. We call them *2-manifolds* for short. The Hamiltonian vector field X_θ is of degree 1 and square 0; graded supermanifolds with a vector field like this are called Q-manifolds, and the vector field is generally denoted as Q. An example of a 2-manifold (corresponding to the Courant algebroid $TM \oplus T^*M$) is $T^*[2](T[1]M)$ with the standard symplectic form on T^*. There is a canonical vector field on $T[1]M$ (the de Rham differential) and it becomes the function θ on $T^*[2](T[1]M)$. Another example is $\mathfrak{g}[1]$, where $\mathfrak{g}$ is a Lie algebra with invariant inner product.

Now let Σ be a surface, $\mathcal{M}$ a 2-manifold, and let us consider maps $\phi : T[1]\Sigma \to \mathcal{M}$. We call such a ϕ an *extremal surface* provided it is homogeneous (i.e., preserves the grading) and Q-equivariant. A *D-brane* is a homogeneous Lagrangian submanifold $\mathcal{D} \subset \mathcal{M}$ on which θ vanishes (i.e., which is Q-invariant). If Σ has boundary, ϕ is supposed to send $T[1]\partial\Sigma$ to $\mathcal{D}$ for some fixed $\mathcal{D}$ (though we may have different $\mathcal{D}$'s for different boundary components, if we wish).

This somewhat surprising definition is equivalent to the standard one (if $\mathcal{M} = T^*[2](T[1]M)$), although it makes no reference to any Lagrangian. If we want it, we choose a trivialization of $T\Sigma$, so that $T[1]\Sigma$ becomes $\mathbb{R}^{0|2}\times\Sigma$ and ϕ a map $\Sigma \to \mathcal{M}^{\mathbb{R}^{0|2}}$. Now $\mathcal{M}^{\mathbb{R}^{0|2}}$ is a graded symplectic manifold (symplectic form is the integral over $\mathbb{R}^{0|2}$ of the form on $\mathcal{M}$), and a Lagrangian is simply a homogeneous Lagrangian submanifold $\Lambda \subset \mathcal{M}^{\mathbb{R}^{0|2}}$. Then we only consider ϕ's that send Σ to Λ.

In this language, Poisson–Lie T-duality is just symplectic reduction.

5. Conclusion

The topics discussed up to now have been fairly elementary. Yet several interesting open problems remain. We may put them together: is it possible to put all this stuff together? Just to mention two of the missing links: According to [9] a 2-manifold (as defined in §4) is the basic ingredient of 3-dimensional topological field theories; this is to be connected with the TFT's of §1. And the second, one can verify that a Courant algebroid is indeed an infinitesimal version of a $U(1)$-gerbe; however, a true 2-dimensional analogue of the geometrical picture of §2 is still missing.

References

[1] P. Ševera, Duality and TFT: A suggestion based on $d=2+1$, hep-th/9811136.

[2] P. Ševera, Contact reductions, mechanics and duality, math.DS/9903168.

[3] M. Kontsevich, Geometry of formulae, lecture given at *Colloque scientifique du 40ème anniversaire*, IHES, Oct. 8, 1998.

[4] F. Klein, *Vorlesungen über die Entwicklung der Mathematik im 19. Jahrhundert*, Springer, Berlin, 1926.

[5] C. Klimčík, P. Ševera, Dual Non-Abelian Duality and the Drinfeld Double, *Phys.Lett.* **B351** (1995), 455–462, hep-th/9502122

[6] J.L. Brylinski, *Loop spaces, characteristic classes and geometric quantization.* Progress in Mathematics, 107. Birkhäuser, Boston, 1993.

[7] Zhang-Ju Liu, Alan Weinstein, Ping Xu, Manin Triples for Lie Bialgebroids, *J. Differential Geom.* **45** (1997), no. 3, 547–574.

[8] Dmitry Roytenberg, *Courant algebroids, derived brackets and even symplectic supermanifolds*, UC Berkeley Ph.D. thesis, math.DG/9910078.

[9] M. Alexandrov, M. Kontsevich, A. Schwarz, O. Zaboronsky, The Geometry of the Master Equation and Topological Quantum Field Theory, *Int.J.Mod.Phys.* **A12** (1997), 1405–1430.

WEYL CALCULUS AND WIGNER TRANSFORM ON THE POINCARÉ DISK

Tatsuya Tate
Department of Mathematics, Faculty of Science and Technology, Keio University
3–14–1, Hiyoshi, Kohoku-ku, Yokohama, 223–8522, JAPAN
tate@math.keio.ac.jp

Abstract The aim of this paper is to establish an analogue of Weyl calculus on the Poincaré disk. The Weyl calculus is defined by functional calculus with respect to the Schrödinger representation of the Heisenberg group, and the Wigner transform is defined to be the Fourier transform of the matrix coefficients of this representation. As in [1], [3], the Weyl calculus and the Wigner transform are closely related to each other. In this paper we will define an analogue of the Weyl calculus on the Poincaré disk by using the middle point $m(z, w)$ on the geodesic through z and w. We will also define the Wigner transform by taking into account a formula describing a relationship between the matrix coefficients of a Weyl pseudo-differential operator and the Wigner transform on Euclidean space.

Mathematics Subject Classification (2000): 43A85, 58J40, 44A05

Keywords: Weyl calculus, Wigner transform, non-Euclidean Fourier transform

1. Introduction

The Weyl calculus on Euclidean spaces was originally defined by functional calculus with respect to the Schrödinger representation, that is, an infinite-dimensional irreducible unitary representation, of the Heisenberg group. The Wigner transform is defined to be the Fourier transform of the matrix coefficients of this representation. There are many remarkable formulas which describe the relation between the Weyl calculus and the Wigner transform.

Y. Maeda, et al. (eds.), Noncommutative Differential Geometry and Its Applications to Physics, 227–243.

For example, the Weyl symbol of a rank 1 projection is given by the Wigner transform. Furthermore, the Weyl symbol of any Weyl operator is given by the Wigner transform of the kernel of the operator. These are important and interesting facts; for other facts and properties of the Wigner transform and the Weyl calculus, we refer the reader to [1], [3].

The purpose of this paper is to formulate analogues of Weyl pseudo-differential operators and Wigner transforms and to investigate relationships between them. In the Euclidean case the Weyl pseudo-differential operator $a^{\mathrm{w}}(x, D)$ with symbol $a(x, \xi)$ has the following form:

$$a^{\mathrm{w}}(x, D)u = (2\pi\hbar)^{-n} \int_{\mathbb{R}^n} e^{i(x-y)\cdot\xi/\hbar} a\left(\frac{x+y}{2}, \xi\right) u(y)\, dy d\xi,$$

where $\hbar$ is a positive parameter. Our idea for defining Weyl pseudo-differential operators on the Poincaré disk is that we regard the term $(x+y)/2$ in the above expression as the middle point on the geodesic through x and y. Also we replace the variable z in the symbol $a(z, b, \nu)$ of the Kohn–Nirenberg pseudo-differential operator defined by Zelditch [5] (see section 3) by the middle point $m(z, w)$ on the geodesic through z, w (w is an integral variable). Moreover, we define, in this paper, the Wigner transform so as to satisfy a similar formula to that of the Wigner transform and the Weyl calculus on $\mathbb{R}^n$ (see Lemma 6).

The product formula for the operators, unfortunately, is not proved at present. However, an analogue of a formula for the Wigner transform on the Euclidean space is given (see Theorem 2). We will also investigate the relationships between the Wigner transform and the kernel of the Weyl pseudo-differential operator. However, as in the Euclidean case, whether or not there is a representation of a group in the background of the Weyl operator defined in this paper is not made clear. Furthermore, Weinstein [4] investigated the relation between the area of a geodesic triangle and the phase function of the (Schwartz) kernel of a product on the space of smooth functions on a hermitian symmetric space. The relations between his work and ours will be discussed elsewhere.

2. Preliminaries

Let $\mathbf{D}$ be the unit disk in the complex plane with Poincaré metric $ds^2 = \dfrac{4}{(1-|z|^2)^2} dz\overline{dz}$. For two point $z, w \in \mathbf{D}$, let $m(z, w)$ be the middle point of z and w on the unique geodesic through z and w, and for simplicity, we set $m_0(z) = m(0, z)$. Clearly the map $m : \mathbf{D} \times \mathbf{D} \to \mathbf{D}$ is $SU(1,1)$-equivariant, symmetric map. It is not difficult to see that an $SU(1,1)$-equivariant symmetric map $m : \mathbf{D} \times \mathbf{D} \to \mathbf{D}$ satisfying $m(z, -z) = 0$ for every z assigns the middle point to two points.

We will describe the middle point $m(z,w)$ of z,w more concretely. A direct computation, by using the polar coordinate on $\mathbf{D}$, gives

$$m_0(z) = \frac{z}{1+\sqrt{1-|z|^2}}.$$

For $z \in \mathbf{D}$, we set

$$g(z) = \frac{1}{\sqrt{1-|z|^2}} \begin{pmatrix} 1 & -z \\ -\bar{z} & 1 \end{pmatrix}, \quad g(z)^{-1} = \frac{1}{\sqrt{1-|z|^2}} \begin{pmatrix} 1 & z \\ \bar{z} & 1 \end{pmatrix}. \quad (1)$$

The matrix $g(z)$ is an element of $SU(1,1)$ and satisfies $g(z)z = 0$. By the $SU(1,1)$-equivariance of m, we have

$$m(z,w) = g(z)^{-1} \frac{g(z)w}{1+\sqrt{1-|g(z)w|^2}}.$$

Note that, for fixed $w \in \mathbf{D}$, the map $z \mapsto m(z,w)$ is a diffeomorphism of $\mathbf{D}$ onto itself, and we will denote its inverse by $p(w,x)$ for $x = m(z,w)$. The map $p : \mathbf{D} \times \mathbf{D} \to \mathbf{D}$ is also $SU(1,1)$-equivariant, but not symmetric. The map $w \mapsto p(w,x)$ is nothing but the reflection with respect to $x \in \mathbf{D}$.

Lemma 1. *Let u be a continuous function on $\mathbf{D}$ with compact support. For any $z \in \mathbf{D}$, we have*

$$\int_{\mathbf{D}} u(m(z,w))\,đw = \int_{\mathbf{D}} u(x) J(x,z)\,đx,$$

where $đz$ denotes the volume measure with respect to Poincaré metric and $J(x,z)$ denotes the function defined by

$$J(x,z) = 4\frac{1+|g(z)x|^2}{1-|g(z)x|^2} = 4\frac{|1-\bar{z}x|^2 + |z-x|^2}{(1-|z|^2)(1-|x|^2)}.$$

For every $z \in \mathbf{D}$ we also have

$$\int_{\mathbf{D}} u(p(z,x))\,đz = \int_{\mathbf{D}} u(z)\,đz.$$

This lemma can be proved by a direct computation. Since $|g(z)w|^2 = \tanh^2 d(z,w)/2$, we have $J(z,w) = 4\cosh d(z,w)$. The group $SU(1,1)$ can be viewed as an $U(1)$ bundle over $\mathbf{D}$, and hence the map $g : \mathbf{D} \ni z \mapsto g(z) \in SU(1,1)$ is a section of this bundle. The action of $g(z)$ for $z \in \mathbf{D}$ is an analogue of the translation on the Euclidean plane. We will summarize the properties for the section g and the maps m, p.

Lemma 2. *For every $z, w \in \mathbf{D}$, we have the following.*

$$g(z)0 = -z, \quad g(z)(-w) = -g(z)^{-1}w, \quad p(z,w) = -g(w)^2 z,$$
$$g(w)^2 = g(p(0,w)), \quad g(m_0(z))^2 = g(z), \quad g(m_0(z))^{-1}m_0(z) = z.$$

Proof. The last equation of the first line is obtained by the equivariance of the map p and the second one. The first equation of the second line is obtained by a direct computation using (1). The another equations of the second line is follows from first one. ∎

We now review on the non-Euclidean Fourier transform, which is introduced by Helgason [2].

Let $P(z,b)$ be the Poisson kernel on $\mathbf{D}$, where we denote a point on $B = \partial\mathbf{D}$ by $b = e^{i\theta}$. Explicitly, the function $P(z,b)$ has the form

$$P(z,b) = \frac{1-|z|^2}{|z-b|^2} = e^{\langle z,b \rangle},$$

where $\langle z,b \rangle$ is the (signed) hyperbolic distance between origin 0 and the horocycle through z tangent to B at b. For $\nu \in \mathbb{R}$ we set

$$F_\nu(z,b) = P(z,b)^{1/2+i\nu}.$$

Note that the function $F_\nu(z,b)$ is an eigenfunction (but not L^2–function) of the Laplacian Δ on the variable z;

$$\Delta F_\nu = (1/4 + \nu^2)F_\nu.$$

Then the non-Euclidean Fourier transform is defined to be a linear map from $C_0^\infty(\mathbf{D})$ to $C^\infty(B \times \mathbb{R}^+)$ by the identity

$$\hat{u}(b,\nu) = \frac{1}{2\pi}\int_{\mathbf{D}} F_{-\nu}(z,b)u(z)đz, \quad u \in C_0^\infty(\mathbf{D})$$

The following theorem is an analogue of the Fourier inversion formula on the Euclidean spaces. See [2] for the proof.

Theorem 1. *The linear map $u \mapsto \hat{u}$ extends to a unitary operator from $L^2(\mathbf{D}, đz)$ onto $L^2(B \times \mathbb{R}^+, db\, đ\nu)$, where db is the Haar measure on the unit circle B with total volume 2π, and $đ\nu = \nu\tanh(\pi\nu)d\nu$. The inverse of this map is given by the following identity;*

$$u(z) = \frac{1}{2\pi}\int_{B\times\mathbb{R}^+} F_\nu(z,b)\hat{u}(b,\nu)\, db đ\nu.$$

Moreover, if the function u *has the compact support, then for each* $N > 0$ *there is a positive constant* $C > 0$ *such that for every* $\nu \in \mathbb{R}^+$, $b \in B$ *we have*

$$|\hat{u}(b, \nu)| \le C\langle \nu \rangle^{-N},$$

where $\langle \nu \rangle = (\frac{1}{4} + \nu^2)^{1/2}$.

3. Weyl pseudo-differential operators

Zelditch [5] has introduced an analogue of the Kohn–Nirenberg type pseudo-differential operator on $\mathbf{D}$ by using the non-Euclidean Fourier transform (Theorem 1). Namely we set the operator $a(x, D)$ in the following form;

$$\begin{aligned} a(z, D)u &= \frac{1}{(2\pi)} \int_{B\times\mathbb{R}^+} F_\nu(z, b) a(z, b, \nu) \hat{u}(b, \nu)\, db\, đ\nu \\ &= \frac{1}{(2\pi)^2} \int F_\nu(z, b) F_{-\nu}(w, b) a(z, b, \nu) u(w)\, đw\, db\, đ\nu, \end{aligned} \tag{2}$$

where the function $a(z, b, \nu)$ is a symbol (see Definition 1). In this section we will define the Weyl calculus $a^{\mathrm{w}}_{\hbar}(z, D)$ associated with a function $a(z, b, \nu)$ by replacing the variable z in (2) by the middle point $m(z, w)$. Before defining the Weyl calculus we will review the formulation of [5]. Let

$$X_+ = \frac{1}{2}\begin{pmatrix} i & -i \\ i & -i \end{pmatrix}, \quad H = \frac{1}{2}\begin{pmatrix} 0 & 1 \\ 1 & 0 \end{pmatrix}, \quad W = \frac{1}{2}\begin{pmatrix} i & 0 \\ 0 & -i \end{pmatrix}. \tag{3}$$

Then X_+, H, W form a basis of the Lie algebra $\mathfrak{su}(1,1)$. We note that the group $PSU(1,1) = SU(1,1)/\{\pm 1\}$ is diffeomorphic to $\mathbf{D} \times B$ via the map $g : \mathbf{D} \times B \to PSU(1,1)$ defined so as to satisfy $g(z,b)0 = z$, $g(z,b)1 = b$. $\mathbf{D} \times B$ is identified with the unit tangent bundle $S\mathbf{D}$ of the Poincaré disk by the map $v : \mathbf{D} \times B \to S\mathbf{D}$ which assigns to $(z, b) \in \mathbf{D} \times B$ the unit tangent vector to the geodesic through z with forward endpoint $b \in B$. Then we can identify $PSU(1,1)$ with the unit cotangent bundle $S^*\mathbf{D}$ on $\mathbf{D}$. Thus we can consider X_+, H, W as the vector fields on $S^*\mathbf{D}$. We also identify $\mathbf{D} \times B \times \mathbb{R}^+$ with the punctured cotangent bundle $T^*\mathbf{D} \setminus 0$ by regarding $\nu \in \mathbb{R}^+$ with the hyperbolic norm $\|\xi\|$ of the covector $\xi \in T^*\mathbf{D} \setminus 0$. X_+, H, W are extended to homogeneous vector fields on $T^*\mathbf{D} \setminus 0$ of degree zero. We recall the definition of symbols made by Zelditch [5].

Definition 1. (1) *A function* $a(z, b, \nu) \in C^\infty(\mathbf{D} \times B \times \mathbb{R}^+)$ *is said to be a symbol of order* m, $a \in S^m$, *if for every positive integers* $k, \alpha_i \in \mathbb{Z}$ $(i = 1, 2, 3)$ *and compact set* $K \subset \mathbf{D}$, *there is a positive constant* $C > 0$ *such that we have*

$$\left| \left(\frac{\partial}{\partial \nu}\right)^k X_+^{\alpha_1} H^{\alpha_2} W^{\alpha_3} a(z, b, \nu) \right| \le C\langle \nu \rangle^{m-k}.$$

(2) *A symbol $a \in S^m$ is said to be classical, $a \in S^m_{cl}$, if*

$$a(z,b,\nu) \sim \sum_{j=0}^{\infty} a_j(z,b,\nu),$$

where a_j is positive homogeneous of degree $m-j$,

$$a_j(z,b,\nu) = \nu^{m-j} a_j(z,b,1) \text{ for } \nu \geq 1.$$

Now we define the Weyl operator on the Poincaré disk.

Definition 2. *For suitable smooth function $a(z,b,\nu)$ on $\mathbf{D} \times B \times \mathbb{R}^+$, we set*

$$\begin{aligned} a^{\mathrm{w}}_{\hbar}(z,D)u &= (2\pi\hbar)^{-2} \int F_{\nu/\hbar}(z,b) F_{-\nu/\hbar}(w,b) a(m(z,w),b,\nu) u(w) đw db đ\nu_{\hbar} \\ &= (2\pi\hbar)^{-2} \int F_{\nu/\hbar}(z,b) F_{-\nu/\hbar}(p(z,w),b) a(w,b,\nu) u(p(z,w)) \\ &\qquad \times J(z,w) đw db đ\nu_{\hbar}, \end{aligned}$$

where $\hbar$ is a positive parameter and $đ\nu_{\hbar} = \nu \tanh(\pi\nu/\hbar)\, d\nu$. We call the operator $a^{\mathrm{w}}_{\hbar}(z,D)$ the Weyl operator with the symbol $a(z,b,\nu)$. We denote $a^{\mathrm{w}}_{1}(z,D)$ by $a^{\mathrm{w}}(z,D)$.

Example 1. For $a(z,b,\nu) = \langle \nu \rangle^2 \in S^2_{\mathrm{cl}}$, we have $a^{\mathrm{w}}_{\hbar}(z,D) = \hbar^2\Delta + (1-\hbar^2)/4$.

Example 2. Let $a(z,b,\nu) = a(z) \in C^{\infty}(\mathbf{D})$. Then we have $a \in S^0_{\mathrm{cl}}$ and the operator $a^{\mathrm{w}}_{\hbar}(z,D)$ is the multiplication by the function $a(z)$ for all $\hbar > 0$.

The following lemma guarantees that the Definition 2 makes sense for symbols in the sense of Definition 1.

Lemma 3. *Let $a(z,b,\nu)$ be a smooth function which satisfies that, there exists a constant $m \in \mathbb{R}$ and $\delta < 1$ such that for every compact set $K \subset \mathbf{D}$ and every non negative integer k, α_1, α_2, α_3, there exists a constant $C > 0$ such that we have*

$$\left| \left(\frac{\partial}{\partial \nu}\right)^k X_+^{\alpha_1} H^{\alpha_2} W^{\alpha_3} a(z,b,\nu) \right| \leq C \langle \nu \rangle^{m+\delta(\alpha_1+\alpha_2+\alpha_3+k)} \tag{4}$$

for $z \in K$. Then the map $a^{\mathrm{w}}_{\hbar}(z,D)$ is a continuous operator from $C_0^{\infty}(\mathbf{D})$ to $C^{\infty}(\mathbf{D})$. Especially, for each symbol $a \in S^m$, we have a continuous linear operator $a^{\mathrm{w}}_{\hbar}(z,D) : C_0^{\infty}(\mathbf{D}) \to C^{\infty}(\mathbf{D})$.

Proof. We give the proof of the statement only for $\hbar = 1$. We set

$$A(z,b,\nu) = (2\pi)^{-1} \int_{\mathbf{D}} F_{-\nu}(w,b) a(m(z,w),b,\nu) u(w) đw,$$

then we have

$$a^{\mathrm{w}}(z,D)u = (2\pi)^{-1} \int_{\mathbf{D}} F_{\nu}(z,b) A(z,b,\nu)\, db đ\nu.$$

Integration by parts enables us to obtain

$$\langle \nu \rangle^{2N} A(z,b,\nu) = (2\pi)^{-1} \int_{\mathbf{D}} F_{-\nu}(w,b) \Delta^N (a \circ m \cdot u) đw.$$

The term of the integrand involving the Laplacian is then estimated locally uniformly in z as follows by the assumption (4) on a;

$$|\Delta^N (a \circ m \cdot u)| \leq C \langle \nu \rangle^{m+2\delta N} \|u\|_{2N},$$

where $\|u\|_{2N}$ is the sum of the sup–norm of u and its derivatives up to $2N$ order and the constant $C > 0$ depends on the volume of the support of u but independent of ν. If δ is negative then the right hand side of the above is replaced by $C\langle \nu \rangle^m \|u\|_{2N}$. Therefore the function $A(z,b,\nu)$ has the following estimate;

$$|A(z,b,\nu)| \leq C_K \|u\|_{2N} \langle \nu \rangle^{m-2N(1-\delta)}, \tag{5}$$

for every $N \in \mathbf{N}$, where K is a compact subset containing the support of u. Since $\delta < 1$, taking N large enough the integral in Definition 2 is absolutely convergent locally uniformly in z. The derivative of the function $A(z,b,\nu)$ has a similar estimate to (5). Hence we can differentiate $a^{\mathrm{w}}(z,D)u$ under the integral sign. Therefore the function $a^{\mathrm{w}}(z,D)u$ is smooth, and also the estimate (5) and that of the derivatives of $A(z,b,\nu)$ shows that the operator $a^{\mathrm{w}}(z,D)$ is continuous. ▌

First of all, we consider the formal adjoint operator of a Weyl operator.

Lemma 4. *Let $a \in S^m$. Then for every $u, v \in C_0^\infty(\mathbf{D})$, we have*

$$\langle a_\hbar^{\mathrm{w}}(z,D)u, v \rangle = \langle u, \bar{a}_\hbar^{\mathrm{w}}(z,D)v \rangle. \tag{6}$$

Namely, the formal adjoint of the operator $a_\hbar^{\mathrm{w}}(z,D)$ is also Weyl operator with symbol $\bar{a}$. Moreover, the operator $a_\hbar^{\mathrm{w}}(z,D)$ extends to a continuous operator from the space $\mathcal{E}'(\mathbf{D})$ of all compactly supported distribution on $\mathbf{D}$ to the space $\mathcal{D}'(\mathbf{D})$ of all distribution on $\mathbf{D}$.

Proof. The equation (6) is obvious. The latter assertion follows from (6) and Lemma 3. ▌

The Weyl operators defined in this section relates to pseudo-differential operators defined by (2).

Lemma 5. *We define an operator* $S : C_0^\infty(\mathbf{D}\times B\times\mathbb{R}^+) \to C^\infty(\mathbf{D}\times B\times\mathbb{R}^+)$ *by*

$$\begin{aligned} &Sa(z,b,\nu) \\ &\quad = \frac{1}{(2\pi)^2} F_\nu(z,b)^{-1}\int F_\mu(z,c)F_{-\mu}(w,c)F_\nu(w,b) \\ &\qquad\qquad \times a(m(z,w),c,\mu)đwdcđ\mu. \qquad (7) \end{aligned}$$

Then for every compactly supported symbol $a(z,b,\nu)$*, we have*

$$a^{\mathrm{w}}(z,D) = (Sa)(z,D), \qquad (8)$$

where the right hand side of the above is the operator defined by (2).

Proof. Note that, If $a \in C_0^\infty(\mathbf{D}\times B\times\mathbb{R}^+)$, then for every $u \in C^\infty(\mathbf{D})$, which is not necessarily compactly supported, the function $a^{\mathrm{w}}(z,D)u$ is well defined as a smooth function. Therefore the function

$$Sa(z,b,\nu) = F_\nu(z,b)^{-1}a^{\mathrm{w}}(z,D)F_\nu(\cdot,b)$$

is smooth on $\mathbf{D}\times B\times\mathbb{R}^+$. By Theorem 1 we have

$$u(z) = (2\pi)^{-2}\int F_\nu(z,b)F_{-\nu}(w,b)u(w)đwdbđ\nu, \qquad (9)$$

for $u \in C_0^\infty(\mathbf{D})$. By the continuity of the operator $a^{\mathrm{w}}(z,D)$, we can apply this operator to u under the integral sign of (9), and hence the assertion follows. ∎

For further investigation of the Weyl operators, one may need to find the inverse of the operator S, if it exists. Note that the operator S can be represented by the operator U in [5]. In fact one has $Sa(z,b,\nu) = U_w Ma(z,w,b,\nu)|_{w=z}$, where $Ma(z,w,b,\nu) = a(m(z,w),b,\nu)$. The operator U is an unitary operator on $L^2(\mathbf{D}\times B\times\mathbb{R}^+, P(z,b)đzdbđ\nu)$ (note that the measure $P(z,b)đzdb$ is the Haar measure on $PSU(1,1)$). It seems difficult to find the inverse of the operator S from this form.

4. Wigner transform

In this section we will define an analogue of the Wigner transform by taking a formula describing a relation between the Weyl calculus and the Wigner transform on the Euclidean space into account. For simplicity we will restrict our attention to Weyl operators associated with compactly supported symbols.

Definition 3. *For every compactly supported function $u,\ v \in C_0^\infty(\mathbf{D})$, we define*

$$\begin{aligned} W_\hbar(u,v)(x,b,\nu) &= (2\pi\hbar)^{-2}P(x,b)^{-1}\int F_{\nu/\hbar}(z,b)F_{-\nu/\hbar}(p(z,x),b) \\ &\qquad \times u(p(z,x))\overline{v(z)}J(z,x)\textit{đ}z \\ &= (2\pi\hbar)^{-2}P(x,b)^{-1}\int F_{\nu/\hbar}(p(z,x),b)F_{-\nu/\hbar}(z,b) \\ &\qquad \times u(z)\overline{v(p(z,x))}J(z,x)\textit{đ}z. \end{aligned}$$

For simplicity we denote $W_\hbar(u,u)$ by $W_\hbar u$. We call the function $W_\hbar(u,v)$ the Wigner transform of u and v.

Note that the equality of the second line of the above is directly obtained by the last assertion of Lemma 1. The reason why we have defined the Wigner transform as Definition 3 is made clear by the following lemma.

Lemma 6. *For a compactly supported symbol a and for $u,\ v \in C_0^\infty(\mathbf{D})$, we have*

$$\langle\, a_\hbar^w(z,D)u, v\,\rangle = \int a(x,b,\nu)W_\hbar(u,v)(x,b,\nu)P(x,b)\textit{đ}x\,db\,\textit{đ}\nu_\hbar. \tag{10}$$

This lemma can be proved by a direct computation. The statement of this lemma is an analogue of the formula (2.5) Proposition in [1].

Lemma 7. *For every $u, v \in C_0^\infty(\mathbf{D})$, we have*

$$\overline{W_\hbar(u,v)} = W_\hbar(v,u) \tag{11}$$

In particular, the function $W_\hbar u$ is real-valued for every $u \in C_0^\infty(\mathbf{D})$.

This lemma also can be shown by a direct computation.

Originally the Wigner transform on the Euclidean spaces was defined to be the Fourier transform of the matrix coefficient $\langle\, \rho(x,\xi)u, v\,\rangle$ of the Schrödinger representation ρ on the Heisenberg group. We do not know whether or not the Wigner transform defined in Definition 3 can be represented by the matrix coefficient of some representation of a group. However, we can prove the following theorem, which is an analogue of (1.96) Proposition in [1].

Theorem 2. *For every $u,\ v \in C_0^\infty(\mathbf{D})$, we have*

$$\int W_\hbar(u,v)(x,b,\nu)P(x,b)\textit{đ}x = \hbar^{-2}\hat{u}(b,\nu/\hbar)\overline{\hat{v}(b,\nu/\hbar)}, \tag{12}$$

$$\int W_\hbar(u,v)(x,b,\nu)P(x,b)\,db\textit{đ}\nu_\hbar = u(x)\overline{v(x)}. \tag{13}$$

Proof. By the definition of the Wigner transform $W_\hbar(u,v)$, we have

$$\int W_\hbar(u,v)(x,b,\nu)P(x,b)\text{đ}x$$
$$= (2\pi\hbar)^{-2}\int F_{\nu/\hbar}(z,b)F_{-\nu/\hbar}(p(z,w),b)$$
$$\times u(p(z,w))v(z)J(z,w)\text{đ}z\text{đ}w. \qquad (14)$$

Since u and v have compact supports, we can change the order of the integral of (14), and by Lemma 1 we have (12).

Next we will show (13). If we set

$$A_\hbar(z,x,\nu) = (2\pi\hbar)^{-1}\int_B F_{\nu/\hbar}(z,b)F_{-\nu/\hbar}(p(z,x),b)\,db, \qquad (15)$$

then we have

$$\int W_\hbar(u,v)(x,b,\nu)P(x,b)\,db\text{đ}\nu_\hbar$$
$$= (2\pi\hbar)^{-1}\int A_\hbar(z,x,\nu)u(p(z,x))\overline{v(z)}J(z,x)\text{đ}z\text{đ}\nu_\hbar. \qquad (16)$$

Since

$$F_{\nu/\hbar}(z,b)F_{-\nu/\hbar}(z,b) = P(z,b),$$

we have

$$A_\hbar(z,x,\nu) = (2\pi\hbar)^{-1}\int F_{-\nu/\hbar}(p(z,x),b)F_{-\nu/\hbar}(z,b)^{-1}P(z,b)\,db. \quad (17)$$

Note that for every $g \in SU(1,1)$ we have the following formula (see [2]);

$$\frac{dg^{-1}b}{db} = P(g0,b). \qquad (18)$$

Changing the variable to $c = g(z)b$ and using (18), we obtain

$$A_\hbar(z,x,\nu)$$
$$= (2\pi\hbar)^{-1}\int F_{-\nu/\hbar}(p(z,x),g(z)^{-1}c)F_{-\nu/\hbar}(z,g(z)^{-1}c)\,dc. \quad (19)$$

To rewrite this form of the function $A_\hbar(z,x,\nu)$, we remark the following (see p. 83 [2]);

$$P(gz,b) = P(z,g^{-1}b)P(g0,b), \quad \forall g \in SU(1,1). \qquad (20)$$

From (20), we easily see

$$F_{-\nu/\hbar}(g(z)w, c)F_{-\nu/\hbar}(z, g(z)^{-1}c) = F_{-\nu/\hbar}(w, g(z)^{-1}c),$$

for every $z, w \in \mathbf{D}$. In particular, taking $w \in \mathbf{D}$ as 0, we have

$$F_{-\nu/\hbar}(z, g(z)^{-1}c)^{-1} = F_{-\nu/\hbar}(g(z)0, c).$$

Combining this with the formula (20) we obtain

$$\begin{aligned} A_\hbar(z, x, \nu) &= (2\pi\hbar)^{-1} \int F_{-\nu/\hbar}(g(z)p(z,x), c)\, dc \\ &= \hbar^{-1}\rho_{\nu/\hbar}(g(z)p(z,x)), \end{aligned}$$

where the function ρ_ν is the spherical function defined by

$$\rho_\nu(z) = (2\pi)^{-1} \int F_\nu(z, b)\, db.$$

(Notice that the function ρ_ν satisfies $\rho_\nu = \rho_{-\nu}$.) By the $SU(1,1)$–equivariance of the map p, we have $g(z)p(z,x) = p_0(g(z)x)$. Note that the function $\rho_\nu \circ p_0$ is invariant under S^1–action (as the rotation around the origin) and $|g(z)x| = |g(x)z|$ for any z, $x \in \mathbf{D}$. Combining this with Lemma 1, we have

$$\begin{aligned} &\int W_\hbar(u,v)(x,b,\nu)P(x,b)\, db đ\nu_\hbar \\ &= (2\pi\hbar)^{-1}\hbar^{-1} \int \rho_{\nu/\hbar}(z)u(p(m(x, g(x)^{-1}z), x)) \\ &\qquad\qquad \times \overline{v(m(x, g(x)^{-1}z))} đz đ\nu_\hbar \\ &= (2\pi\hbar)^{-1}\hbar^{-1} \int \rho_{\nu/\hbar}(g(w)x)\tilde{u}(w) đw đ\nu_\hbar \\ &= (2\pi)^{-1} \int \rho_\nu(g(w)x)\tilde{u}(w) đw đ\nu, \end{aligned} \tag{21}$$

where

$$\tilde{u}(w) = u(p(m(x,w),x))\overline{v(m(x,w))}. \tag{22}$$

The function $\tilde{u}$ has a compact support. The right hand side of (21) is nothing but the integral of the convolution $\tilde{u} * \rho_\nu(x)$ on $\mathbb{R}^+$ with respect to the measure $đ\nu$. By the standard argument as in [2], we have

$$\int W_\hbar(u,v)(x,b,\nu)P(x,b)\, db đ\nu_\hbar = (2\pi)^{-1} \int \tilde{u} * \rho_\nu(x) đ\nu = \tilde{u}(x). \tag{23}$$

Since $\tilde{u}(x) = u(x)\overline{v(x)}$, the assertion follows. ∎

5. Invariance for $SU(1,1)$-action

In this section, we investigate the invariance of the Weyl operators and Wigner transform for the action of $SU(1,1)$. We denote the action of $SU(1,1)$ on $L^2(\mathbf{D})$ by T, that is $T_g u(z) = u(g^{-1}z)$ for $u \in L^2(\mathbf{D})$, $g \in SU(1,1)$. We also denotes by T the action of $SU(1,1)$ on the space of functions on $\mathbf{D} \times B \times \mathbb{R}^+$, that is $T_g a(z,b,\nu) = a(g^{-1}z, g^{-1}b, \nu)$, $g \in SU(1,1)$.

First, we mention the invariance of the Weyl operator $a^{\mathrm{w}}(z,D)$.

Proposition 1. *For every symbol a and every $g \in SU(1,1)$, we have*

$$T_g a^w_\hbar(z,D) T_g^{-1} = (T_g a)^w_\hbar(z,D). \tag{24}$$

Proof. We calculate the operator $T_g a^{\mathrm{w}}_\hbar(x,D) T_{g^{-1}}$ as follows;

$$\begin{aligned} &T_g a^{\mathrm{w}}_\hbar(x,D) T_{g^{-1}} u(x) \\ &\quad = (2\pi\hbar)^{-2} \int F_{\nu/\hbar}(g^{-1}x, b) F_{-\nu/\hbar}(w,b) \\ &\qquad\qquad \times a(m(g^{-1}x, w), b, \nu)(T_{g^{-1}}u)(w) \,đw\,db\,đ\nu_\hbar \\ &\quad = (2\pi\hbar)^{-2} \int F_{\nu/\hbar}(g^{-1}x, b) F_{-\nu/\hbar}(g^{-1}z, b) \\ &\qquad\qquad \times a(g^{-1}m(x,z), b, \nu) u(z) \,đz\,db\,đ\nu_\hbar . \end{aligned}$$

Here we have changed the variable w to gw to deduce the last line from the second one. By (20) we have

$$F_{\nu/\hbar}(g^{-1}x, b) F_{-\nu/\hbar}(g^{-1}z, b) = F_{\nu/\hbar}(x, gb) F_{-\nu/\hbar}(z, gb) P(g^{-1}0, b).$$

Therefore we obtain

$$\begin{aligned} &T_g a^{\mathrm{w}}_\hbar(x,D) T_{g^{-1}} u(x) \\ &\quad = (2\pi\hbar)^{-2} \int F_{\nu/\hbar}(x, gb) F_{-\nu/\hbar}(z, gb) P(g^{-1}0, b) \\ &\qquad\qquad \times a(g^{-1}m(x,z), b, \nu) u(z) \,đz\,db\,đ\nu_\hbar . \end{aligned}$$

Using (18), we obtain

$$\begin{aligned} &T_g a^{\mathrm{w}}_\hbar(x,D) T_{g^{-1}} u(x) \\ &\quad = (2\pi\hbar)^{-2} \int F_{\nu/\hbar}(x, c) F_{-\nu/\hbar}(z, c) \\ &\qquad\qquad \times a(g^{-1}m(x,z), g^{-1}c, \nu) u(z) \,đz\,db\,đ\nu_\hbar \\ &\quad = (2\pi\hbar)^{-2} \int F_{\nu/\hbar}(x, c) F_{-\nu/\hbar}(z, c) (T_g a)(m(x,z), c, \nu) u(z) \,đz\,db\,đ\nu_\hbar , \end{aligned}$$

which completes the proof. ∎

Next we will investigate the invariance of the Wigner transform.

Lemma 8. *For every $g \in SU(1,1)$ and every $u,\ v \in C_0^\infty(\mathbf{D})$, we have*

$$T_g W_\hbar(u,v) = W_\hbar(T_g u, T_g v). \tag{25}$$

Proof. A direct computation leads us to obtain

$$\begin{aligned}(T_g W_\hbar(u,v))(x,b,\nu)& \\ = (2\pi\hbar)^{-2} P(g^{-1}x, g^{-1}b)^{-1}& \\ \times \int F_{\nu/\hbar}(g^{-1}p(w,x), g^{-1}b) F_{-\nu/\hbar}(g^{-1}w, g^{-1}b)& \\ \times u(g^{-1}w)\overline{v(g^{-1}p(gz,x))} J(w,x) đw.&\end{aligned} \tag{26}$$

By (20) we have

$$\begin{aligned} F_{\nu/\hbar}(g^{-1}p(w,x), g^{-1}b) &= F_{\nu/\hbar}(p(w,x),b) F_{\nu/\hbar}(g^{-1}0, g^{-1}b), \\ F_{-\nu/\hbar}(g^{-1}w, g^{-1}b) &= F_{-\nu/\hbar}(w,b) F_{-\nu/\hbar}(g^{-1}0, g^{-1}b), \\ P(g^{-1}x, g^{-1}b)^{-1} &= P(x,b)^{-1} P(g^{-1}0, g^{-1}b)^{-1}. \end{aligned}$$

Substituting this in (26), we conclude the assertion. ∎

6. Kernel of Weyl operators and Wigner transform

As in the case of the Euclidean spaces, one may expect that the kernel of the Weyl calculus and the Wigner transform are closely related. In this section we will describe the kernel of the Weyl operator and reformulate the Wigner transform. This makes us be aware of an analogy of the form of the kernel and the Wigner transform.

Let a be a compactly supported smooth function on $\mathbf{D} \times B \times \mathbb{R}^+$. Then the kernel $K_\hbar^{\mathrm{w}}$ of the operator $a_\hbar^{\mathrm{w}}(z,D)$ is a smooth function on $\mathbf{D} \times \mathbf{D}$ which is given by the following;

$$\begin{aligned} K_\hbar^{\mathrm{w}}(z,w)& \\ = (2\pi\hbar)^{-2} \int F_{\nu/\hbar}(z,b) F_{-\nu/\hbar}(w,b) a(m(z,w),b,\nu)\, db đ\nu_\hbar.& \end{aligned} \tag{27}$$

To rewrite (27) to more symmetric form, we prepare the following lemma. Recall that, for $z, w \in \mathbf{D}$, we denote the middle point between z and w by $m(z,w)$, and denote the middle point between 0 and z by $m_0(z)$.

Lemma 9. *We define the maps* $\alpha, \beta : \mathbf{D} \times \mathbf{D} \to \mathbf{D}$ *by*

$$\alpha(x,t) = g(x)^{-1} m_0(t), \quad \beta(x,t) = g(x)^{-1} m_0(-t), \tag{28}$$

for $(x,t) \in \mathbf{D} \times \mathbf{D}$. *Then we have*

$$m(\alpha(x,t), \beta(x,t)) = x. \tag{29}$$

Conversely, if two maps $\alpha, \beta : \mathbf{D} \times \mathbf{D} \to \mathbf{D}$ *satisfying*

$$m(\alpha(x,t), \beta(x,t)) = x, \quad g(x)\alpha(x,t) = m_0(t) \tag{30}$$

for all $(x,t) \in \mathbf{D} \times \mathbf{D}$ *is given by (28).*

Proof. First assertion (29) follows from the equivariance of m. Note that, if the two points $z, w \in \mathbf{D}$ satisfy $m(z,w) = 0$, then we have $w = -z$. Therefore by (30) we have

$$g(x)\alpha(x,t) = -g(x)\beta(x,t) = m_0(t).$$

The second assertion follows from this and Lemma 2. ∎

Proposition 2. *The kernel* $K_\hbar^{\mathrm{w}}$ *of the operator* $a_\hbar^{\mathrm{w}}(z,D)$ *for* $a \in C_0^\infty(\mathbf{D} \times B \times \mathbb{R}^+)$ *has the following expression;*

$$\begin{aligned} &K_\hbar^{\mathrm{w}}(\alpha(x,z), \beta(x,z)) \\ &\quad = (2\pi\hbar)^{-2} \int F_{\nu/\hbar}(m_0(z), b) F_{-\nu/\hbar}(-m_0(z), b) \\ &\qquad\qquad\qquad \times a(x, g(x)^{-1}b, \nu)\, db\, đ\nu. \end{aligned}$$

Proof. For $t \in \mathbf{D}$, we set $s = m_0(z)$. By the formula (20), we have

$$\begin{aligned} &F_{\nu/\hbar}(g(x)^{-1}s, b) F_{-\nu/\hbar}(g(x)^{-1}(-s), b) \\ &\qquad = F_{\nu/\hbar}(s, g(x)b) F_{-\nu/\hbar}(-s, g(x)b) P(x,b). \end{aligned}$$

Then the assertion follows from a direct computation by using this fact and Lemma 9. ∎

Corollary *The trace* $\mathrm{Tr}(a_\hbar^{\mathrm{w}}(z,D))$ *of the operator* $a_\hbar^{\mathrm{w}}(z,D)$ *is given by*

$$\mathrm{Tr}(a_\hbar^{\mathrm{w}}(z,D)) = (2\pi)^{-2} \int a(x, c, \hbar\nu) P(x,c)\, đx\, dc\, đ\nu.$$

Proof. We note that the trace $\mathrm{Tr}(a_\hbar^{\mathrm{w}}(z,D))$ is given by the integral of the function $K_\hbar^{\mathrm{w}}(x,x)$ on $\mathbf{D}$. Note also that $\alpha(x,0)=\beta(x,0)=x$. Therefore the assertion follows from these facts and the formula (18). ∎

Before rewriting the Wigner transform, we will define a transform $\widetilde{W}_\hbar$ of smooth functions on $\mathbf{D}\times\mathbf{D}$. For $K\in C^\infty(\mathbf{D}\times\mathbf{D})$, we set

$$\begin{aligned}\widetilde{W}_\hbar K(x,b,\nu) &\\ = (2\pi\hbar)^{-2}P(x,b)^{-1}\int & F_{\nu/\hbar}(z,b)F_{-\nu/\hbar}(p(z,x),b)\\ &\times K(p(z,x),z)J(z,x)\,đz, \qquad (31)\end{aligned}$$

if the integral exists. Note that we have $W_\hbar(u,v)=\widetilde{W}_\hbar K$ for $K(z,w)=u(z)\overline{v(w)}$.

Proposition 3. *For* $K\in C_0^\infty(\mathbf{D}\times\mathbf{D})$, *we have*

$$\begin{aligned}\widetilde{W}_\hbar K(x,g(x)^{-1}b,\nu) &\\ = (2\pi\hbar)^{-2}\int & F_{\nu/\hbar}(-m_0(z),b)F_{-\nu/\hbar}(m_0(z),b)\\ &\times K(\alpha(x,z),\beta(x,z))\,đz, \qquad (32)\end{aligned}$$

where $\alpha(x,z)$ *and* $\beta(x,z)$ *is the maps defined in Lemma 9.*

Proof. We note that by Lemma 2 we have

$$p(z,x)=-g(x)^2z=g(x)^{-1}(-g(x)z).$$

By (20) we have

$$\begin{aligned}F_{\nu/\hbar}(p(z,x),b) &= F_{\nu/\hbar}(-g(x)z,g(x)b)F_{\nu/\hbar}(x,b),\\ F_{-\nu/\hbar}(z,b) &= F_{-\nu/\hbar}(g(x)z,g(x)b)F_{-\nu/\hbar}(x,b).\end{aligned}$$

Hence we obtain

$$\begin{aligned}&F_{\nu/\hbar}(p(z,x),b)F_{-\nu/\hbar}(z,b)\\ &\quad = F_{\nu/\hbar}(-g(x)z,g(x)b)F_{-\nu/\hbar}(g(x)z,g(x)b)P(x,b).\end{aligned}$$

From this the assertion follows. ∎

The above two propositions make us to consider the operator A from $C_0^\infty(\mathbf{D})$ to $C^\infty(B\times\mathbb{R}^+)$ defined by

$$Au(b,\nu)=(2\pi)^{-1}\int F_\nu(-m_0(z),b)F_{-\nu}(m_0(z),b)u(z)\,đz. \qquad (33)$$

In the Euclidean case the operator corresponding to the operator A is nothing but the Fourier transform, since $e^{-ix\xi/2}e^{-ix\xi/2} = e^{-ix\xi}$. But in the non-Euclidean case the operator A might be different from the Fourier transform. However, the first term of the Fourier coefficient of Au with respect to b coincides with that of the Fourier transform $\hat{u}$. Namely, we have the following.

Proposition 4. *For every $u \in C_0^\infty(\mathbf{D})$ we have*

$$\int_B Au(b,\nu)\,db = \int_B \hat{u}(b,\nu)\,db. \tag{34}$$

Proof. As in the proof of Theorem 2 we set

$$\rho_\nu(z) = (2\pi)^{-1}\int_B F_\nu(z,b)\,db,$$

which is a spherical function on $\mathbf{D}$ satisfying $\rho_\nu = \rho_{-\nu}$. For every $g \in SU(1,1)$, we have the following formula ([2]);

$$\rho_\nu(g^{-1}z) = (2\pi)^{-1}\int_B F_{-\nu}(z,b)F_\nu(g0,b)\,db.$$

Taking $g(z)$ as g in the above formula we obtain

$$\rho_\nu(g(z)^{-1}z) = \rho_\nu(p(0,z)) = (2\pi)^{-1}\int_B F_{-\nu}(z,b)F_\nu(-z,b)\,db.$$

Replacing z by $m_0(z)$ in this expression we have

$$\rho_\nu(z) = (2\pi)^{-1}\int_B F_\nu(-m_0(z),b)F_{-\nu}(m_0(z),b)\,db.$$

Therefore for every $u \in C_0^\infty(\mathbf{D})$ we obtain

$$\begin{aligned}
&\int_B Au(b,\nu)\,db \\
&\quad = (2\pi)^{-1}\int F_\nu(-m_0(z),b)F_{-\nu}(m_0(z),b)u(z)\,đz\,db \\
&\quad = \int \rho_\nu(z)u(z)\,đz \\
&\quad = \int \hat{u}(b,\nu)\,db,
\end{aligned}$$

which completes the proof. ▌

References

[1] G. Folland, *Harmonic analysis in phase space,* Princeton Univ. Press, 1989.

[2] S. Helgason, *Topics in Harmonic Analysis on Homogeneous Spaces,* Birkhauser, Boston, 1981.

[3] M. Taylor, *Noncommutative harmonic analysis,* Mathematical surveys and monograph, No. 22, AMS.

[4] A. Weinstein, Traces and Triangles in Symmetric Symplectic Spaces, *Contemporary Math.* **179** (1994), 261–270.

[5] S. Zelditch, Pseudo-differential Analysis on Hyperbolic Surfaces, *J. Funct. Anal.* **68** (1986), 72–105.

LECTURES ON GRADED DIFFERENTIAL ALGEBRAS AND NONCOMMUTATIVE GEOMETRY

Michel Dubois-Violette
L.P.T.-ORSAY, Bâtiment 210, Université Paris XI,
91405 ORSAY Cedex, FRANCE
patricia@osiris.th.u-psud.fr

Abstract These notes contain a survey of some aspects of the theory of graded differential algebras and of noncommutative differential calculi as well as of some applications connected with physics. They also give a description of several new developments.

Mathematics Subject Classification (2000): 81T20, 81T70, 81T75

Keywords: graded differential algebras; categories of algebras; bimodules; noncommutative differential calculus; noncommutative symplectic geometry.

1. Introduction

The correspondence between 'spaces' and 'commutative algebras' is by now familiar in mathematics and in theoretical physics. This correspondence allows an algebraic translation of various geometrical concepts on spaces in terms of the appropriate algebras of functions on these spaces. Replacing these commutative algebras by noncommutative algebras, i.e., forgetting commutativity, leads then to noncommutative generalizations of geometries where notions of 'spaces of points' are not involved. Such a noncommutative generalization of geometry was a need in physics for the formulation of quantum theory and the understanding of its relations with classical physics. In fact, the relation

Y. Maeda, et al. (eds.), Noncommutative Differential Geometry and Its Applications to Physics, 245–306.

between spectral theory and geometry was implicitly understood very early in physics.

Gel'fand's transformation associates to each compact topological space X the algebra $C(X)$ of complex continuous functions on X. Equipped with the sup norm, $C(X)$ is a commutative unital C^*-algebra. One of the main points of Gel'fand theory is that *the correspondence $X \mapsto C(X)$ defines an equivalence between the category of compact topological spaces and the category of commutative unital C^*-algebras.* The compact space X is then identified to the spectrum of $C(X)$, (i.e., to the set of homomorphisms of unital $*$-algebras of $C(X)$ into $\mathbb{C}$ equipped with the weak topology). Let X be a compact space and let $\mathcal{E}(X)$ denote the category of finite rank complex vector bundles over X. To any vector bundle E of $\mathcal{E}(X)$ one can associate the $C(X)$-module $\Gamma(E)$ of all continuous sections of E. The module $\Gamma(E)$ is a finite projective $C(X)$ module and the Serre–Swan theorem asserts that *the correspondence $E \mapsto \Gamma(E)$ defines an equivalence between the category $\mathcal{E}(X)$ and the category $\mathcal{P}(C(X))$ of finite projective $C(X)$-modules.* Thus the compact spaces and the complex vector bundles over them can be replaced by the commutative unital C^*-algebras and the finite projective modules over them. In this sense noncommutative unital C^*-algebras provide 'noncommutative generalizations' of compact spaces whereas the notion of finite projective right module over them is a corresponding generalization of the notion of complex vector bundle. It is worth noticing here that for the latter generalization one can use as well left modules but these are not the only possibilities (see below) and that something else has to be used for the generalization of the notion of real vector bundle.

Remark 1. Let X be an arbitrary topological space, then the algebra $C^{\mathrm{b}}(X)$ of complex continuous bounded functions on X is a C^*-algebra if one equips it with the sup norm. In view of Gel'fand theory one has $C^{\mathrm{b}}(X) = C(\widehat{X})$ (as C^*-algebras), where $\widehat{X}$ denotes the spectrum of $C^{\mathrm{b}}(X)$. The spectrum $\widehat{X}$ is a compact space and the evaluation defines a continuous mapping $e : X \to \widehat{X}$ with dense image ($\overline{e(X)} = \widehat{X}$). The compact space $\widehat{X}$ is called *the Stone-Čech compactification of* X and the pair $(e, \widehat{X})$ is characterized (uniquely up to an isomorphism) by the following universal property: *For any continuous mapping $f : X \to Y$ of X into a compact space Y there is a unique continuous mapping $\hat{f} : \widehat{X} \to Y$ such that $f = \hat{f} \circ e$.* Notice that $e : X \to \widehat{X}$ is generally not injective and that it is an isomorphism, i.e., $X = \widehat{X}$, if and only if X is compact. The above universal property means that $\widehat{X}$ is the biggest compactification of X . For instance if X is locally compact then e is injective, i.e., $X \subset \widehat{X}$ canonically, but $\widehat{X}$ is generally much bigger than the one point compactification $X \cup \{\infty\}$ of X, (e.g., for $X = \mathbb{R}$ the canonical projection $\widehat{\mathbb{R}} \to \mathbb{R} \cup \{\infty\}$ has a huge inverse image of ∞).

If instead of (compact) topological spaces one is interested in the geometry of measure spaces, what replaces algebras of continuous functions are of course algebras of measurable functions. In this case the class of algebras is the class of commutative W^*-algebras (or von Neumann algebras). The noncommutative generalizations are therefore provided by general (noncommutative) W^*-algebras. It has been shown by A. Connes that the corresponding noncommutative measure theory (i.e., the theory of von Neumann algebras) has a very rich structure with no classical (i.e., commutative) counterpart (e.g., the occurrence of a canonical dynamical system) [12].

In the case of differential geometry, it is more or less obvious that the appropriate class of commutative algebras are algebras of smooth functions. Indeed if X is a smooth manifold and if $\mathcal{C}$ is the algebra of complex smooth function on X, ($\mathcal{C} = C^\infty(X)$), one can reconstruct X with its smooth structure and the objects attached to X (differential forms, etc.) by starting from $\mathcal{C}$ considered as an abstract (commutative) unital $*$-algebra. As a set X can be identified with the set of characters of $\mathcal{C}$, i.e., with the set of homomorphisms of unital $*$-algebras of $\mathcal{C}$ into $\mathbb{C}$; its differential structure is connected with the abundance of derivations of $\mathcal{C}$ which identify with the smooth vector fields on X as well known. In fact, in [50] J.L. Koszul gave a powerful algebraic generalization of differential geometry in terms of a commutative (associative) algebra $\mathcal{C}$, of $\mathcal{C}$-modules and connections (called derivation laws there) on these modules. For the applications to differential geometry, $\mathcal{C}$ is of course the algebra of smooth functions on a smooth manifold and the $\mathcal{C}$-modules are modules of smooth sections of smooth vector bundles over the manifold.

In this approach what generalizes the vector fields are the derivations of $\mathcal{C}$ (into itself). The space $\mathrm{Der}(\mathcal{C})$ of all derivations of $\mathcal{C}$ is a Lie algebra and a $\mathcal{C}$-module, both structures being connected by $[X, fY] = f[X,Y] + X(f)Y$ for $X, Y \in \mathrm{Der}(\mathcal{C})$ and $f \in \mathcal{C}$. Using the latter property one can extract (by $\mathcal{C}$-multilinearity) a graded differential algebra generalizing the algebra of differential forms, from the graded differential algebra $C_\wedge(\mathrm{Der}(\mathcal{C}), \mathcal{C})$ of $\mathcal{C}$-valued Chevalley–Eilenberg cochains of the Lie algebra $\mathrm{Der}(\mathcal{C})$ (with its canonical action on $\mathcal{C}$). This construction admits a generalization to the noncommutative case; it is the *derivation-based* differential calculus ([25], [26], [27],[34] [35]) which will be described below. As will be explained (see also [26] and [27]) this is the right differential calculus for quantum mechanics, in particular we shall show that the corresponding noncommutative symplectic geometry is exactly what is needed there.

For commutative algebras there is another well known generalization of the calculus of differential forms which is the Kähler differential calculus [6], [43], [52], [58]. This differential calculus is 'universal' and consequently functorial for the category of (associate unital) commutative algebras. In these lectures we shall give a generalization of the Kähler differential calculus for the noncommutative algebras. By its very construction this differential calculus will be functorial for the algebra-homomorphisms mapping the centers into the centers. More precisely this differential calculus will be shown to be the universal differential calculus for the category of algebra $\mathbf{Alg}_Z$ whose objects are the unital associative $\mathbb{C}$-algebras and whose morphisms are the homomorphisms of unital algebras mapping the centers into the centers. This differential calculus generalizes the Kähler differential calculus in the sense that it reduces to it for a commutative (unital associative $\mathbb{C}$) algebra. This latter property is in contrast with what happens for the so called *universal differential calculus*, which is universal for the category $\mathbf{Alg}$ of unital associative $\mathbb{C}$-algebras and of *all* unital algebra-homomorphisms, the construction of which will be recalled in these lectures.

Concerning the generalizations of the notion of module over a commutative algebra $\mathcal{C}$ when one replaces it by a noncommutative algebra $\mathcal{A}$, there are the notion of right $\mathcal{A}$-module and the dual notion of left $\mathcal{A}$-module, but since a module over a commutative algebra is also canonically a bimodule (of a certain kind) and since a commutative algebra coincides with its center, there is a notion of bimodule over $\mathcal{A}$ and also the notion of module over the center $Z(\mathcal{A})$ of $\mathcal{A}$ which are natural. The 'good choices' depend on the kind of problems involved. Again categorical notions can be of some help. As will be explained in these lectures, for each category of algebras there is a notion of bimodule over the objects of the category. Furthermore, for the category $\mathbf{Algcom}$ of unital commutative associative $\mathbb{C}$-algebras the notion of bimodule just reduces to the notion of module. Again, as for the universal differential calculus, for the notion of bimodule it is immaterial for a commutative algebra $\mathcal{C}$ whether one considers $\mathcal{C}$ as an object of $\mathbf{Algcom}$ or of $\mathbf{Alg}_Z$ whereas the notion of bimodule over $\mathcal{C}$ in $\mathbf{Alg}$ is much wider.

This problem of the choice of the generalization of the notion of module over a commutative algebra $\mathcal{C}$ when $\mathcal{C}$ is replaced by a noncommutative algebra $\mathcal{A}$ is closely connected with the problem of the noncommutative generalization of the classical notion of reality. If $\mathcal{C}$ is the algebra of complex continuous functions on a topological space or the algebra of complex smooth functions on a smooth manifold, then it is a $*$-algebra and the (real) algebra of real functions is the real subspace $\mathcal{C}^h$ of Hermitian (i.e., $*$-invariant) elements of $\mathcal{C}$. More generally if $\mathcal{C}$ is a commutative associative complex $*$-algebra the set $\mathcal{C}^h$ of Hermitian elements of $\mathcal{C}$ is a commutative associative real algebra. Conversely if $\mathcal{C}_{\mathbb{R}}$ is a commutative associative real algebra then its complexification $\mathcal{C}$ is canonically a commutative associative complex $*$-algebra, and one has $\mathcal{C}^h = \mathcal{C}_{\mathbb{R}}$. In fact, the correspondence $\mathcal{C} \mapsto \mathcal{C}^h$ defines an equivalence between the category of commutative associative complex $*$-algebras and the category of commutative associative real algebras (the morphisms of the first category being the $*$-homomorphisms). This is in contrast with what happens for noncommutative algebras. Recall that an associative complex $*$-algebra is an associative complex algebra $\mathcal{A}$ equipped with an antilinear involution $x \mapsto x^*$ such that $(xy)^* = y^*x^*$, $(\forall x, y \in \mathcal{A})$. From the fact that the involution reverses the order of the product it follows that the real subspace $\mathcal{A}^h$ of Hermitian elements of a complex associative $*$-algebra is generally not stable under the product but only by the symmetrized Jordan product $x \circ y = \frac{1}{2}(xy + yx)$. Thus $\mathcal{A}^h$ is not (generally) an associative algebra but is a real *Jordan algebra*. Therefore, one has two natural choices for the generalization of an algebra of real functions : either the real Jordan algebra $\mathcal{A}^h$ of Hermitian elements of a complex associative $*$-algebra $\mathcal{A}$ which plays the role of the algebra of complex functions or a real associative algebra. In these lectures we take the first choice which is dictated by quantum theory (and spectral theory). This choice has important consequences on the possible generalizations of real vector bundles and, more generally, of modules over commutative real algebras.

Let $\mathcal{C}$ be a commutative associative $*$-algebra and let $\mathcal{M}^h$ be a $\mathcal{C}^h$-module. The complexified $\mathcal{M} = \mathcal{M}^h \oplus i\mathcal{M}^h = \mathcal{M}^h \otimes_{\mathbb{R}} \mathbb{C}$ of $\mathcal{M}^h$ is canonically a $\mathcal{C}$-module. Furthermore there is a canonical antilinear involution $(\Phi + i\Psi) \mapsto (\Phi + i\Psi)^* = \Phi - i\Psi$ $(\Phi, \Psi \in \mathcal{M}^h)$ for which $\mathcal{M}^h$ is the set of $*$-invariant elements. This involution is compatible with the one of $\mathcal{C}$ in the sense that one has $(x\Phi)^* = x^*\Phi^*$ for $x \in \mathcal{C}$ and $\Phi \in \mathcal{M}$; $\mathcal{M}$ will be said to be a $*$-*module over the commutative* $*$-*algebra* $\mathcal{C}$. In view of the above discussion what generalizes $\mathcal{C}$ is a noncommutative $*$-algebra $\mathcal{A}$ and we have to generalize the $*$-module $\mathcal{M}$ and its 'real part' $\mathcal{M}^h$. However, it is clear that there is no noncommutative generalization of a $*$-module over $\mathcal{A}$ as right or left module. The reason is that since the involution of $\mathcal{A}$ reverses the order in products, it intertwines between actions of $\mathcal{A}$ and actions of the opposite algebra $\mathcal{A}^0$, i.e., between a structure of right (resp., left) module and a structure of left (resp., right) module. Fortunately, as already mentioned, a $\mathcal{C}$-module is canonically a bimodule (of a certain kind) and the above compatibility condition can be equivalently written $(x\Phi)^* = \Phi^* x^*$. This latter condition immediately generalizes for $\mathcal{A}$, namely a $*$-*bimodule over the* $*$-*algebra* $\mathcal{A}$ is a bimodule $\mathcal{M}$ over $\mathcal{A}$ equipped with an antilinear involution $\Phi \mapsto \Phi^*$ such that $(x\Phi y)^* = y^*\Phi^*x^*$, $(\forall x, y \in \mathcal{A}$,

$\forall \Phi \in \mathcal{M}$). The real subspace $\mathcal{M}^h = \{\Phi \in \mathcal{M} | \Phi^* = \Phi\}$ of the $*$-invariant element of $\mathcal{M}$ can play the role of the sections of a real vector bundle (for some specific kind of $*$-bimodule $\mathcal{M}$). Since a commutative algebra is its center, one can also generalize $*$-modules over $\mathcal{C}$ by $*$-modules over the center $Z(\mathcal{A})$ of $\mathcal{A}$ and modules over $\mathcal{C}^h$ by modules over $Z(\mathcal{A})^h$. In a sense these two types of generalizations of the reality (for modules) are dual ([34], [27]) as we shall see later. The main message of this little discussion is that notions of reality force us to consider bimodules and not only right or left modules as generalization of vector bundles, [34], [27], [18], [61].

Remark 2. One can be more radical. Instead of generalizing an associative commutative $\mathbb{R}$-algebra $\mathcal{C}_\mathbb{R}$ by the Jordan algebra $\mathcal{A}^h$ of Hermitian elements of an associative complex $*$-algebra $\mathcal{A}$, one can more generally choose to generalize $\mathcal{C}_\mathbb{R}$ by a real Jordan algebra $\mathcal{J}_\mathbb{R}$ (not *a priori* a special one). The corresponding generalization of a $\mathcal{C}_\mathbb{R}$-module could then be a *Jordan bimodule over* $\mathcal{J}_\mathbb{R}$ [44] instead of the real subspace of a $*$-bimodule over $\mathcal{A}$, (what is a Jordan bimodule will be explained later). We, however, refrain from doing that because it is relatively complicated technically for a slight generalization practically.

In these lectures we shall be interested in noncommutative versions of differential geometry where the algebra of smooth complex functions on a smooth manifold is replaced by a noncommutative associative unital complex $*$-algebra $\mathcal{A}$. Since there are commutative $*$-algebras of this sort which are not (and cannot be) algebras of smooth functions on smooth manifolds, one cannot expect that an arbitrary $*$-algebra as above is a good noncommutative generalization of an algebra of smooth functions. What is involved here is the generalization of the notion of smoothness. It is possible to characterize among the unital commutative associative complex $*$-algebras the ones which are isomorphic to algebras of smooth functions; however, there are several inequivalent noncommutative generalizations of this characterization and no one is universally accepted. Thus although it is an interesting subject on which work is currently in progress [30], we shall not discuss it here. This means that if the algebra $\mathcal{A}$ is not 'good enough', some of our constructions can become a little trivial.

The plan of these notes is the following. After this introduction, in Section 2 we recall the definition of graded differential algebras and of various concepts related to them; we state in particular the result of D. Sullivan concerning the structure of connected finitely generated free graded commutative differential algebras and we review H. Cartan's notion of operation of a Lie algebra in a graded differential algebra. In Section 3 we explain the equivalence between the category of finite-dimensional Lie algebras and the category of the free connected graded commutative differential algebras which are finitely generated in degree 1 (i.e., exterior algebras of finite-dimensional spaces equipped with differentials); we describe several examples related to Lie algebras such as the Chevalley–Eilenberg complexes, the Weil algebra (and we state the result defining the Weil homomorphism) and we introduce the graded differential algebras of the derivation-based calculus. In Section 4 we start in an analogous way as in Section 3, that is, we explain the equivalence between the category of finite-dimensional associative algebras and the category of free connected graded differential algebras which are generated in degree 1 (i.e., tensor algebras of finite-dimensional spaces equipped with differentials); we describe examples related to associative algebras such as Hochschild complexes. In Section 5, we introduce categories of algebras and we define the associated notions of bimodules which we follow on several relevant examples. In Section 6 we recall the notion of first order differential calculus over an algebra, and we introduce our generalization of the module of Kähler differentials and discuss its functorial properties; we also recall in this section the definition and properties of the universal first order calculus. In Section 7 we introduce the higher order differential calculi and discuss, in particular, the universal one as well as our generalization of Kähler exterior forms; we give, in particular, their universal properties and study their functorial properties. In Section 8 we introduce another new differential calculus, the diagonal calculus, which, although not functorial, is characterized by a universal property, and we compare it with the other differential calculi attached to an algebra. In Section 9 we define and study noncommutative Poisson and symplectic structures and show their relation with quantum theory. In Section 10 we describe the theory of connections on modules and on bimodules; in the latter case we recall, in particular, the generalization of the proposal of J. Mourad (concerning linear connections) and describe its basic properties and its relations with the theory of first-order operators in bimodules. In Section 11 we discuss in some examples the relations between connections in the noncommutative setting and classical Yang–Mills–Higgs models. Section 12, which serves as the conclusion, contains some further remarks concerning, in particular, the differential calculus on quantum groups.

Apart from in §5, an *algebra* without other specification will always mean a unital associative complex algebra, and by a $*$-*algebra* without other specification we shall mean a unital associative complex $*$-algebra. Given two algebras $\mathcal{A}$ and $\mathcal{B}$ in this sense, a $(\mathcal{A},\mathcal{B})$-*bimodule* is a vector space $\mathcal{M}$ equipped with linear maps $\mathcal{A}\otimes\mathcal{M}\rightarrow\mathcal{M}$ and $\mathcal{M}\otimes\mathcal{B}\rightarrow\mathcal{M}$ denoted by $a\otimes m\mapsto am$ and $m\otimes b\mapsto mb$, respectively, such that $(aa')m=a(a'm)$, $m(bb')=(mb)b'$, $(am)b=a(mb)$, $\mathbb{1}m=m$ and $m\mathbb{1}=m$, $\forall a,a'\in\mathcal{A}$, $\forall b,b'\in\mathcal{B}$, $\forall m\in\mathcal{M}$ where $\mathbb{1}$ denotes the unit of $\mathcal{A}$ as well as the one of $\mathcal{B}$. In Section 5 we shall define for a more general algebra $\mathcal{A}$ a notion of $\mathcal{A}$-bimodule which is relative to a category of algebras; the notion of $(\mathcal{A}, \mathcal{A})$-bimodule as above is the notion of $\mathcal{A}$-bimodule for the category **Alg** of unital associative complex algebras. A *complex* $\mathfrak{C}$ will be a $\mathbb{Z}$-graded vector space (over $\mathbb{C}$) equipped with a homogeneous endomorphism d of degree ± 1 and such that $d^2 = 0$. If d is of degree -1, $\mathfrak{C}$ is said to be a *chain complex*, its elements are called *chains*, and d is called the *boundary*; if d is of degree $+1$, $\mathfrak{C}$ is said to be a *cochain complex*, its elements are called *cochains*, and d is called the *coboundary*. The graded vector space $H(\mathfrak{C}) = \mathrm{Ker}(d)/\mathrm{Im}(d)$ is called the *homology* of $\mathfrak{C}$ if $\mathfrak{C}$ is a chain complex and the *cohomology* of $\mathfrak{C}$ if $\mathfrak{C}$ is a cochain complex.

2. Graded differential algebras

A *graded algebra* will here be a unital associative complex algebra $\mathfrak{A}$ which is a $\mathbb{Z}$-graded vector space $\mathfrak{A} = \oplus_{n\in\mathbb{Z}} \mathfrak{A}^n$ such that $\mathfrak{A}^m.\mathfrak{A}^n \subset \mathfrak{A}^{m+n}$. A *homomorphism of graded algebras* will be a homomorphism of the corresponding graded vector spaces (i.e., a homogeneous linear mapping of degree 0) which is also a homomorphism of unital algebras. A graded algebra $\mathfrak{A}$ is said to be *graded commutative* if one has $xy = (-1)^{mn}yx$, $\forall x \in \mathfrak{A}^m$ and $\forall y \in \mathfrak{A}^n$. Most graded algebras involved in these lectures will be $\mathbb{N}$-graded, i.e., $\mathfrak{A}^n = 0$ for $n \leq -1$. A graded algebra $\mathfrak{A}$ is said to be 0-*connected* or *connected* if it is $\mathbb{N}$-graded with $\mathfrak{A}^0 = \mathbb{C}\mathbf{1}$, where $\mathbf{1}$ denotes the unit of $\mathfrak{A}$. An example of connected graded algebra is the tensor algebra over $\mathbb{C}$ of a complex vector space E which will be denoted by $T(E)$. In this example the graduation is the tensorial degree, which means that the degree 1 is given to the elements of E. The exterior algebra $\bigwedge(E)$ of E is an example of connected graded commutative algebra (the graduation being again induced by the tensorial degree).

More generally, let $C = \oplus_n C^n$ be a $\mathbb{Z}$-graded complex vector space and let $T(C)$ be the tensor algebra of C. One has $C \subset T(C)$ and we equip the algebra $T(C)$ with the unique grading of algebra which induces on C the original grading. Since this is not the usual grading of the tensor algebra we shall denote the corresponding graded algebra by $\mathfrak{T}(C)$. The graded algebra $\mathfrak{T}(C)$ is characterized (uniquely up to an isomorphism) by the following universal property: *Any homomorphism of graded vector spaces $\alpha : C \to \mathfrak{A}$ of the graded vector space C into a graded algebra $\mathfrak{A}$ extends uniquely as a homomorphism of graded algebras $\mathfrak{T}(\alpha) : \mathfrak{T}(C) \to \mathfrak{A}$.* Let $\mathcal{I}$ be the graded two-sided ideal of $\mathfrak{T}(C)$ generated by the graded commutators $\psi_r \otimes \varphi_s - (-1)^{rs}\varphi_s \otimes \psi_r$ with $\psi_n, \varphi_n \in C^n$ and let $\mathfrak{F}(C)$ denote the quotient graded algebra $\mathfrak{T}(C)/\mathcal{I}$. Then $\mathfrak{F}(C)$ is a graded commutative algebra which contains again C as graded subspace. The graded commutative algebra $\mathfrak{F}(C)$ is characterized (uniquely up to an isomorphism) by the following universal property, (which is the graded commutative counterpart of the above one): *Any homomorphism of graded vector spaces* $\alpha : C \to \mathfrak{A}$ *of the graded vector space* C *into a graded commutative algebra* $\mathfrak{A}$ *extends uniquely as a homomorphism of graded commutative algebras* $\mathfrak{F}(\alpha) : \mathfrak{F}(C) \to \mathfrak{A}$. Notice that $\mathfrak{T}(C)$ (resp., $\mathfrak{F}(C)$) is connected if and only if $C^n = 0$ for $n \leq 0$ and that $\mathfrak{T}(C) = T(C)$ (resp., $\mathfrak{F}(C) = \bigwedge(C)$) as graded algebras if and only if $C^n = 0$ for $n \neq 1$. Notice also that, as algebra $\mathfrak{F}(C) = \bigwedge(\oplus_r C^{2r+1}) \otimes S(\oplus_s C^{2s})$ where $S(E)$ denotes the symmetric algebra of the vector space E. The graded algebra $\mathfrak{T}(C)$ will be referred to as *the free graded algebra generated by the graded vector space* C, whereas the graded algebra $\mathfrak{F}(C)$ will be referred to as *the free graded commutative algebra generated by the graded vector space* C. Finally, a *finitely generated free graded algebra* will be a graded algebra of the form $\mathfrak{T}(C)$ for some finite-dimensional graded vector space C, whereas an algebra of the form $\mathfrak{F}(C)$ for some finite-dimensional graded vector space C will be called a *finitely generated free graded commutative algebra*.

If $\mathfrak{A}$ and $\mathfrak{A}'$ are two graded algebras their *tensor product*, $\mathfrak{A} \otimes \mathfrak{A}'$ will here be their *skew tensor product* which means that the product in $\mathfrak{A} \otimes \mathfrak{A}'$ is defined by $(x \otimes x')(y \otimes y') = (-1)^{m'n} xy \otimes x'y'$ for $x' \in \mathfrak{A}'^{m'}$, $y \in \mathfrak{A}^n$, $x \in \mathfrak{A}$ and $y' \in \mathfrak{A}'$. With this convention the tensor product of two (or more) graded commutative algebras is again a graded commutative algebra. If C and C' are $\mathbb{Z}$-graded complex vector spaces one has $\mathfrak{F}(C \oplus C') = \mathfrak{F}(C) \otimes \mathfrak{F}(C')$.

By a *graded* $*$*-algebra* we here mean a graded algebra $\mathfrak{A} = \oplus_n \mathfrak{A}^n$ equipped with an involution $x \mapsto x^*$ satisfying:

(i) $x \in \mathfrak{A}^n \Rightarrow x^* \in \mathfrak{A}^n$ (homogeneity of degree = 0)

(ii) $(\lambda x + y)^* = \bar{\lambda} x^* + y^*$, $\forall x, y \in \mathfrak{A}$ and $\forall \lambda \in \mathbb{C}$ (antilinearity)

(iii) $(xy)^* = (-1)^{mn} y^* x^*$, $\forall x \in \mathfrak{A}^m$ and $\forall y \in \mathfrak{A}^n$.

Notice that Property (iii) implies that if $\mathfrak{A}$ is graded commutative then one has $(xy)^* = x^* y^*$, ($\forall x, y \in \mathfrak{A}$).

For a graded algebra $\mathfrak{A}$ there is, besides the notion of derivation, the notion of antiderivation: A linear mapping $\theta : \mathfrak{A} \to \mathfrak{A}$ is called an *antiderivation of* $\mathfrak{A}$ if it satisfies $\theta(xy) = \theta(x)y + (-1)^m x\theta(y)$ for any $x \in \mathfrak{A}^m$ and $y \in \mathfrak{A}$. However, the best generalizations of the notions of center and of derivations are the following graded generalizations. The *graded center* $Z_{\text{gr}}(\mathfrak{A})$ of $\mathfrak{A}$ is the graded subspace of $\mathfrak{A}$ generated by the homogeneous elements $x \in \mathfrak{A}^m$ ($m \in \mathbb{Z}$) satisfying $xy = (-1)^{mn} yx$, $\forall y \in \mathfrak{A}^n$ and $\forall n \in \mathbb{Z}$, (i.e., $Z_{\text{gr}}(\mathfrak{A})$ is the graded commutant of $\mathfrak{A}$ in $\mathfrak{A}$). The graded center is a graded subalgebra of $\mathfrak{A}$ which is graded commutative. A *graded derivation of degree k of $\mathfrak{A}$*, ($k \in \mathbb{Z}$), is a homogeneous linear mapping $X : \mathfrak{A} \to \mathfrak{A}$ which is of degree k and satisfies $X(xy) = X(x)y + (-1)^{km} xX(y)$ for $x \in \mathfrak{A}^m$ and $y \in \mathfrak{A}$. Thus a homogeneous graded derivation of even (resp., odd) degree is a derivation (resp., antiderivation). The vector space of all these graded derivations of degree k will be denoted by $\mathrm{Der}^k_{\mathrm{gr}}(\mathfrak{A})$ and the graded vector space $\mathrm{Der}_{\mathrm{gr}}(\mathfrak{A}) = \oplus_{k\in\mathbb{Z}} \mathrm{Der}^k_{\mathrm{gr}}(\mathfrak{A})$ of all graded derivations is a graded Lie algebra for the *graded commutator* $[X,Y]_{\mathrm{gr}} = XY - (-1)^{k\ell} YX$, $X \in \mathrm{Der}^k_{\mathrm{gr}}(\mathfrak{A})$, $Y \in \mathrm{Der}^\ell_{\mathrm{gr}}(\mathfrak{A})$. If $x \in \mathfrak{A}^m$, one defines a graded derivation of degree m of $\mathfrak{A}$, denoted by $\mathrm{ad}_{\mathrm{gr}}(x)$, by setting $\mathrm{ad}_{\mathrm{gr}}(x)y = xy - (-1)^{mn} yx = [x,y]_{\mathrm{gr}}$ for $y \in \mathfrak{A}^n$. The graded subspace of $\mathrm{Der}_{\mathrm{gr}}(\mathfrak{A})$ generated by these $\mathrm{ad}(x)$ (when x runs over $\mathfrak{A}^m$ and m runs over $\mathbb{Z}$) is denoted by $\mathrm{Int}_{\mathrm{gr}}(\mathfrak{A})$ and its elements are called *inner graded derivations of* $\mathfrak{A}$. It is an ideal of the graded Lie algebra $\mathrm{Der}_{\mathrm{gr}}(\mathfrak{A})$ and the quotient graded Lie algebra will be denoted by $\mathrm{Out}_{\mathrm{gr}}(\mathfrak{A})$. Notice that the graded center $Z_{\mathrm{gr}}(\mathfrak{A})$ is stable under the graded derivations of $\mathfrak{A}$ and that this leads to a canonical homomorphism $\mathrm{Out}_{\mathrm{gr}}(\mathfrak{A}) \to \mathrm{Der}_{\mathrm{gr}}(Z_{\mathrm{gr}}(\mathfrak{A}))$, since the inner graded derivations vanish on $Z_{\mathrm{gr}}(\mathfrak{A})$. If $\mathfrak{A}$ is a graded $*$-algebra then $Z_{\mathrm{gr}}(\mathfrak{A})$ is stable by the involution, (i.e., it is a graded $*$-subalgebra of $\mathfrak{A}$), one defines in the obvious manner an involution on $\mathrm{Der}_{\mathrm{gr}}(\mathfrak{A})$ and one has then $(\mathrm{ad}_{\mathrm{gr}}(x))^* = -\mathrm{ad}_{\mathrm{gr}}(x^*)$ for $x \in \mathfrak{A}$. One recovers the usual ungraded notions for an ordinary (ungraded) algebra $\mathcal{A}$ by considering $\mathcal{A}$ as a graded algebra which has non-zero elements only in degree 0.

Finally a *graded differential algebra* is a graded algebra $\mathfrak{A} = \oplus_n \mathfrak{A}^n$ equipped with an antiderivation d of degree 1 satisfying $d^2 = 0$, (i.e., d is linear, $d(xy) = d(x)y + (-1)^m xd(y)$ $\forall x \in \mathfrak{A}^m$ and $\forall y \in \mathfrak{A}$, $d(\mathfrak{A}^n) \subset \mathfrak{A}^{n+1}$ and $d^2 = 0$); d is the *differential* of the graded differential algebra. Notice that then the graded center $Z_{\mathrm{gr}}(\mathfrak{A})$ of $\mathfrak{A}$ is stable by the differential d and that it is therefore a graded differential subalgebra of $\mathfrak{A}$ which is graded commutative. A *graded differential $*$-algebra* will be a graded differential algebra $\mathfrak{A}$ which is also a graded $*$-algebra such that $d(x^*) = (d(x))^*$, $\forall x \in \mathfrak{A}$.

Given a graded differential algebra $\mathfrak{A}$ its *cohomology* $H(\mathfrak{A})$ is a graded algebra. Indeed, the antiderivation property of d implies that $\mathrm{Ker}(d)$ is a subalgebra of $\mathfrak{A}$ and that $\mathrm{Im}(d)$ is a two-sided ideal of $\mathrm{Ker}(d)$ and the homogeneity of d implies that they are graded. If $\mathfrak{A}$ is graded commutative then $H(\mathfrak{A})$ is also graded commutative and if $\mathfrak{A}$ is a graded differential $*$-algebra then $H(\mathfrak{A})$ is a graded $*$-algebra.

If $\mathfrak{A}'$ and $\mathfrak{A}''$ are two graded differential algebras their tensor product $\mathfrak{A}' \otimes \mathfrak{A}''$ will be the tensor product of the graded algebras equipped with the differential

d defined by

$$d(x' \otimes x'') = d(x') \otimes x'' + (-1)^{n'} x' \otimes dx'', \ \forall x' \in \mathfrak{A}'^{n'} \text{ and } \forall x'' \in \mathfrak{A}''.$$

For the cohomology one has the Künneth formula [60]

$$H(\mathfrak{A}' \otimes \mathfrak{A}'') = H(\mathfrak{A}') \otimes H(\mathfrak{A}'')$$

for the corresponding graded algebra.

Remark 3. More generally if $\mathfrak{A}'$ and $\mathfrak{A}''$ are (co)chain complexes of vector spaces with (co)boundaries denoted by d, then one defines a (co)boundary d on the graded vector space $\mathfrak{A}' \otimes \mathfrak{A}''$ by the same formula as above and one has the Künneth formula $H(\mathfrak{A}' \otimes \mathfrak{A}'') = H(\mathfrak{A}') \otimes H(\mathfrak{A}'')$ for the corresponding graded vector spaces of (co)homologies [60].

Let $\mathfrak{A}$ be a graded differential algebra which is connected, i.e., $\mathfrak{A} = \mathbb{C}\mathbb{1} \oplus \mathfrak{A}^+$ where $\mathfrak{A}^+$ is the ideal of elements of strictly positive degrees. Then $\mathfrak{A}$ will be said to be *minimal* or to be a *minimal graded differential algebra* if it satisfies the *condition of minimality* [59]:

$$d\mathfrak{A} \subset \mathfrak{A}^+ . \mathfrak{A}^+ \quad \text{(minimal condition)}.$$

A *free graded differential algebra* is a graded differential algebra which is of the form $\mathfrak{T}(C)$ for some graded vector space C as a graded algebra whereas a *free graded commutative differential algebra* is a graded differential algebra which is of the form $\mathfrak{F}(C)$ as a graded algebra.

For instance, if $\mathfrak{C}$ is a cochain complex its coboundary extends uniquely as a differential of $\mathfrak{T}(\mathfrak{C})$ and also as a differential of $\mathfrak{F}(\mathfrak{C})$. The corresponding graded differential algebra, which will be again denoted by $\mathfrak{T}(\mathfrak{C})$ and $\mathfrak{F}(\mathfrak{C})$ when no confusion arises, will be referred to respectively as *the free graded differential algebra generated by the complex* $\mathfrak{C}$ and *the free graded commutative differential algebra generated by the complex* $\mathfrak{C}$. One can show (by using the Künneth formula) that one has in cohomology $H(\mathfrak{T}(\mathfrak{C})) = \mathfrak{T}(H(\mathfrak{C}))$ and $H(\mathfrak{F}(\mathfrak{C})) = \mathfrak{F}(H(\mathfrak{C}))$. We let the reader guess the universal properties which characterize $\mathfrak{T}(\mathfrak{C})$ and $\mathfrak{F}(\mathfrak{C})$, and deduce from these the functorial character of the construction. A free graded (resp., graded commutative) differential algebra will be said to be *contractible* if it is of the form $\mathfrak{T}(\mathfrak{C})$ (resp., $\mathfrak{F}(\mathfrak{C})$) for a cochain complex (of vector spaces) $\mathfrak{C}$ such that $H(\mathfrak{C}) = 0$ (trivial cohomology). In Theorem 1 below we shall be interested in free graded commutative contractible differential algebras which are connected and finitely generated; such a differential algebra is a finite tensor product $\otimes_\alpha \mathfrak{F}(\mathbb{C}e_\alpha \oplus \mathbb{C}de_\alpha)$ with the e_α of degrees ≥ 1 (connected property).

Concerning the structure of connected finitely generated free graded commutative differential algebras, one has the following result [59].

THEOREM 1 *Every connected finitely generated free graded commutative differential algebra is the tensor product of a unique minimal one and a unique contractible one.*

This result has been for instance an important constructive ingredient in the computation of the local B.R.S. cohomology of gauge theory [37], [24].

There is probably a similar statement for the non-graded commutative case (i.e., for connected finitely generated free graded differential algebras) in which the tensor product is replaced by the free product of unital algebras.

An *operation of a Lie algebra* $\mathfrak{g}$ *in a graded differential algebra* $\mathfrak{A}$ [9], [41] is a linear mapping $X \mapsto i_X$ of $\mathfrak{g}$ into the space of antiderivations of degree -1 of $\mathfrak{A}$ such that one has $(\forall X, Y \in \mathfrak{g})$:

(i) $i_X i_Y + i_Y i_X = 0$ i.e., $[i_X, i_Y]_{\text{gr}} = 0$

$(ii) L_X i_Y - i_Y L_X = i_{[X,Y]}$ i.e., $[L_X, i_Y]_{\text{gr}} = i_{[X,Y]}$

where L_X denotes the derivation of degree 0 of $\mathfrak{A}$ defined by $L_X = i_X d + d i_X = [d, i_X]_{\text{gr}}$ for $X \in \mathfrak{A}$. Property (ii) above implies

(iii) $L_X L_Y - L_Y L_X = L_{[X,Y]}$, $(\forall X, Y \in \mathfrak{g})$

which means that $X \mapsto L_X$ is a Lie algebra-homomorphism of $\mathfrak{g}$ into the Lie algebra of derivations of degree 0 of $\mathfrak{A}$. The definition implies that L_X commutes with the differential d for any $X \in \mathfrak{g}$.

Given an operation of $\mathfrak{g}$ in $\mathfrak{A}$ as above, an element x of $\mathfrak{A}$ is said to be *horizontal* if $i_X(x) = 0$ $(\forall X \in \mathfrak{g})$, *invariant* if $L_X(x) = 0$ $(\forall X \in \mathfrak{g})$, and *basic* if it is both horizontal and invariant, i.e., if $i_X(x) = 0 = L_X(x)$ $(\forall X \in \mathfrak{g})$. The set $\mathfrak{A}_H$ of horizontal elements is a graded subalgebra of $\mathfrak{A}$ stable by the representation $X \mapsto L_X$ of $\mathfrak{g}$. The set $\mathfrak{A}_I$ of invariant elements is a graded differential subalgebra of $\mathfrak{A}$ and the set $\mathfrak{A}_B$ of basic elements is a graded differential subalgebra of $\mathfrak{A}_I$ (and therefore also of $\mathfrak{A}$). The cohomologies of $\mathfrak{A}_I$ and $\mathfrak{A}_B$ are called respectively *invariant cohomology* and *basic cohomology of* $\mathfrak{A}$ and are denoted by $H_I(\mathfrak{A})$ and $H_B(\mathfrak{A})$.

A prototype of a graded differential algebra is the graded differential algebra $\Omega(M)$ of differential forms on a smooth manifold M. We shall discuss various generalizations of it in these lectures. Let P be a smooth principal bundle with structure group G and with basis M. One defines an operation $X \mapsto i_X$ of the

Lie algebra $\mathfrak{g}$ of G in the graded differential algebra $\Omega(P)$ of differential forms on P by letting i_X be the contraction by the vertical vector field corresponding to $X \in \mathfrak{g}$. Then the elements of $\Omega(P)_H$ are the horizontal forms in the usual sense, $\Omega(P)_I$ is the differential algebra of the differential forms which are invariant by the action of G on P, whereas the graded differential algebra $\Omega(P)_B$ is canonically isomorphic to the graded differential algebra $\Omega(M)$ of differential forms on the basis. The terminology adopted above for operations comes from this fundamental example. In [24], [25] very different kinds of operations of Lie algebras in graded differential algebras have been considered.

3. Examples related to Lie algebras

Let $\mathfrak{g}$ be a finite-dimensional complex vector space with dual space $\mathfrak{g}^*$. Let $X, Y \mapsto [X,Y]$ be an antisymmetric bilinear product on $\mathfrak{g}$, i.e., a linear mapping $[\cdot,\cdot] : \bigwedge^2 \mathfrak{g} \to \mathfrak{g}$ of the second exterior power of $\mathfrak{g}$ into $\mathfrak{g}$. The dual of the bracket $[\cdot,\cdot]$ is a linear mapping of $\mathfrak{g}^*$ into $\bigwedge^2 \mathfrak{g}^* (= (\bigwedge^2 \mathfrak{g})^*)$, and such a linear mapping of $\mathfrak{g}^*$ into $\bigwedge^2 \mathfrak{g}^*$ has a unique extension as a graded derivation δ of degree 1 of the exterior algebra $\bigwedge \mathfrak{g}^*$. Conversely, given a graded derivation δ of degree 1 of $\bigwedge \mathfrak{g}^*$, the dual of $\delta : \mathfrak{g}^* \to \bigwedge^2 \mathfrak{g}^*$ is a bilinear antisymmetric product on $\mathfrak{g}(= (\mathfrak{g}^*)^*)$ and δ is the unique graded derivation of degree 1 of $\bigwedge \mathfrak{g}^*$ which extends the dual of this antisymmetric product. Thus to give an antisymmetric product $[\cdot,\cdot]$ on $\mathfrak{g}$ is the same thing as to give a graded derivation δ of degree 1 of the exterior algebra $\bigwedge \mathfrak{g}^*$. For notational reasons one usually introduces the antiderivation $d = -\delta$, i.e., the unique antiderivation of $\bigwedge \mathfrak{g}^*$ such that

$$d(\omega)(X,Y) = -\omega([X,Y])$$

for $\omega \in \mathfrak{g}^*$ and $X, Y \in \mathfrak{g}$. We shall call d the antiderivation of $\bigwedge \mathfrak{g}^*$ corresponding to the bilinear antisymmetric product on $\mathfrak{g}$.

LEMMA 1 *The bilinear antisymmetric product $[\cdot,\cdot]$ on $\mathfrak{g}$ satisfies the Jacobi identity if and only if the corresponding antiderivation d of $\bigwedge \mathfrak{g}^*$ satisfies $d^2 = 0$.*

i.e., $\mathfrak{g}$ is a Lie algebra if and only if $\bigwedge \mathfrak{g}^*$ is a graded differential algebra (for the d corresponding to the bracket of $\mathfrak{g}$).

Proof. One has $d^2 = \frac{1}{2}[d,d]_{\text{gr}}$ so d^2 is a derivation (a graded derivation of degree 2) of $\bigwedge \mathfrak{g}^*$. Since, as unital algebra $\bigwedge \mathfrak{g}^*$ is generated by $\mathfrak{g}^*$, $d^2 = 0$ is equivalent to $d^2(\mathfrak{g}^*) = 0$. On the other hand by definition one has $d(\omega)(X,Y) = -\omega([X,Y])$, for $\omega \in \mathfrak{g}^*$ and $X, Y \in \mathfrak{g}$, and, by the antiderivation property one has for $X, Y, Z \in \mathfrak{g}$

$$3!d^2(\omega)(X,Y,Z) = (d(\omega)(X,[Y,Z]) - d(\omega)([X,Y],Z)) + \text{ cycl }(X,Y,Z)$$

i.e.,

$$d^2(\omega)(X,Y,Z) = \omega([[X,Y],Z] + [[Y,Z],X] + [[Z,X],Y]).$$

Therefore $d^2(\omega) = 0\ \forall\omega \in \mathfrak{g}^*$ is equivalent to the Jacobi identity for $[\cdot,\cdot]$. □

Thus to give a finite-dimensional Lie algebra is the same thing as to give the exterior algebra of a finite-dimensional vector space equipped with a differential, that is, to give a finitely generated free graded commutative differential algebra which is generated in degree 1. Such a graded differential algebra is automatically connected and minimal. This is why, as pointed out in [59], the connected finitely generated free graded commutative differential algebras which are minimal constitute a natural categorical closure of finite-dimensional Lie algebras. In fact, such generalizations of Lie algebras occur in some physical models [5].

Let $\mathfrak{g}$ be a finite-dimensional Lie algebra, then *the cohomology* $H(\mathfrak{g})$ *of* $\mathfrak{g}$ is the cohomology of $\bigwedge\mathfrak{g}^*$. More generally, $\bigwedge\mathfrak{g}^*$ is the basic building block to construct the cochain complexes for the cohomology of $\mathfrak{g}$ with values in representations.

Assume that $\mathfrak{g}$ is the Lie algebra of a Lie group G. Then by identifying $\mathfrak{g}$ with the Lie algebra of left invariant vector fields on G one defines a canonical homomorphism of $\Lambda\mathfrak{g}^*$ into the graded differential algebra $\Omega(G)$ of differential forms on G, (in fact onto the algebra of left invariant forms). This induces a homomorphism of $H(\mathfrak{g})$ into the cohomology $H(G)$ of differential forms on G which is an isomorphism when G is compact.

In the following we consider the symmetric algebra $S\mathfrak{g}^*$ (i.e., the algebra of polynomials on $\mathfrak{g}$) to be evenly graded by giving the degree two to its generators, i.e., by writing $(S\mathfrak{g}^*)^{2n} = S^n\mathfrak{g}^*$ and $(S\mathfrak{g}^*)^{2n+1} = 0$. With this convention $S\mathfrak{g}^*$ is graded commutative and one defines the graded commutative algebra $W(\mathfrak{g})$ by $W(\mathfrak{g}) = \Lambda\mathfrak{g}^* \otimes S\mathfrak{g}^*$. Let (E_α) be a basis of $\mathfrak{g}$ with dual basis (E^α) and let us define correspondingly generators A^α and F^α of $W(\mathfrak{g})$ by $A^\alpha = E^\alpha \otimes 1$ and $F^\alpha = 1 \otimes E^\alpha$ so that $W(\mathfrak{g})$ is just the free connected graded commutative algebra (freely) generated by the A^α's in degree 1 and the F^α's in degree 2. It is convenient to introduce the elements A and F of $\mathfrak{g} \otimes W(\mathfrak{g})$ defined by $A = E_\alpha \otimes A^\alpha$ and $F = E_\alpha \otimes F^\alpha$. One then defines the elements dA^α and dF^α of $W(\mathfrak{g})$ by setting

$$\begin{aligned} dA &= E_\alpha \otimes dA^\alpha &= -\tfrac{1}{2}[A,A] + F, \\ dF &= E_\alpha \otimes dF^\alpha &= -[A,F], \end{aligned}$$

where the bracket is the graded Lie bracket obtained by combining the bracket of $\mathfrak{g}$ with the graded commutative product of $W(\mathfrak{g})$. One then extends d as an antiderivation of $W(\mathfrak{g})$ of degree 1. One has $d^2 = 0$, and since an alternative free system of homogeneous generators of $W(\mathfrak{g})$ is provided by the A^α's and the dA^α's, $W(\mathfrak{g})$ is a connected free graded commutative differential algebra which is contractible and which is referred to as the *Weil algebra* of the Lie algebra $\mathfrak{g}$ [9], [41]. It is straightforward to verify that one defines an operation of $\mathfrak{g}$ in $W(\mathfrak{g})$ by setting $i_X(A^\alpha) = X^\alpha$ and $i_X(F^\alpha) = 0$ for $X = X^\alpha E_\alpha \in \mathfrak{g}$ and by extending i_X as an antiderivation of $W(\mathfrak{g})$. Since $W(\mathfrak{g})$ is contractible, its cohomology is trivial; the same is true for the invariant cohomology $H_I(W(\mathfrak{g}))$ of $W(\mathfrak{g})$, i.e., one has $H^0_I(W(\mathfrak{g})) = \mathbb{C}$ and $H^n_I(W(\mathfrak{g})) = 0$ for $n \geq 1$ [9] (see also in [24]). The graded subalgebra of horizontal elements of $W(\mathfrak{g})$ is obviously $\mathbb{1}\otimes S\mathfrak{g}^*$ so it follows that the graded subalgebra of basis elements of $W(\mathfrak{g})$ is just $\mathbb{1}\otimes \mathcal{I}_S(\mathfrak{g})$, where $\mathcal{I}_S(\mathfrak{g})$ denotes the algebra of invariant polynomials on $\mathfrak{g}$ (with the degree $2n$ given in $W(\mathfrak{g})$ to the homogeneous polynomials of degree n). On the other hand one has $d(\mathbb{1} \otimes \mathcal{I}_S(\mathfrak{g})) = 0$, and it is easily seen that the corresponding homomorphism $\mathbb{1} \otimes \mathcal{I}_S(\mathfrak{g}) \to H_B(W(\mathfrak{g}))$ onto the basic cohomology of $W(\mathfrak{g})$ is an isomorphism. Therefore, one has $H^{2n}_B(W(\mathfrak{g})) = \mathcal{I}^n_S(\mathfrak{g})$ and $H^{2n+1}_B(W(\mathfrak{g})) = 0$, where $\mathcal{I}^n_S(\mathfrak{g})$ denotes the space of invariant homogeneous polynomials of degree n on $\mathfrak{g}$. Let now P be a smooth principal bundle with basis M and with structure group G such that its Lie algebra is $\mathfrak{g}$. One has the canonical operation $X \mapsto i_X$ of $\mathfrak{g}$ in $\Omega(P)$ defined at the end of last section. Given a connection $\omega = E_\alpha \otimes \omega^\alpha \in \mathfrak{g} \otimes \Omega^1(P)$ on P, there is a unique homomorphism of graded differential algebras $\Psi : W(\mathfrak{g}) \to \Omega(P)$ such that $\Psi(A^\alpha) = \omega^\alpha$. This homomorphism satisfies $\Psi(i_X(w)) = i_X(\Psi(w))$ for any $X \in \mathfrak{g}$ and $w \in W(\mathfrak{g})$. It follows that it induces a homomorphism in basic cohomology $\varphi : H_B(W(\mathfrak{g})) \to H_B(P)$, i.e., a homomorphism of $\mathcal{I}_S(\mathfrak{g})$ into the cohomology $H(M)$ of the basis M of P, such that $\varphi(\mathcal{I}^n_S(\mathfrak{g})) \subset H^{2n}(M)$, (it is an homomorphism of commutative algebras). One has $\mathrm{Im}(\varphi) \subset H^{ev}(M) = \oplus_p H^{2p}(M)$.

THEOREM 2 *The above homomorphism $\varphi : \mathcal{I}_S(\mathfrak{g}) \to H^{ev}(M)$ does not depend on the choice of the connection ω on P.*

That is, φ only depends on P; it is called the *Weil homomorphism* of the principal bundle P. Before leaving this subject it is worth noticing here that there is a very interesting noncommutative (or quantized) version of the Weil algebra of $\mathfrak{g}$ in the case where $\mathfrak{g}$ admits a non-degenerate invariant symmetric bilinear form, i.e., for $\mathfrak{g}$ reductive, where $S\mathfrak{g}^*$ is replaced by the enveloping algebra $U(\mathfrak{g})$ and where $\Lambda\mathfrak{g}^*$ is replaced by the Clifford algebra $C\ell(\mathfrak{g})$ of the bilinear form, which has been introduced and studied in [1].

In these lectures the Lie algebras involved will be generally not finite-dimensional and some care must be taken with respect to duality and tensor products. For instance, if $\mathfrak{g}$ is not finite-dimensional then the dual of the Lie bracket $[\cdot,\cdot] : \bigwedge^2 \mathfrak{g} \to \mathfrak{g}$ is a linear mapping $\delta : \mathfrak{g}^* \to (\bigwedge^2 \mathfrak{g})^*$ and one only has an inclusion $\bigwedge^2 \mathfrak{g}^* \subset (\bigwedge^2 \mathfrak{g})^*$. In the following we give the formulation adapted to this more general situation.

Let $\mathfrak{g}$ be a Lie algebra, let E be a representation space of $\mathfrak{g}$ (i.e., a $\mathfrak{g}$-module or, as will be explained in Section 5, a $\mathfrak{g}$-bimodule for the category **Lie** of Lie algebras) and let $X \mapsto \pi(X) \in \mathrm{End}(E)$ denote the action of $\mathfrak{g}$ on E. An *E-valued (Lie algebra) n-cochain of* $\mathfrak{g}$ is a linear mapping $X_1 \wedge \cdots \wedge X_n \mapsto \omega(X_1,\dots,X_n)$ of $\bigwedge^n \mathfrak{g}$ into E. The vector space of these n-cochains will be denoted by $C^n_\wedge(\mathfrak{g},E)$. One defines a homogeneous endomorphism d of degree 1 of the $\mathbb{N}$-graded vector space $C_\wedge(\mathfrak{g},E) = \oplus_n C^n_\wedge(\mathfrak{g},E)$ of all E-valued cochains of $\mathfrak{g}$ by setting

$$\begin{aligned} d(\omega)(X_0,\dots,X_n) &= \textstyle\sum_{k=0}^n (-1)^k \pi(X_k)\omega(X_0,\overset{k}{\dots\vee\dots},X_n) \\ &\quad + \textstyle\sum_{0\leq r<s\leq n} (-1)^{r+s} \omega([X_r,X_s],X_0 \overset{r}{\dots\vee}\overset{s}{\dots\vee\dots} X_n) \end{aligned}$$

for $\omega \in C^n_\wedge(\mathfrak{g},E)$ and $X_i \in \mathfrak{g}$. It follows from the Jacobi identity and from $\pi(X)\pi(Y) - \pi(Y)\pi(X) = \pi([X,Y])$ that $d^2 = 0$. Thus, equipped with d, $C_\wedge(\mathfrak{g},E)$ is a cochain complex and its cohomology, denoted by $H(\mathfrak{g},E)$, is called the *E-valued cohomology of* $\mathfrak{g}$. When $E = \mathbb{C}$ and π is the trivial representation $\pi = 0$, it is the *cohomology* $H(\mathfrak{g})$ of $\mathfrak{g}$. One verifies that if $\mathfrak{g}$ is finite-dimensional, it is the same as the cohomology of $\bigwedge \mathfrak{g}^*$; in fact, in this case one has $C_\wedge(\mathfrak{g},E) = E \otimes \bigwedge \mathfrak{g}^*$.

Assume now that E is an algebra $\mathcal{A}$ (unital, associative, complex) and that $\mathfrak{g}$ acts on $\mathcal{A}$ by derivations, i.e., that one has $\pi(X)(xy) = \pi(X)(x)y + x\pi(X)(y)$ for $X \in \mathfrak{g}$ and $x,y \in \mathcal{A}$. Then $C_\wedge(\mathfrak{g},\mathcal{A})$ is canonically a graded differential algebra. Indeed the product is obtained by taking the product in $\mathcal{A}$ after evaluation and then antisymmetrizing whereas, the derivation property of the action of $\mathfrak{g}$ implies that d is an antiderivation. The trivial representation $\pi = 0$ in $\mathbb{C}$ is of this kind, this is why $H(\mathfrak{g})$ is a graded algebra.

In particular, the vector space $\mathrm{Der}(\mathcal{A})$ of all derivations of $\mathcal{A}$ into itself is a Lie algebra, and therefore $C_\wedge(\mathrm{Der}(\mathcal{A}),\mathcal{A})$ is a graded differential algebra. Furthermore, $\mathrm{Der}(\mathcal{A})$ is also a module over the center $Z(\mathcal{A})$ of $\mathcal{A}$ and one has $[X,zY] = z[X,Y] + X(z)Y$ from which it follows that the graded subalgebra $\underline{\Omega}_{\mathrm{Der}}(\mathcal{A})$ of $C_\wedge(\mathrm{Der}(\mathcal{A}),\mathcal{A})$ which consists of $Z(\mathcal{A})$-multilinear cochains is stable by the differential and is therefore a graded differential subalgebra of $C_\wedge(\mathrm{Der}(\mathcal{A}),\mathcal{A})$. Since $\underline{\Omega}^0_{\mathrm{Der}}(\mathcal{A}) = \mathcal{A}$, a smaller differential subalgebra is the smallest differential subalgebra $\Omega_{\text{Der}}(\mathcal{A})$ of $C_\wedge(\text{Der}(\mathcal{A}),\mathcal{A})$ containing $\mathcal{A}$. When M is a 'good' smooth manifold (finite-dimensional, paracompact, etc.) and $\mathcal{A} = C^\infty(M)$ then $\underline{\Omega}_{\text{Der}}(\mathcal{A})$ and $\Omega_{\text{Der}}(\mathcal{A})$ both coincide with the graded differential algebra $\Omega(M)$ of differential forms on M. In general, the inclusion $\Omega_{\text{Der}}(\mathcal{A}) \subset \underline{\Omega}_{\text{Der}}(\mathcal{A})$ is a strict one, even when $\mathcal{A}$ is commutative (e.g., for the smooth functions on a ∞-dimensional manifold). The differential calculus over $\mathcal{A}$ (see in Sections 7, 8) using $\underline{\Omega}_{\text{Der}}(\mathcal{A})$ (or $\Omega_{\text{Der}}(\mathcal{A})$) as generalization of differential forms will be referred to as the *derivation-based calculus*, [25], [26], [27], [28], [29], [33], [34], [35], [36]. If $\mathcal{A}$ is a $*$-algebra one defines an involution $X \mapsto X^*$ on $\text{Der}(\mathcal{A})$ by setting $X^*(a) = (X(a^*))^*$ and an involution $\omega \mapsto \omega^*$ on $C_\wedge(\text{Der}(\mathcal{A}),\mathcal{A})$ by setting $\omega^*(X_1,\dots,X_n) = (\omega(X_1^*,\dots,X_n^*))^*$. So equipped $C_\wedge(\text{Der}(\mathcal{A}),\mathcal{A})$ is a graded differential $*$-algebra and $\Omega_{\text{Der}}(\mathcal{A})$, as well as $\underline{\Omega}_{\text{Der}}(\mathcal{A})$, are stable by the involution and are therefore also graded differential $*$-algebras.

One defines a linear mapping $X \mapsto i_X$ of $\mathfrak{g}$ into the homogeneous endomorphisms of degree -1 of $C_\wedge(\mathfrak{g},E)$ by setting $i_X(\omega)(X_1,\dots,X_{n-1}) = \omega(X,X_1,\dots,X_{n-1})$ for $\omega \in C^n_\wedge(\mathfrak{g},E)$ and $X_i \in \mathfrak{g}$. Then $X \mapsto L_X = i_X d + d i_X$ is a representation of $\mathfrak{g}$ in $C_\wedge(\mathfrak{g},E)$ by homogeneous endomorphisms of degree 0 which extends the original representation π in $E = C^0_\wedge(\mathfrak{g},E)$, i.e., $L_X \restriction E = \pi(X)$ for $X \in \mathfrak{g}$. In the case where E is an algebra $\mathcal{A}$ and where $\mathfrak{g}$ acts by derivations on $\mathcal{A}$, we have seen that $C_\wedge(\mathfrak{g},\mathcal{A})$ is a graded differential algebra and it is easy to show that $X \mapsto i_X$ is an operation of the Lie algebra $\mathfrak{g}$ in the graded differential algebra $C_\wedge(\mathfrak{g},\mathcal{A})$; in fact, properties (i) and (ii) of operations (see last section) hold already in $C_\wedge(\mathfrak{g},E)$ for any $\mathfrak{g}$-module E.

In particular, one has the operation $X \mapsto i_X$ of the Lie algebra $\text{Der}(\mathcal{A})$ in the graded differential algebra $C_\wedge(\text{Der}(\mathcal{A}),\mathcal{A})$ defined as above. It is not hard to verify that the graded differential subalgebras $\underline{\Omega}_{\text{Der}}(\mathcal{A})$ and $\Omega_{\text{Der}}(\mathcal{A})$ are stable by the i_X $(X \in \text{Der}(\mathcal{A}))$. The corresponding operations will be referred to as *the canonical operations of* $\text{Der}(\mathcal{A})$ *in* $\underline{\Omega}_{\text{Der}}(\mathcal{A})$ *and in* $\Omega_{\text{Der}}(\mathcal{A})$.

4. Examples related to associative algebras

Let $\mathcal{A}$ be a finite-dimensional complex vector space with dual space $\mathcal{A}^*$ and let $x, y \mapsto xy$ be an arbitrary bilinear product on $\mathcal{A}$, i.e., a linear mapping $\otimes^2\mathcal{A} \to \mathcal{A}$ where $\otimes^2\mathcal{A}$ denotes the second tensor power of $\mathcal{A}$. The dual of the product is a linear mapping of $\mathcal{A}^*$ into $\otimes^2\mathcal{A}^*$ and again such a linear mapping uniquely extends as a graded derivation δ of degree 1 of the tensor algebra $T(\mathcal{A}^*) = \oplus_{n\geq 0} \otimes^n \mathcal{A}^*$. Conversely, given such a graded derivation δ of degree 1 (i.e., an antiderivation of degree 1) of $T(\mathcal{A}^*)$, the dual mapping of the restriction $\delta : \mathcal{A}^* \to \otimes^2\mathcal{A}^*$ of δ to $\mathcal{A}^*$ is a bilinear product on $\mathcal{A}$ which is such that δ is obtained from it by the above construction. Thus, to give a bilinear product on $\mathcal{A}$ is the same thing as to give an antiderivation of degree 1 of $T(\mathcal{A}^*)$. Again, for notational reasons it is usual to consider the antiderivation $d = -\delta$, i.e., the unique antiderivation of $T(\mathcal{A}^*)$ such that

$$d(\omega)(x, y) = -\omega(xy)$$

for $\omega \in \mathcal{A}^*$ and $x, y \in \mathcal{A}$. We shall call this d the antiderivation of $T(\mathcal{A}^*)$ corresponding to the bilinear product of $\mathcal{A}$.

LEMMA 2 *The bilinear product on $\mathcal{A}$ is associative if and only if the corresponding antiderivation of $T(\mathcal{A}^*)$ satisfies $d^2 = 0$.*

i.e., $\mathcal{A}$ is an associative algebra if and only if $T(\mathcal{A}^*)$ is a graded differential algebra (for the d corresponding to the product of $\mathcal{A}$).

Proof. By definition, one has for $\omega \in \mathcal{A}^*$ and $x, y, z \in \mathcal{A}$

$$d(d(\omega))(x, y.z) = d(\omega)(x, yz) - d(\omega)(xy, z) = \omega((xy)z - x(yz)).$$

Therefore the product of $\mathcal{A}$ is associative if and only if d^2 vanishes on $\mathcal{A}^*$, but this is equivalent to $d^2 = 0$ since d^2 is a derivation and since the (unital) graded algebra $T(\mathcal{A}^*)$ is generated by $\mathcal{A}^*$. □

Therefore to give a finite-dimensional associative algebra is the same thing as to give a finitely generated free graded differential algebra which is generated in degree 1. Again, such a graded differential algebra is automatically connected and minimal. The situation is very similar to the one of last section except that here one has not graded commutativity. So one can consider in particular that the connected finitely generated free graded differential algebras which are minimal constitute a natural categorical closure of finite-dimensional associative algebras, i.e., a natural generalization of the notion of associative algebra.

Let $\mathcal{A}$ be a finite-dimensional associative algebra; we shall see that if $\mathcal{A}$ has a unit then the cohomology of the graded differential algebra $T(\mathcal{A}^*)$ is trivial. Nevertheless, $T(\mathcal{A}^*)$ is the basic building block for constructing the Hochschild cochain complexes. Namely, if $\mathcal{M}$ is a $(\mathcal{A}, \mathcal{A})$-bimodule then the graded vector space of $\mathcal{M}$-valued Hochschild cochains of $\mathcal{A}$ is the graded space $\mathcal{M} \otimes T(\mathcal{A}^*)$ and the Hochschild coboundary d_H is given by

$$\begin{aligned} d_H(\omega)(x_0, \dots, x_n) &= x_0\omega(x_1, \dots, x_n) + (I_{\mathcal{M}} \otimes d)(\omega)(x_0, \dots, x_n) \\ &\quad + (-1)^{n+1}\omega(x_0, \dots, x_{n-1})x_n \end{aligned}$$

for $\omega \in \mathcal{M} \otimes (\otimes^n \mathcal{A}^*)$ and $x_i \in \mathcal{A}$.

In these lectures we shall have to deal with infinite dimensional algebras as algebras of smooth functions and their generalizations, so again (as in last section) one has to take some care of duality and tensor products.

Let $\mathcal{A}$ be now an arbitrary associative algebra and let $C(\mathcal{A})$ denote the graded vector space of multilinear forms on $\mathcal{A}$, i.e., $C(\mathcal{A}) = \oplus_n C^n(\mathcal{A})$ where $C^n(\mathcal{A}) = (\otimes^n \mathcal{A})^*$ is the dual of the n-th tensor power of $\mathcal{A}$. One has $T(\mathcal{A}^*) \subset C(\mathcal{A})$ and the equality $T(\mathcal{A}^*) = C(\mathcal{A})$ holds if and only if $\mathcal{A}$ is finite-dimensional. The product of $T(\mathcal{A}^*)$ (i.e., the tensor product) canonically extends to $C(\mathcal{A})$ which so equipped is a graded algebra. Furthermore minus the dual of the product of $\mathcal{A}$ is a linear mapping of $C^1(\mathcal{A}) = \mathcal{A}^*$ into $C^2(\mathcal{A}) = (\mathcal{A} \otimes \mathcal{A})^*$ which also canonically extends as an antiderivation d of $C(\mathcal{A})$ which is a differential as a consequence of the associativity of the product of $\mathcal{A}$. It is given by:

$$d\omega(x_0, \dots, x_n) = \sum_{k=1}^{n} (-1)^k \omega(x_0, \dots, x_{i-1}x_i, \dots, x_n)$$

for $\omega \in C^n(\mathcal{A})$ and $x_i \in \mathcal{A}$. The graded differential algebra $C(\mathcal{A})$ is the generalization of the above $T(\mathcal{A}^*)$ for an infinite dimensional algebra $\mathcal{A}$. As announced before the cohomology of $C(\mathcal{A})$ is trivial whenever $\mathcal{A}$ has a unit.

LEMMA 3 *Let $\mathcal{A}$ be a unital associative algebra (over $\mathbb{C}$). Then the cohomology $H(C(\mathcal{A}))$ of $C(\mathcal{A})$ is trivial in the sense that one has:*

$$H^0(C(\mathcal{A})) = \mathbb{C} \quad \text{and} \quad H^n(C(\mathcal{A})) = 0 \quad \text{for} \quad n \geq 1.$$

Proof. By definition $C(\mathcal{A})$ is connected so $H^0(C(\mathcal{A})) = \mathbb{C}$ is obvious. For $\omega \in C^n(\mathcal{A})$ with $n \geq 1$ let us define $h(\omega) \in C^{n-1}(\mathcal{A})$ by $h(\omega)(x_1, \dots, x_{n-1}) = \omega(\mathbf{1}, x_1, \dots, x_{n-1})$, $\forall x_i \in \mathcal{A}$. One has

$$d(h(\omega)) + h(d(\omega)) = \omega \text{ for any } \omega \in C^n(\mathcal{A}) \text{ with } n \geq 1$$

which implies $H^n(C(\mathcal{A})) = 0$ for $n \geq 1$. □

If $\mathcal{M}$ is a $(\mathcal{A}, \mathcal{A})$-bimodule then the graded vector space of *$\mathcal{M}$-valued Hochschild cochains of $\mathcal{A}$* is the graded vector space $C(\mathcal{A}, \mathcal{M})$ of multilinear mappings of $\mathcal{A}$ into $\mathcal{M}$, i.e., $C^n(\mathcal{A}, \mathcal{M})$ is the space of linear mappings of $\otimes^n \mathcal{A}$ into $\mathcal{M}$, equipped with the Hochschild coboundary d_H defined by

$$\begin{aligned} d_H(\omega)(x_0, \dots, x_n) \quad &= \quad x_0\omega(x_1, \dots, x_n) + d(\omega)(x_0, \dots, x_n) \\ &\quad +(-1)^{n+1}\omega(x_0, \dots, x_{n-1})x_n \end{aligned}$$

for $\omega \in C^n(\mathcal{A}, \mathcal{M})$, $x_i \in \mathcal{A}$ and where d is 'the obvious extension' to $C(\mathcal{A}, \mathcal{M})$ of the differential d of $C(\mathcal{A})$. When $\mathcal{A}$ is finite-dimensional all this reduces to the previous definitions; in particular, in this case one has $C(\mathcal{A}, \mathcal{M}) = \mathcal{M} \otimes T(\mathcal{A}^*)$. The cohomology $H(\mathcal{A}, \mathcal{M})$ of $C(\mathcal{A}, \mathcal{M})$ is the *$\mathcal{M}$-valued Hochschild cohomology of $\mathcal{A}$* or the *Hochschild cohomology of $\mathcal{A}$ with coefficients in $\mathcal{M}$*. The $\mathcal{M}$-valued Hochschild cochains of $\mathcal{A}$ which vanishes whenever one of their arguments is the unit $\mathbf{1}$ of $\mathcal{A}$ are said to be *normalized Hochschild cochains*. The graded vector space $C_0(\mathcal{A}, \mathcal{M})$ of $\mathcal{M}$-valued normalized Hochschild cochains is stable under the Hochschild coboundary d_H and it is well known and easy to show that the injection of $C_0(\mathcal{A}, \mathcal{M})$ into $C(\mathcal{A}, \mathcal{M})$ induces an isomorphism in cohomology, i.e., the cohomology of $C_0(\mathcal{A}, \mathcal{M})$ is again $H(\mathcal{A}, \mathcal{M})$. Notice that an $\mathcal{M}$-valued Hochschild 1-cocycle (i.e., an element of $C^1(\mathcal{A}, \mathcal{M})$ in $\mathrm{Ker}(d_H)$) is a derivation δ of $\mathcal{A}$ in $\mathcal{M}$, and that it is automatically normalized. If $\mathcal{N}$ is another $(\mathcal{A}, \mathcal{A})$-bimodule then the tensor product over $\mathcal{A}$ of $\mathcal{M}$ and $\mathcal{N}$, $(\mathcal{M}, \mathcal{N}) \mapsto \mathcal{M} \otimes_{\mathcal{A}} \mathcal{N}$, induces a product $(\alpha, \beta) \mapsto \alpha \cup \beta$, *the cup product* $\cup : C(\mathcal{A}, \mathcal{M}) \otimes C(\mathcal{A}, \mathcal{N}) \to C(\mathcal{A}, \mathcal{M} \otimes_{\mathcal{A}} \mathcal{N})$ such that $C^m(\mathcal{A}, \mathcal{M}) \cup C^n(\mathcal{A}, \mathcal{N}) \subset C^{m+n}(\mathcal{A}, \mathcal{M} \otimes_{\mathcal{A}} \mathcal{N})$ defined by

$$(\alpha \cup \beta)(x_1, \dots, x_{m+n}) = \alpha(x_1, \dots, x_m) \underset{\mathcal{A}}{\otimes} \beta(x_{m+1}, \dots, x_{m+n})$$

for $\alpha \in C^m(\mathcal{A}, \mathcal{M}), \beta \in C^n(\mathcal{A}, \mathcal{N})$ and $x_i \in \mathcal{A}$. If $\mathcal{P}$ is another $(\mathcal{A}, \mathcal{A})$-bimodule and if $\gamma \in C^p(\mathcal{A}, \mathcal{P})$, one has:$(\alpha \cup \beta) \cup \gamma = \alpha \cup (\beta \cup \gamma)$. Furthermore one has $d_H(\alpha \cup \beta) = d_H(\alpha) \cup \beta + (-1)^m \alpha \cup d(\beta)$ for $\alpha \in C^m(\mathcal{A}, \mathcal{M})$, $\beta \in C(\mathcal{A}, \mathcal{N})$. This implies, in particular, that $C(\mathcal{A}, \mathcal{A})$ is a graded differential algebra (when equipped with the cup product and with d_H). In fact, $C(\mathcal{A}, \mathcal{A})$ has a very rich structure which was first described in [40]. As pointed out in [40], its cohomology $H(\mathcal{A}, \mathcal{A})$ which it inherits from this structure is graded commutative (as graded algebra for the cup product). The cohomology $H(\mathcal{A}, \mathcal{A})$ is a sort of graded commutative Poisson algebra.

A unital associative algebra $\mathcal{A}$ is said to be of *Hochschild dimension* n if n is the smaller integer such that $H^k(\mathcal{A}, \mathcal{M}) = 0$ for any $k \geq n+1$ and any $(\mathcal{A}, \mathcal{A})$-bimodule $\mathcal{M}$. The Hochschild dimension of the algebra $\mathbb{C}[X_1, \dots, X_n]$ of complex polynomials with n indeterminates is n. If one considers $\mathcal{A}$ as the generalization of the algebra of smooth functions on a noncommutative space then its Hochschild dimension n is the analog of the dimension of the noncommutative space.

In spite of the triviality of the cohomology of $C(\mathcal{A})$, several complexes with non-trivial cohomologies can be extracted from it. Let $S : C(\mathcal{A}) \to C(\mathcal{A})$ and

$\mathcal{C} : C(\mathcal{A}) \to C(\mathcal{A})$ be linear mappings defined by

$$\mathcal{S}(\omega)(x_1, \dots, x_n) = \sum_{\pi \in \mathcal{S}_n} \varepsilon(\pi)\omega(x_{\pi(1)}, \dots, x_{\pi(n)})$$

and

$$\mathcal{C}(\omega)(x_1, \dots, x_n) = \sum_{\gamma \in \mathcal{C}_n} \varepsilon(\gamma)\omega(x_{\gamma(1)}, \dots, x_{\gamma(n)})$$

for $\omega \in C^n(\mathcal{A})$, $x_i \in \mathcal{A}$ and where $\mathcal{S}_n$ is the group of permutations of $\{1, \dots, n\}$ and $\mathcal{C}_n$ is the subgroup of cyclic permutations, ($\varepsilon(\pi)$ denoting the signature of the permutation π). The mapping $C(\mathcal{A}) \overset{\mathcal{S}}{\to} \mathcal{S}(C(\mathcal{A}))$ is a homomorphism of graded differential algebras of $C(\mathcal{A})$ onto the graded differential algebra $C_\wedge(\mathcal{A}_{\text{Lie}})$ of Lie algebra cochains of the underlying Lie algebra $\mathcal{A}_{\text{Lie}}$ with values in the trivial representation of $\mathcal{A}_{\text{Lie}}$ in $\mathbb{C}$; (notice that the product of $C_\wedge(\mathcal{A}_{\text{Lie}})$ is not induced by the inclusion $C_\wedge(\mathcal{A}_{\text{Lie}}) \subset C(\mathcal{A})$). The cohomology of $\text{Im}(\mathcal{S}) = C_\wedge(\mathcal{A}_{\text{Lie}})$ is therefore the Lie algebra cohomology of $\mathcal{A}_{\text{Lie}}$. On the other hand (see Lemma 3 in [13] part II), one has $\mathcal{C} \circ d = d_H \circ \mathcal{C}$, where d_H is the Hochschild coboundary of $C(\mathcal{A}, \mathcal{A}^*)$, and therefore $(\text{Im}(\mathcal{C}), d_H)$ is a complex the cohomology of which coincides with *the cyclic cohomology* $H_\lambda(\mathcal{A})$ *of* $\mathcal{A}$ up to a shift of -1 in degree [13].

Let us define for $a \in \mathcal{A}$ the homogeneous linear mapping i_a of degree -1 of $C(\mathcal{A})$ into itself by setting

$$i_a(\omega)(x_1, \dots, x_{n-1}) = \sum_{k=0}^{n-1} (-1)^k \omega(x_1, \dots, x_k, a, x_{k+1}, \dots, x_{n-1})$$

for $\omega \in C^n(\mathcal{A})$ with $n \geq 1$ and $x_i \in \mathcal{A}$, and by setting $i_a(C^0(\mathcal{A})) = 0$. For each $a \in \mathcal{A}$, i_a is an antiderivation of $C(\mathcal{A})$ and it is easy to verify that $a \mapsto i_a$ is an operation of the Lie algebra $\mathcal{A}_{\text{Lie}}$ in the graded differential algebra $C(\mathcal{A})$. The homotopy h used in the proof of Lemma 3 commutes with the L_a's, which implies that the invariant cohomology $H_I(C(\mathcal{A}))$ of $C(\mathcal{A})$ is also trivial. The basic cohomology of $C(\mathcal{A})$ for this operation has been called *basic cohomology of* $\mathcal{A}$ and denoted by $H_B(\mathcal{A})$ in [31]. It is given by the following theorem [31]

THEOREM 3 *The basic cohomology $H_B(\mathcal{A})$ of $\mathcal{A}$ identifies with the algebra $\mathcal{I}_S(\mathcal{A}_{\text{Lie}})$ of invariant polynomials on the Lie algebra $\mathcal{A}_{\text{Lie}}$, where the degree $2n$ is given to the homogeneous polynomials of degree n, that is, $H_B^{2n}(\mathcal{A}) = \mathcal{I}_S^n(\mathcal{A}_{\text{Lie}})$ and $H_B^{2n+1}(\mathcal{A}) = 0$.*

The proof of this theorem, which is not straightforward, uses a familiar trick in equivariant cohomology to convert the operation i of $\mathcal{A}_{\text{Lie}}$ into a differential.

Two algebras $\mathcal{A}$ and $\mathcal{B}$ (associative unital, etc.) are said to be *Morita equivalent* if there is a $(\mathcal{A},\mathcal{B})$-bimodule $\mathcal{U}$ and a $(\mathcal{B},\mathcal{A})$-bimodule $\mathcal{V}$ such that one has an isomorphism of $(\mathcal{A},\mathcal{A})$-bimodules $\mathcal{U}\otimes_{\mathcal{B}}\mathcal{V}\simeq\mathcal{A}$ and an isomorphism of $(\mathcal{B},\mathcal{B})$-bimodules $\mathcal{V}\otimes_{\mathcal{A}}\mathcal{U}\simeq\mathcal{B}$. This is an equivalence relation and this induces an equivalence between the category of right $\mathcal{A}$-modules (resp., left $\mathcal{A}$-modules, $(\mathcal{A},\mathcal{A})$-bimodules) and the category of right $\mathcal{B}$-modules (resp., left $\mathcal{B}$-modules, $(\mathcal{B},\mathcal{B})$-bimodules). The algebras $M_m(\mathcal{A})$ and $M_n(\mathcal{A})$ of $m\times m$ matrices and of $n\times n$ matrices with entries in $\mathcal{A}$ are Morita equivalent for any $m,n\in\mathbb{N}$; in fact, the $(M_m(\mathcal{A}),M_n(\mathcal{A}))$-bimodule $M_{mn}(\mathcal{A})$ of rectangular $m\times n$ matrices and the $(M_n(\mathcal{A}),M_m(\mathcal{A}))$-bimodule $M_{nm}(\mathcal{A})$ of rectangular $n\times m$ matrices with entries in $\mathcal{A}$ are such that $M_n(\mathcal{A}) = M_{nm}(\mathcal{A})\otimes_{M_m(\mathcal{A})}M_{mn}(\mathcal{A})$ and $M_m(\mathcal{A}) = M_{mn}(\mathcal{A})\otimes_{M_n(\mathcal{A})}M_{nm}(\mathcal{A})$, (the tensor products over $M_m(\mathcal{A})$ and $M_n(\mathcal{A})$ being canonically the usual matricial products).

An important property of Hochschild cohomology and cyclic cohomology (and of the corresponding homologies) is their Morita invariance [45], [52], [60]. More precisely if $\mathcal{A}$ and $\mathcal{B}$ are Morita equivalent with $\mathcal{U}$ and $\mathcal{V}$ as above and if $\mathcal{M}$ is a $(\mathcal{A},\mathcal{A})$-bimodule (resp., $\mathcal{N}$ is a $(\mathcal{B},\mathcal{B})$-bimodule) one has a canonical isomorphism $H(\mathcal{A},\mathcal{M})\simeq H(\mathcal{B},\mathcal{V}\otimes_{\mathcal{A}}\mathcal{M}\otimes_{\mathcal{A}}\mathcal{U})$, (resp., $H(\mathcal{B},\mathcal{N})\simeq H(\mathcal{A},\mathcal{U}\otimes_{\mathcal{B}}\mathcal{N}\otimes_{\mathcal{B}}\mathcal{V})$) in Hochschild cohomology and also $H_\lambda(\mathcal{A})\simeq H_\lambda(\mathcal{B})$ in cyclic cohomology. In contrast, the Lie algebra cohomology $H(\mathcal{A}_{\text{Lie}})$ and the basic cohomology $H_B(\mathcal{A})$ are not Morita invariant, since, for instance, for $\mathcal{A}=M_n(\mathbb{C})$ they depend on the number $n\in\mathbb{N}$, whereas $M_n(\mathbb{C})$ is Morita equivalent to $\mathbb{C}$.

5. Categories of algebras

In this section we consider general algebras over $\mathbb{C}$. That is, by an *algebra* we here mean a complex vector space $\mathcal{A}$ equipped with a bilinear product $m:\mathcal{A}\otimes\mathcal{A}\to\mathcal{A}$. Given two such algebras $\mathcal{A}$ and $\mathcal{B}$, an *algebra homomorphism* of $\mathcal{A}$ into $\mathcal{B}$ is a linear mapping $\varphi:\mathcal{A}\to\mathcal{B}$ such that $\varphi(m(x\otimes y)) = m(\varphi(x)\otimes\varphi(y))$, $(\forall x,y\in\mathcal{A})$, i.e., $\varphi\circ m = m\circ(\varphi\otimes\varphi)$.

Let us define the category $\mathbf{A}$ to be the category such that the class $\text{Ob}(\mathbf{A})$ of its objects is the class of all algebras (in the above sense) and such that for any $\mathcal{A},\mathcal{B}\in\text{Ob}(\mathbf{A})$ the set $\text{Hom}_{\mathbf{A}}(\mathcal{A},\mathcal{B})$ of morphisms from $\mathcal{A}$ to $\mathcal{B}$ is the set of all algebra homomorphisms of $\mathcal{A}$ into $\mathcal{B}$.

A subcategory of $\mathbf{A}$ will be called a *category of algebras*. Thus a category $\mathbf{C}$ is a category of algebras if $\text{Ob}(\mathbf{C})$ is a subclass of $\text{Ob}(\mathbf{A})$ and if, for any $\mathcal{A},\mathcal{B}\in\text{Ob}(\mathbf{C})$, one has $\text{Hom}_{\mathbf{C}}(\mathcal{A},\mathcal{B})\subset\text{Hom}_{\mathbf{A}}(\mathcal{A},\mathcal{B})$. We now list some categories of algebras which will be used later.

1. The category $\mathbf{Alg}$ of unital associative algebras: $\mathrm{Ob}(\mathbf{Alg})$ is the class of all complex unital associative algebras and for any $\mathcal{A},\mathcal{B}\in\mathrm{Ob}(\mathbf{Alg})$, $\mathrm{Hom}_{\mathbf{Alg}}(\mathcal{A},\mathcal{B})$ is the set of all algebra homomorphisms mapping the unit of $\mathcal{A}$ onto the unit of $\mathcal{B}$.

2. The category $\mathbf{Alg}_Z$ is the subcategory of $\mathbf{Alg}$ defined by $\mathrm{Ob}(\mathbf{Alg}_Z)=\mathrm{Ob}(\mathbf{Alg})$ and for any $\mathcal{A},\mathcal{B}\in\mathrm{Ob}(\mathbf{Alg}_Z)$, $\mathrm{Hom}_{\mathbf{Alg}_Z}(\mathcal{A},\mathcal{B})$ is the set of all $\varphi\in\mathrm{Hom}_{\mathbf{Alg}}(\mathcal{A},\mathcal{B})$ mapping the center $Z(\mathcal{A})$ of $\mathcal{A}$ into the center $Z(\mathcal{B})$ of $\mathcal{B}$, i.e., such that $\varphi(Z(\mathcal{A}))\subset Z(\mathcal{B})$.

3. The category $\mathbf{Jord}$ of complex unital Jordan algebras: $\mathrm{Ob}(\mathbf{Jord})$ is the class of all complex unital Jordan algebras and for any $\mathcal{A},\mathcal{B}\in\mathrm{Ob}(\mathbf{Jord})$, $\mathrm{Hom}_{\mathbf{Jord}}(\mathcal{A},\mathcal{B})$ is the set of all algebra homomorphisms mapping the unit of $\mathcal{A}$ onto the unit of $\mathcal{B}$.

4. The category $\mathbf{Algcom}$ of unital associative commutative algebras: $\mathrm{Ob}(\mathbf{Algcom})$ is the class of all complex unital associative commutative algebras and for any $\mathcal{A},\mathcal{B}\in\mathrm{Ob}(\mathbf{Algcom})$, $\mathrm{Hom}_{\mathbf{Algcom}}(\mathcal{A},\mathcal{B})=\mathrm{Hom}_{\mathbf{Alg}}(\mathcal{A},\mathcal{B})$.

5. The category $\mathbf{Lie}$ of Lie algebras: $\mathrm{Ob}(\mathbf{Lie})$ is the class of all complex Lie algebras and for any $\mathcal{A},\mathcal{B}\in\mathrm{Ob}(\mathbf{Lie})$, $\mathrm{Hom}_{\mathbf{Lie}}(\mathcal{A},\mathcal{B})=\mathrm{Hom}_{\mathbf{A}}(\mathcal{A},\mathcal{B})$.

Remark 4. If $\mathcal{A}\in\mathrm{Ob}(\mathbf{Alg})$ and $\mathcal{B}\in\mathrm{Ob}(\mathbf{Algcom})$, one has $\mathrm{Hom}_{\mathbf{Alg}}(\mathcal{A},\mathcal{B})=\mathrm{Hom}_{\mathbf{Alg}_Z}(\mathcal{A},\mathcal{B})$.

On the other hand if $\mathcal{A}$ and $\mathcal{B}$ are objects of $\mathbf{Algcom}$ then

$$\mathrm{Hom}_{\mathbf{Algcom}}(\mathcal{A},\mathcal{B})=\mathrm{Hom}_{\mathbf{Jord}}(\mathcal{A},\mathcal{B}).$$

Thus $\mathbf{Algcom}$ is *a full subcategory* of $\mathbf{Alg}$, of $\mathbf{Alg}_Z$ and of $\mathbf{Jord}$, i.e., for any $\mathcal{A},\mathcal{B}\in\mathrm{Ob}(\mathbf{Algcom})$ one has :

$$\begin{aligned}\mathrm{Hom}_{\mathbf{Algcom}}(\mathcal{A},\mathcal{B}) &= \mathrm{Hom}_{\mathbf{Alg}}(\mathcal{A},\mathcal{B})=\mathrm{Hom}_{\mathbf{Alg}_Z}(\mathcal{A},\mathcal{B})\\ &= \mathrm{Hom}_{\mathbf{Jord}}(\mathcal{A},\mathcal{B}).\end{aligned}$$

In order to discuss reality conditions we shall also need categories of $*$-algebras. By a $*$-algebra we here mean a general complex algebra $\mathcal{A}$ as above equipped with an antilinear involution $x\mapsto x^*$ such that $m(x\otimes y)^*=m(y^*\otimes x^*)$, (i.e., such that it reverses the order in the product). If $\mathcal{A}$ and $\mathcal{B}$ are $*$-algebras, a $*$-algebra homomorphism of $\mathcal{A}$ into $\mathcal{B}$ is an algebra homomorphism φ of $\mathcal{A}$ into $\mathcal{B}$ which preserves the involutions, i.e., $\varphi(x^*)=\varphi(x)^*$ for $x\in\mathcal{A}$. One defines the category of algebras $*$-$\mathbf{A}$ to be the category such that

Ob($*$-$\mathbf{A}$) is the class of all $*$-algebras and such that for any $\mathcal{A}, \mathcal{B} \in$ Ob($*$-$\mathbf{A}$), $\mathrm{Hom}_{*\text{-}\mathbf{A}}(\mathcal{A}, \mathcal{B})$ is the set of $*$-algebra homomorphisms of $\mathcal{A}$ into $\mathcal{B}$. A subcategory of $*$-$\mathbf{A}$ will be called a *category of $*$-algebras* and one defines in the obvious manner the categories of $*$-algebras $*$-$\mathbf{Alg}$, $*$-$\mathbf{Alg}_Z$, $*$-$\mathbf{Jord}$, $*$-$\mathbf{Algcom}$, $*$-$\mathbf{Lie}$ corresponding to the above examples 1, 2, 3, 4, 5.

Let $\mathbf{C}$ be a category of algebras and let $\mathcal{A}$ be an object of $\mathbf{C}$ with product denoted by $a \otimes a' \mapsto aa'$ $(a, a' \in a)$. A complex vector space $\mathcal{E}$ will be said to be a *$\mathcal{A}$-bimodule for* $\mathbf{C}$ if there are linear mappings $\mathcal{A} \otimes \mathcal{E} \to \mathcal{E}$ and $\mathcal{E} \otimes \mathcal{A} \to \mathcal{E}$, denoted by $a \otimes e \mapsto ae$ and $e \otimes a \mapsto ea$ $(a \in \mathcal{A}, e \in \mathcal{E})$ respectively, such that the direct sum $\mathcal{A} \oplus \mathcal{E}$ equipped with the product

$$(a \oplus e) \otimes (a' \oplus e') \mapsto aa' \oplus (ae' + ea')$$

is an object of $\mathbf{C}$ and such that the canonical linear mappings

$$i : \mathcal{A} \to \mathcal{A} \oplus \mathcal{E} \quad \text{and} \quad p : \mathcal{A} \oplus \mathcal{E} \to \mathcal{A}$$

defined by $i(a) = a \oplus 0$ and $p(a \oplus e) = a$ $(\forall a \in \mathcal{A}$ and $\forall e \in \mathcal{E})$ are morphisms of $\mathbf{C}$. In other words $\mathcal{E}$ is a $\mathcal{A}$-bimodule for $\mathbf{C}$ if $\mathcal{A} \oplus \mathcal{E}$ is equipped with a bilinear product vanishing on $\mathcal{E} \otimes \mathcal{E}$ and such that $\mathcal{A} \oplus \mathcal{E} \in \mathrm{Ob}(\mathbf{C})$, $i \in \mathrm{Hom}_{\mathbf{C}}(\mathcal{A}, \mathcal{A} \oplus \mathcal{E})$ and $p \in \mathrm{Hom}_{\mathbf{C}}(\mathcal{A} \oplus \mathcal{E}, \mathcal{A})$.

For the category $\mathbf{A}$ this notion of bimodule is not very restrictive. In fact, if $\mathcal{A}$ is an algebra (i.e., $\mathcal{A} \in \mathrm{Ob}(\mathbf{A})$) then a $\mathcal{A}$-bimodule for $\mathbf{A}$ is simply a complex vector space $\mathcal{E}$ with two bilinear mappings corresponding to linear mappings $\mathcal{A} \otimes \mathcal{E} \to \mathcal{E}$ and $\mathcal{E} \otimes \mathcal{A} \to \mathcal{E}$ as above. These two linear mappings will be always denoted by $a \otimes e \mapsto ae$ and $e \otimes a \mapsto ea$ and called *left* and *right action of $\mathcal{A}$ on $\mathcal{E}$*. Let us describe what restrictions occur for the categories of algebras of examples 1, 2, 3, 4, 5.

1. Let $\mathcal{A}$ be a unital associative complex algebra with product denoted by $a \otimes a' \mapsto aa'$ and unit denoted by $\mathbf{1}$. Then $\mathcal{E}$ is an $\mathcal{A}$-bimodule for $\mathbf{Alg}$ if and only if one has

$$\begin{aligned} &(i) \qquad (aa')e = a(a'e) \quad \text{and} \quad \mathbf{1}e = e, \\ &(ii) \qquad e(aa') = (ea)a' \quad \text{and} \quad e\mathbf{1} = e, \\ &(iii) \qquad (ae)a' = a(ea'), \end{aligned}$$

for any $a, a' \in \mathcal{A}$ and $e \in \mathcal{E}$. Conditions (i) express that $\mathcal{E}$ is a left $\mathcal{A}$-module in the usual sense; conditions (ii) express that $\mathcal{E}$ is a right $\mathcal{A}$-module in the usual sense; whereas, completed with the compatibility condition (iii), all these conditions express that $\mathcal{E}$ is a $(\mathcal{A}, \mathcal{A})$-bimodule in the usual sense for unital associative algebras.

2. Let $\mathcal{A}$ be as in 1 above. Then $\mathcal{E}$ is a $\mathcal{A}$-bimodule for $\mathbf{Alg}_Z$ if and only if it is a $\mathcal{A}$-bimodule for $\mathbf{Alg}$ such that one has

$$ze = ez$$

for any element z of the center $Z(\mathcal{A})$ of $\mathcal{A}$ and $e \in \mathcal{E}$. This condition expresses that, as a $(Z(\mathcal{A}), Z(\mathcal{A}))$-bimodule, $\mathcal{E}$ is the underlying bimodule of a $Z(\mathcal{A})$-module. Such $(\mathcal{A}, \mathcal{A})$-bimodules were called *central bimodules* over $\mathcal{A}$ in [34], [35] (see also in [27]). We shall keep this terminology here and call central bimodule a bimodule for $\mathbf{Alg}_Z$.

Let $\mathcal{E}$ be a $\mathcal{A}$-bimodule for $\mathbf{Alg}$ (i.e., a $(\mathcal{A}, \mathcal{A})$-bimodule). One can associate to $\mathcal{E}$ two $\mathcal{A}$-bimodules for $\mathbf{Alg}_Z$ (i.e., two central bimodules) $\mathcal{E}^Z$ and $\mathcal{E}_Z$. The bimodule $\mathcal{E}^Z$ is the biggest $(\mathcal{A}, \mathcal{A})$-sub-bimodule of $\mathcal{E}$ which is central and we denote by i^Z the canonical inclusion of $\mathcal{E}^Z$ into $\mathcal{E}$ whereas $\mathcal{E}_Z$ is the quotient of $\mathcal{E}$ by the $(\mathcal{A}, \mathcal{A})$-sub-bimodule $[Z(\mathcal{A}), \mathcal{E}]$ generated by the $ze - ez$ where z is in the center $Z(\mathcal{A})$ of $\mathcal{A}$, $e \in \mathcal{E}$ and we denote by p_Z the canonical projection of $\mathcal{E}$ onto $\mathcal{E}_Z$. The pair $(\mathcal{E}^Z, i^Z)$ is characterized by the following universal property: *For any $(\mathcal{A}, \mathcal{A})$-bimodule homomorphism $\Phi : \mathcal{N} \to \mathcal{E}$ of a central bimodule $\mathcal{N}$ into $\mathcal{E}$, there is a unique $(\mathcal{A}, \mathcal{A})$-bimodule homomorphism $\Phi^Z : \mathcal{N} \to \mathcal{E}^Z$ such that $\Phi = i^Z \circ \Phi^Z$.* The pair $(\mathcal{E}_Z, p_Z)$ is characterized by the following universal property: *For any $(\mathcal{A}, \mathcal{A})$-bimodule homomorphism $\varphi : \mathcal{E} \to \mathcal{M}$ of $\mathcal{E}$ into a central bimodule $\mathcal{M}$ there is a unique $(\mathcal{A}, \mathcal{A})$-bimodule homomorphism $\varphi_Z : \mathcal{E}_Z \to \mathcal{M}$ such that $\varphi = \varphi_Z \circ p_Z$.* In functorial language this means that $\mathcal{E} \mapsto \mathcal{E}^Z$ is a right adjoint and that $\mathcal{E} \mapsto \mathcal{E}_Z$ is a left adjoint of the canonical functor I_Z from the category of $\mathcal{A}$-bimodules for $\mathbf{Alg}_Z$ in the category of $\mathcal{A}$-bimodules for $\mathbf{Alg}$. Notice also that $\mathcal{E}$ is central if and only if $\mathcal{E} = \mathcal{E}_Z$, which is equivalent to $\mathcal{E} = \mathcal{E}^Z$, and that if $\mathcal{M}$ and $\mathcal{N}$ are two $\mathcal{A}$-bimodules for $\mathbf{Alg}_Z$ (i.e., two central bimodules) then one has $(\mathcal{M} \otimes \mathcal{N})_Z = \mathcal{M} \otimes_{Z(\mathcal{A})} \mathcal{N}$. One has the further following stability properties for the $\mathcal{A}$-bimodules for $\mathbf{Alg}_Z$: Every sub-bimodule of a central bimodule is central, every quotient of a central bimodule is central and any product of central bimodules is central. For all this we refer to [35].

3. Let $\mathcal{J}$ be a complex unital Jordan algebra with product denoted by $x \otimes y \mapsto x \bullet y$ $(x, y \in \mathcal{J})$ and unit $\mathbb{1}$. Then $\mathcal{E}$ is a $\mathcal{J}$-bimodule for $\mathbf{Jord}$ if and only if one has

$$(i) \qquad xe = ex \quad \text{and} \quad \mathbb{1}e = e,$$

$$(ii) \qquad x((x \bullet x)e) = (x \bullet x)(xe),$$

$$(iii) \qquad ((x \bullet x) \bullet y)e - (x \bullet x)(ye) = 2((x \bullet y)(xe) - x(y(xe))),$$

for any $x, y \in \mathcal{J}$ and $e \in \mathcal{E}$. Such a bimodule for **Jord** is called a *Jordan module* over $\mathcal{J}$ [44] which is natural since, in view of (i), there is only one bilinear mapping of $\mathcal{J} \times \mathcal{E}$ into $\mathcal{E}$.

4. Let $\mathcal{C}$ be a unital associative commutative complex algebra. Then $\mathcal{E}$ is a $\mathcal{C}$-bimodule for **Algcom** if and only if it is a $\mathcal{C}$-bimodule for **Alg** such that one has

$$ce = ec$$

for any $c \in \mathcal{C}$ and $e \in \mathcal{E}$. This means that a $\mathcal{C}$-bimodule for **Algcom** is the same thing as (the underlying bimodule of) a $\mathcal{C}$-module in the usual sense. Since the center of $\mathcal{C}$ coincides with $\mathcal{C}$, $Z(\mathcal{C}) = \mathcal{C}$, this implies that it is also the same thing as a $\mathcal{C}$-bimodule for $\mathbf{Alg}_Z$, as announced in the introduction. Notice that in the case of a $\mathcal{C}$-bimodule for **Alg** one generally has $ce \neq ec$.

5. Let $\mathfrak{g}$ be a complex Lie algebra with product (Lie bracket) denoted by $X \otimes Y \mapsto [X, Y]$ for $X, Y \in \mathfrak{g}$. Then, $\mathcal{E}$ is a $\mathfrak{g}$-bimodule for **Lie** if and only if one has

$$(i) \qquad Xe = -eX,$$

$$(ii) \qquad [X, Y]e = X(Ye) - Y(Xe),$$

for any $X, Y \in \mathfrak{g}$ and $e \in \mathcal{E}$. Condition (i) shows that again there is only one bilinear mapping of $\mathfrak{g} \times \mathcal{E}$ into $\mathcal{E}$ and (ii) means that $\mathcal{E}$ is the space of a linear representation of $\mathfrak{g}$; Thus a $\mathfrak{g}$-bimodule for **Lie** is what is usually called a $\mathfrak{g}$-module (or a linear representation of $\mathfrak{g}$).

Similarly, one has the notion of $*$-bimodule for a category $*$-**C** of $*$-algebras. Namely, if $\mathcal{A} \in \mathrm{Ob}(*\text{-}\mathbf{C})$, a complex vector space $\mathcal{E}$ will be said to be a $\mathcal{A}$-$*$-*bimodule for* $*$-**C** if $\mathcal{A} \oplus \mathcal{E}$ is equipped with a structure of $*$-algebra with product vanishing on $\mathcal{E} \otimes \mathcal{E}$ such that $\mathcal{A} \oplus \mathcal{E} \in \mathrm{Ob}(*\text{-}\mathbf{C})$, $i \in \mathrm{Hom}_{*\text{-}\mathbf{C}}(\mathcal{A}, \mathcal{A} \oplus \mathcal{E})$ and $p \in \mathrm{Hom}_{*\text{-}\mathbf{C}}(\mathcal{A} \oplus \mathcal{E}, \mathcal{A})$.

One can easily describe what is a $*$-bimodule for the various categories of $*$-algebras. If $\mathcal{A}$ is a $*$-algebra we also denote by $\mathcal{A}$ the algebra obtained by 'forgetting the involution'. If $\mathcal{A}$ is an object of $*$-**Alg** then a $\mathcal{A}$-$*$-bimodule for $*$-**Alg** is a $\mathcal{A}$-bimodule $\mathcal{E}$ for **Alg** which is equipped with an antilinear involution $e \mapsto e^*$ such that $(xey)^* = y^* e^* x^*$ for $x, y \in \mathcal{A}$ and $e \in \mathcal{E}$, i.e., it is what has been called in the introduction a $*$-bimodule over the (unital associative complex) $*$-algebra $\mathcal{A}$. A $\mathcal{A}$-$*$-bimodule for $*$-$\mathbf{Alg}_Z$ is then just such a $*$-bimodule over $\mathcal{A}$ which is central. If $\mathcal{C}$ is a unital associative complex commutative $*$-algebra, then a $\mathcal{C}$-$*$-bimodule for $*$-**Algcom** is just what has been called a $*$-module over the (unital associative complex) commutative $*$-algebra $\mathcal{C}$.

One can proceed similarly with real algebras. However, to be in conformity with the point of view of the introduction concerning reality, we shall work with $*$-algebras and, eventually, extract their Hermitian parts as well as the Hermitian parts of the $*$-bimodules over them.

6. First order differential calculi

Throughout the following $\mathcal{A}$ denotes a unital associative complex algebra. A pair (Ω^1, d) where Ω^1 is a $(\mathcal{A}, \mathcal{A})$-bimodule (i.e., a $\mathcal{A}$-bimodule for **Alg**) and where $d : \mathcal{A} \to \Omega^1$ is a derivation of $\mathcal{A}$ into Ω^1, that is a linear mapping which satisfies (the Leibniz rule)

$$d(xy) = d(x)y + xd(y)$$

for any $x, y \in \mathcal{A}$, will be called a *first order differential calculus over* $\mathcal{A}$ *for* **Alg** or simply a *first order differential calculus over* $\mathcal{A}$ [61]. If ,furthermore, Ω^1 is a central bimodule (i.e., a $\mathcal{A}$-bimodule for $\mathbf{Alg}_Z$), we shall say that (Ω^1, d) is a *first order differential calculus over* $\mathcal{A}$ *for* $\mathbf{Alg}_Z$. One can, more generally, define the notion of first order differential calculus over $\mathcal{A}$ for any category $\mathbf{C}$ of algebras such that $\mathcal{A} \in \mathrm{Ob}(\mathbf{C})$.

Remark 5. If Ω^1 is a $\mathcal{A}$-bimodule for $\mathbf{C}$ a derivation $d : \mathcal{A} \to \Omega^1$ can be defined to be a linear mapping such that $a \mapsto a \oplus d(a)$ is in $\mathrm{Hom}_{\mathbf{C}}(\mathcal{A}, \mathcal{A} \oplus \Omega^1)$. However, for the category $\mathbf{Alg}_Z$ this does not impose restrictions on first order differential calculus. Indeed if Ω^1 is a central bimodule and if $d : \mathcal{A} \to \Omega^1$ is a derivation one has $d(z)a + zd(a) = d(za) = d(az) = ad(z) + d(a)z$ for any $a \in \mathcal{A}$ and z in the center $Z(\mathcal{A})$ of $\mathcal{A}$, i.e., $d(z)a = ad(z)$ since, by 'centrality', $zd(a) = d(a)z$; again, by centrality $z\omega = \omega z$, $\forall z \in Z(\mathcal{A})$ and $\forall \omega \in \Omega^1$, which finally implies $(z \oplus d(z))(a \oplus \omega) = (a \oplus \omega)(z \oplus d(z))$ and therefore $z \oplus d(z) \in Z(\mathcal{A} \oplus \Omega^1)$ for any $z \in Z(\mathcal{A})$, which means that the linear mapping $a \mapsto a \oplus d(a)$ is in $\mathrm{Hom}_{\mathbf{Alg}_Z}(\mathcal{A}, \mathcal{A} \oplus \Omega^1)$.

We shall refer to d as the *first order differential*; by definition it is an Ω^1-valued Hochschild cocycle of degree 1 of $\mathcal{A}$, i.e., $d \in Z^1_H(\mathcal{A}, \Omega^1)$. Examples of first order differentials are thus provided by Hochschild coboundaries i.e., given by $d(x) = \tau x - x\tau$ ($\forall x \in \mathcal{A}$) for some $\tau \in \Omega^1$. We shall now explain that there are 'universal first order differential calculi' for **Alg** and for $\mathbf{Alg}_Z$ which respectively define functors from **Alg** and from $\mathbf{Alg}_Z$ in the corresponding categories of first order differential calculi. For the case of a commutative algebra there is also a well known universal first order differential calculus for **Algcom** which is the universal derivation into the module of

Kähler differentials ([6], [52], [58]). We shall see however that it reduces to the universal calculus for $\mathbf{Alg}_Z$ (Corollary 1).

Let m be the product of $\mathcal{A}$, $(x, y) \mapsto m(x \otimes y) = xy$ and let $\Omega^1_u(\mathcal{A})$ be the kernel of m, i.e., one has the short exact sequence

$$0 \to \Omega^1_u(\mathcal{A}) \overset{\subseteq}{\to} \mathcal{A} \otimes \mathcal{A} \overset{m}{\to} \mathcal{A} \to 0$$

of $(\mathcal{A}, \mathcal{A})$-bimodules ($\mathcal{A}$-bimodules for $\mathbf{Alg}$). Define $d_u : \mathcal{A} \to \Omega^1_u(\mathcal{A})$ by $d_u(x) = \mathbb{1} \otimes x - x \otimes \mathbb{1}$, $\forall x \in \mathcal{A}$. One verifies easily that d_u is a derivation. The first order differential calculus $(\Omega^1_u(\mathcal{A}), d_u)$ over $\mathcal{A}$ is characterized uniquely (up to an isomorphism) by the following universal property [10], [6].

PROPOSITION 1 *For any first order differential calculus (Ω^1, d) over $\mathcal{A}$, there is a unique bimodule homomorphism i_d of $\Omega^1_u(\mathcal{A})$ into Ω^1 such that $d = i_d \circ d_u$.*

Proof. $\Omega^1_u(\mathcal{A})$ is generated by $d_u(\mathcal{A})$ as left module since $x^\alpha \otimes y_\alpha$ with $x^\alpha y_\alpha = 0$ is the same thing as $x^\alpha d(y_\alpha)$. On the other hand $d_u(\mathbb{1}) = 0 (= d_u(\mathbb{1}^2) = 2d_u(\mathbb{1}))$. Therefore one has a surjective left $\mathcal{A}$-module homomorphism of $\mathcal{A} \otimes (\mathcal{A}/\mathbb{C}\mathbb{1})$ onto $\Omega^1_u(\mathcal{A})$, $x \otimes \dot{y} \mapsto x d_u(y)$, which is easily shown to be an isomorphism. Then $x d_u(y) \mapsto x d(y)$ defines a left $\mathcal{A}$-module homomorphism i_d of $\Omega^1_u(\mathcal{A})$ into Ω^1 which is easily shown to be a bimodule homomorphism by using the Leibniz rule for d_u and for d. One clearly has $d = i_d \circ d_u$. Uniqueness is straightforward. □

Concerning the image of i_d, let us notice the following easy lemma.

LEMMA 4 *Let (Ω^1, d) be a first order differential calculus over $\mathcal{A}$. The following conditions are equivalent:*

- (i) *Ω^1 is generated by $d\mathcal{A}$ as left $\mathcal{A}$-module;*
- (ii) *Ω^1 is generated by $d\mathcal{A}$ as right $\mathcal{A}$-module;*
- (iii) *Ω^1 is generated by $d\mathcal{A}$ as $(\mathcal{A}, \mathcal{A})$-bimodule;*
- $(iiii)$ *The homomorphism i_d is surjective, i.e., $\Omega^1 = i_d(\Omega^1_u(\mathcal{A}))$.*

Proof. The equivalences $(i) \Leftrightarrow (ii) \Leftrightarrow (iii)$ follows from (Leibniz' rule)

$$ud(v)w = ud(vw) - uvd(w) = d(uv)w - d(u)vw$$

for $u, v, w \in \mathcal{A}$, whereas the equivalence $(iii) \Leftrightarrow (iiii)$ is straightforward from the definitions. □

Remark 6. Proposition 1 claims that there is a unique bimodule homomorphism i_d of $\Omega^1_u(\mathcal{A})$ into Ω^1 mapping the $\Omega^1_u(\mathcal{A})$-valued Hochschild 1-cocycle d_u on the Ω^1-valued Hochschild 1-cocycle d. One can complete the statement by the following: *The Ω^1-valued Hochschild 1-cocycle d is a Hochschild coboundary (i.e., there is a $\tau \in \Omega^1$ such that $d(a) = \tau a - a\tau$ for any $a \in \mathcal{A}$) if and only if i_d has an extension $\tilde{i}_d$ as a bimodule homomorphism of $\mathcal{A} \otimes \mathcal{A}$ into Ω^1*, [8]. In fact τ is then $\tilde{i}_d(\mathbb{1} \otimes \mathbb{1})$, which gives essentially the proof.

The first order differential calculus $(\Omega^1_u(\mathcal{A}), d_u)$ *is universal for* **Alg**, it is usually simply called *the universal first order differential calculus over $\mathcal{A}$*. From Proposition 1 follows the functorial property.

PROPOSITION 2 *Let $\mathcal{A}$ and $\mathcal{B}$ be algebras and let $\varphi : \mathcal{A} \to \mathcal{B}$ be a homomorphism, (i.e., let $\mathcal{A}, \mathcal{B} \in \mathrm{Ob}(\mathbf{Alg})$ and let $\varphi \in \mathrm{Hom}_{\mathbf{Alg}}(\mathcal{A}, \mathcal{B})$), then there is a unique linear mapping $\Omega^1_u(\varphi)$ of $\Omega^1_u(\mathcal{A})$ into $\Omega^1_u(\mathcal{B})$ satisfying $\Omega^1_u(\varphi)(x\omega y) = \varphi(x)\Omega^1_u(\varphi)(\omega)\varphi(y)$ for any $x, y \in \mathcal{A}$ and $\omega \in \Omega^1_u(\mathcal{A})$ and such that $d_u \circ \varphi = \Omega^1_u(\varphi) \circ d_u$.*

Proof. One equips $\Omega^1_u(\mathcal{B})$ of a structure of $(\mathcal{A}, \mathcal{A})$-bimodule by setting $x\lambda y = \varphi(x)\lambda\varphi(y)$ for $x, y \in \mathcal{A}$ and $\lambda \in \Omega^1_u(\mathcal{B})$. Then $d = d_u \circ \varphi$ is a derivation of $\mathcal{A}$ into the $(\mathcal{A}, \mathcal{A})$-bimodule $\Omega^1_u(\mathcal{B})$, i.e., $(\Omega^1_u(\mathcal{B}), d)$ is a first order differential calculus over $\mathcal{A}$, and the result follows from Proposition 1 with $\Omega^1_u(\varphi) = i_d$. □

One can summarize the content of Proposition 2 by the following: *For any homomorphism $\varphi : \mathcal{A} \to \mathcal{B}$ (of unital associative $\mathbb{C}$-algebras) there is a unique $(\mathcal{A}, \mathcal{A})$-bimodule homomorphism $\Omega^1_u(\varphi) : \Omega^1_u(\mathcal{A}) \to \Omega^1_u(\mathcal{B})$ for which the diagram*

$$\begin{array}{ccc} \mathcal{A} & \xrightarrow{\varphi} & \mathcal{B} \\ \downarrow d_u & & \downarrow d_u \\ \Omega^1_u(\mathcal{A}) & \xrightarrow{\Omega^1_u(\varphi)} & \Omega^1_u(\mathcal{B}) \end{array}$$

is commutative. All this was for the category **Alg**, we now pass to $\mathbf{Alg}_Z$.

Let $[Z(\mathcal{A}), \Omega^1_u(\mathcal{A})]$ be the sub-bimodule of $\Omega^1_u(\mathcal{A})$ defined by

$$[Z(\mathcal{A}), \Omega^1_u(\mathcal{A})] = \{z\omega - \omega z | z \in Z(\mathcal{A}), \omega \in \Omega^1_u(\mathcal{A})\}.$$

By definition the quotient $\Omega^1_Z(\mathcal{A}) = \Omega^1_u(\mathcal{A})/[Z(\mathcal{A}), \Omega^1_u(\mathcal{A})]$ is a central bimodule, i.e., a $\mathcal{A}$-bimodule for $\mathbf{Alg}_Z$. Let $p_Z : \Omega^1_u(\mathcal{A}) \to \Omega^1_Z(\mathcal{A})$ be the canonical projection and let $d_Z : \mathcal{A} \to \Omega^1_Z(\mathcal{A})$ be defined by $d_Z = p_Z \circ d_u$. Then d_Z is again a derivation so $(\Omega^1_Z(\mathcal{A}), d_Z)$ is a first order differential calculus over $\mathcal{A}$ for $\mathbf{Alg}_Z$. It is characterized uniquely (up to an isomorphism) among the first order differential calculi over $\mathcal{A}$ for $\mathbf{Alg}_Z$ by the following universal property [35].

PROPOSITION 3 *For any first order differential calculus (Ω^1, d) over $\mathcal{A}$ for $\mathbf{Alg}_Z$, there is a unique bimodule homomorphism i_d of $\Omega^1_Z(\mathcal{A})$ into Ω^1 such that $d = i_d \circ d_Z$; i.e., there is a unique morphism of first order differential calculi over $\mathcal{A}$ for $\mathbf{Alg}_Z$ from $(\Omega^1_Z(\mathcal{A}), d_Z)$ to (Ω^1, d).*

Proof. The unique bimodule homomorphism $i_d : \Omega^1_u(\mathcal{A}) \to \Omega^1$ of Proposition 1 vanishes on $[Z(\mathcal{A}), \Omega^1_u(\mathcal{A})]$ since Ω^1 is central. Therefore it factorizes as $\Omega^1_u(\mathcal{A}) \overset{p_Z}{\to} \Omega^1_Z(\mathcal{A}) \to \Omega^1$ through a unique bimodule homomorphism, again denoted i_d, of $\Omega^1_Z(\mathcal{A})$ into Ω^1 for which one has $d = i_d \circ d_Z$. Again, uniqueness is obvious. □

Remark 7. Proposition 3 can be slightly improved. One can replace the assumption '(Ω^1, d) over $\mathcal{A}$ for $\mathbf{Alg}_Z$' by '(Ω^1, d) over $\mathcal{A}$ such that $zd(a) = d(a)z$ for any $a \in \mathcal{A}$ and $z \in Z(\mathcal{A})$' in the statement. That is to say, what is important is that the sub-bimodule of Ω^1 generated by $d\mathcal{A}$ is central.

The first order differential calculus $(\Omega^1_Z(\mathcal{A}), d_Z)$ will be called *the universal first order differential calculus over $\mathcal{A}$ for* $\mathbf{Alg}_Z$. Concerning the functorial property of this first order differential calculus, Proposition 2 has the following counterpart for $\mathbf{Alg}_Z$.

PROPOSITION 4 *Let $\mathcal{A}$ and $\mathcal{B}$ be algebras as above and let $\varphi : \mathcal{A} \to \mathcal{B}$ be a homomorphism such that $\varphi(Z(\mathcal{A})) \subset Z(\mathcal{B})$, (i.e., let $\mathcal{A}, \mathcal{B} \in \mathrm{Ob}(\mathbf{Alg}_Z)$ and let $\varphi \in \mathrm{Hom}_{\mathbf{Alg}_Z}(\mathcal{A}, \mathcal{B})$), then there is a unique linear mapping $\Omega^1_Z(\varphi)$ of $\Omega^1_Z(\mathcal{A})$ into $\Omega^1_Z(\mathcal{B})$ satisfying $\Omega^1_Z(\varphi)(x\omega y) = \varphi(x)\Omega^1_Z(\varphi)(\omega)\varphi(y)$ for any $x, y \in \mathcal{A}$ and $\omega \in \Omega^1_Z(\mathcal{A})$ and such that $d_Z \circ \varphi = \Omega^1_Z(\varphi) \circ d_Z$.*

Proof. Again, as in the proof of Proposition 2, $\Omega^1_Z(\mathcal{B})$ is an $(\mathcal{A}, \mathcal{A})$-bimodule by setting $x\lambda y = \varphi(x)\lambda\varphi(y)$ for $x, y \in \mathcal{A}$ and $\lambda \in \Omega^1_Z(\mathcal{B})$. Thus Proposition 4 follows from Proposition 3 if one can show that this bimodule is central, i.e., if $\varphi(z)\lambda = \lambda\varphi(z)$ for any $z \in Z(\mathcal{A})$ and $\lambda \in \Omega^1_Z(\mathcal{B})$. This, however, follows from $\Omega^1_Z(\mathcal{B})$ being central over $\mathcal{B}$ and that φ maps the center $Z(\mathcal{A})$ of $\mathcal{A}$ into the center $Z(\mathcal{B})$ of $\mathcal{B}$. □

Again this can be summarized (by identifying $\Omega^1_Z(\mathcal{B})$ with a central bimodule over $\mathcal{A}$ via φ) as : *For any $\varphi \in \mathrm{Hom}_{\mathbf{Alg}_Z}(\mathcal{A}, \mathcal{B})$, there is a unique homomorphism of $\mathcal{A}$-bimodules for $\mathbf{Alg}_Z$, $\Omega^1_Z(\varphi) : \Omega^1_Z(\mathcal{A}) \to \Omega^1_Z(\mathcal{B})$, for*

which the diagram

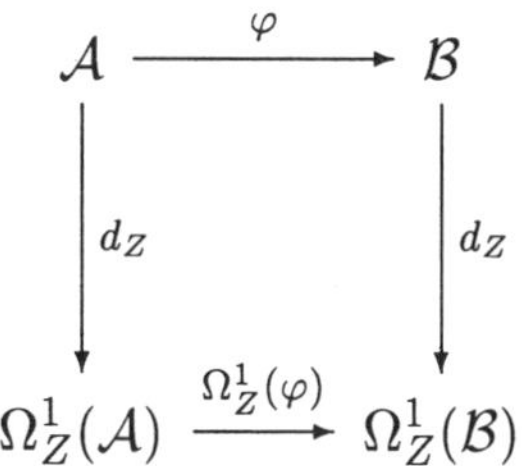

is commutative.

Proposition 3 has the following corollary:

COROLLARY 1 *If $\mathcal{A}$ is commutative $\Omega^1_Z(\mathcal{A})$ identifies canonically with the module of Kähler differentials $\Omega^1_{\mathcal{A}|\mathbb{C}}$ and d_Z identifies with the corresponding universal derivation.*

Proof. The proof is straightforward, since for a commutative algebra $\mathcal{A}$ a central bimodule is just (the underlying bimodule of) a $\mathcal{A}$-module, and Proposition 3 then just reduces to the universal property which characterizes the first order Kähler differential calculus (see e.g., in [6], [52], [58]). □

Remark 8. If $\mathcal{A}$ is commutative the module of Kähler differentials $\Omega^1_{\mathcal{A}|\mathbb{C}}$ is known to be a version of differential 1-forms. There is, however, a little subtlety. In fact $\Omega^1_{\mathcal{A}|\mathbb{C}}$ is the quotient of $\Omega^1_u(\mathcal{A})$ which is a commutative algebra (a subalgebra of $\mathcal{A} \otimes \mathcal{A}$) by the ideal $(\Omega^1_u(\mathcal{A}))^2$. If $\mathcal{A}$ is the algebra of smooth functions $C^\infty(M)$ on a manifold M, this means that $\Omega^1_{\mathcal{A}|\mathbb{C}}$ is the algebra of functions in $\mathcal{A}\otimes\mathcal{A} = C^\infty(M)\otimes C^\infty(M)$ vanishing on the diagonal of $M \times M$ modulo functions vanishing to order one on the diagonal of $M \times M$. On the other hand, it is clear (by using the Taylor expansion around the diagonal) that the ordinary differential 1-forms are smooth functions of $C^\infty(M \times M)$ vanishing on the diagonal of $M \times M$ modulo the functions vanishing to order one on the diagonal of $M \times M$. The subtlety here lies in that without completion of the tensor product the inclusion $C^\infty(M) \otimes C^\infty(M) \subset C^\infty(M \times M)$ is a strict one, so there is generally a slight difference between $\Omega^1_{C^\infty(M)|\mathbb{C}}$ and the module $\Omega^1(M)$ of smooth 1-forms on M. Apart from this one can consider that $(\Omega^1_Z(\mathcal{A}), d_Z)$ generalizes the ordinary first order differential calculus. This is in contrast to what happens for $(\Omega^1_u(\mathcal{A}), d_u)$. Indeed, if $\mathcal{A}$ is an algebra of functions on a set S containing more than one element, $(\text{card}(S) > 1)$, then $\Omega^1_u(\mathcal{A})$ consists of functions on $S \times S$ which vanish on the diagonal, and is therefore *not* the underlying bimodule of a module (non-locality), whereas $(d_u f)(x, y) = f(y) - f(x)$ $(x, y \in S)$ is the finite difference.

7. Higher order differential calculi

Let $\mathcal{A}$ be as before a unital associative complex algebra. A $\mathbb{N}$-graded differential algebra $\Omega = \oplus_{n\geq 0} \Omega^n$ such that the subalgebra Ω^0 of its elements of degree 0 coincides with $\mathcal{A}$, $\Omega^0 = \mathcal{A}$, will be called a *differential calculus over* $\mathcal{A}$ *for* **Alg**, or simply a *differential calculus over* $\mathcal{A}$. If, furthermore, the Ω^n ($n \in \mathbb{N}$) are central bimodules over $\mathcal{A}$ (i.e., $\mathcal{A}$-bimodules for $\mathbf{Alg}_Z$) Ω will be said to be a *differential calculus over* $\mathcal{A}$ *for* $\mathbf{Alg}_Z$.

Let us define the $(\mathcal{A},\mathcal{A})$-bimodules $\Omega^n_u(\mathcal{A})$ for $n \geq 0$ by $\Omega^0_u(\mathcal{A}) = \mathcal{A}$ and by $\Omega^n_u(\mathcal{A}) = \Omega^1_u(\mathcal{A}) \otimes_\mathcal{A} \dots \otimes_\mathcal{A} \Omega^1_u(\mathcal{A})$ (n factors) for $n \geq 1$. The direct sum $\Omega_u(\mathcal{A}) = \oplus_{n\geq 0} \Omega^n_u(\mathcal{A})$ is canonically a graded algebra, it is *the tensor algebra over* $\mathcal{A}$, $T_\mathcal{A}(\Omega^1_u(\mathcal{A}))$, *of the* $(\mathcal{A},\mathcal{A})$*-bimodule* $\Omega^1_u(\mathcal{A})$. The derivation $d_u : \mathcal{A} \to \Omega^1_u(\mathcal{A})$ has a unique extension, again denoted by d_u, as a differential of $\Omega_u(\mathcal{A})$: In fact, it is known on $\mathcal{A} = \Omega^0_u(\mathcal{A})$ and $d_u^2 = 0$ fixes it on $d_u(\mathcal{A})$ to be 0 so it is known on the generators of $\Omega_u(\mathcal{A})$ and the extension by the antiderivation property to the whole $\Omega_u(\mathcal{A})$ is well defined and unique; moreover, d_u^2 is a derivation vanishing on the generators and therefore $d_u^2 = 0$. So equipped, $\Omega_u(\mathcal{A})$ is a graded differential algebra [46] which is characterized uniquely (up to an isomorphism) by the following universal property.

PROPOSITION 5 *Any homomorphism* φ *of unital algebras of* $\mathcal{A}$ *into the subalgebra* Ω^0 *of elements of degree 0 of a graded differential algebra* Ω *has a unique extension* $\tilde{\varphi} : \Omega_u(\mathcal{A}) \to \Omega$ *as a homomorphism of graded differential algebras.*

Proof. The (Ω^0, Ω^0)-bimodule Ω^1 can be considered as a $(\mathcal{A},\mathcal{A})$-bimodule by setting $x\lambda y = \varphi(x)\lambda\varphi(y)$ for $x, y \in \mathcal{A}$ and $\lambda \in \Omega^1$ and then $d \circ \varphi$ defines a derivation of $\mathcal{A}$ into Ω^1. Therefore, by Proposition 1, there is a unique bimodule homomorphism $\varphi^1 : \Omega^1_u(\mathcal{A}) \to \Omega^1$ such that $d\circ\varphi = \varphi^1 \circ d_u : \mathcal{A} \to \Omega^1$ (namely $\varphi^1 = i_{d\circ\varphi}$). The property of $\Omega_u(\mathcal{A})$ to be the tensor algebra $T_\mathcal{A}(\Omega^1_u(\mathcal{A}))$ implies that φ and φ^1 uniquely extend as a homomorphism $\tilde{\varphi} : \Omega_u(\mathcal{A}) \to \Omega$ of graded algebras. By construction one has $\tilde{\varphi} \circ d_u = d \circ \tilde{\varphi}$ on $\mathcal{A}$ and on $d_u\mathcal{A}$ where it vanishes which implies $\tilde{\varphi} \circ d_u = d \circ \tilde{\varphi}$ everywhere by the antiderivation property of d_u and d. □

The graded differential algebra $\Omega_u(\mathcal{A})$ is called (in view of the above universal property) *the universal differential calculus over* $\mathcal{A}$ (it is universal for **Alg**). The functorial property follows immediately: *For any homomorphism* $\varphi : \mathcal{A} \to \mathcal{B}$ *(of unital* $\mathbb{C}$*-algebras), there is a unique homomorphism* $\Omega_u(\varphi) : \Omega_u(\mathcal{A}) \to \Omega_u(\mathcal{B})$ *of graded differential algebra which extends* φ *(i.e.,* $\varphi = \Omega_u(\varphi) \restriction \mathcal{A}$*).* This defines the covariant functor Ω_u from the category **Alg** to the category **Dif** of graded differential algebras (the morphisms being the homomorphisms of graded differential algebras preserving the units).

Proposition 5 is clearly a generalization of Proposition 1. There is another useful generalization of the universality of the Hochschild 1-cocycle $a \mapsto d_u(a)$ (which is the content of Proposition 1) and of Remark 6 which is described in [8] (see also in [22]) and which we now review (Proposition 6 below). First, notice that $(a_1, \dots, a_n) \mapsto d_u(a_1)\dots d_u(a_n)$ is a $\Omega^n_u(\mathcal{A})$-valued Hochschild n-cocycle which is normalized (i.e., which vanishes whenever one of the a_i is the unit **1** of $\mathcal{A}$). Second, notice that the short exact sequence of Section 6 (before Proposition 1) has the following generalization for $n \geq 1$

$$0 \to \Omega^n_u(\mathcal{A}) \overset{\subseteq}{\to} \mathcal{A} \otimes \Omega^{n-1}_u(\mathcal{A}) \overset{m}{\to} \Omega^{n-1}_u(\mathcal{A}) \to 0$$

as a short exact sequence of $(\mathcal{A},\mathcal{A})$-bimodules, where m is the left multiplication by elements of $\mathcal{A}$ of elements of $\Omega^{n-1}_u(\mathcal{A})$ (the inclusion is canonical). One has the following [8].

PROPOSITION 6 *Let $\mathcal{M}$ be a $(\mathcal{A},\mathcal{A})$-bimodule and let $(a_1,\dots,a_n) \mapsto c(a_1,\dots,a_n)$ be a normalized $\mathcal{M}$-valued Hochschild n-cocycle. Then there is a unique bimodule homomorphism $i_c : \Omega^n_u(\mathcal{A}) \to \mathcal{M}$ such that*

$$c(a_1,\dots,a_n) = i_c(d_u(a_1)\dots d_u(a_n)),\ \forall a_i \in \mathcal{A}.$$

Furthermore, c is a Hochschild coboundary if and only if i_c has an extension $\tilde{i}_c$ as a bimodule homomorphism of $\mathcal{A} \otimes \Omega^{n-1}_u(\mathcal{A})$ into $\mathcal{M}$.

Proof. We only give here some indications and refer to [8] for the detailed proof. The proof of the first part proceeds exactly as the proof of Proposition 1: One first shows that the mapping $a_0 \otimes a_1 \otimes \dots \otimes a_n \mapsto a_0 d_u(a_1)\dots d_u(a_n)$ induces a left module isomorphism of $\mathcal{A} \otimes (\otimes^n(\mathcal{A}/\mathbb{C}\mathbf{1}))$ onto $\Omega^n_u(\mathcal{A})$, which implies that $a_0 d_u(a_1)\dots d_u(a_n) \mapsto a_0 c(a_1,\dots,a_n)$ defines a left module homomorphism i_c of $\Omega^n_u(\mathcal{A})$ into $\mathcal{M}$; the cocycle property of c then implies that i_c is a bimodule homomorphism. Again, uniqueness is straightforward. Concerning the last part, if there is an extension $\tilde{i}_c$ to $\mathcal{A} \otimes \Omega^{n-1}_u(\mathcal{A})$ then c is the Hochschild coboundary of $(a_1,\dots,a_{n-1}) \mapsto \tilde{i}_c(\mathbf{1} \otimes d_u(a_1)\dots d_u(a_{n-1}))$ and conversely, if c is the coboundary of a normalized $(n-1)$-cochain c' then one defines an extension $\tilde{i}_c$ by setting $\tilde{i}_c(\mathbf{1} \otimes d_u(a_1)\dots d_u(a_{n-1})) = c'(a_1,\dots,a_{n-1})$. □

Thus for each integer $n \geq 1$ the normalized n-cocycle $d_u^{\cup n}$, defined by $d_u^{\cup n}(a_1,\dots,a_n) = d_u(a_1)\dots d_u(a_n)$ is universal among the normalized Hochschild n-cocycles.

By its very construction $\Omega_u(\mathcal{A})$ is a graded subalgebra of the tensor algebra over $\mathcal{A}$, $T_{\mathcal{A}}(\mathcal{A} \otimes \mathcal{A})$, of the $(\mathcal{A},\mathcal{A})$-bimodule $\mathcal{A} \otimes \mathcal{A}$. Indeed, $T^n_{\mathcal{A}}(\mathcal{A} \otimes \mathcal{A})$ is the

$(n+1)$-th tensor power (over $\mathbb{C}$) $\otimes^{n+1}\mathcal{A}$ of $\mathcal{A}$, whereas $\Omega^n_u(\mathcal{A}) = T^n_{\mathcal{A}}(\Omega^1_u(\mathcal{A}))$ is the intersection of the kernels of the $(\mathcal{A},\mathcal{A})$-bimodule homomorphisms $m_k : \otimes^{n+1}\mathcal{A} \to \otimes^n\mathcal{A}$ defined by

$$\begin{array}{ll} m_1(x_0 \otimes \cdots \otimes x_n) & = x_0x_1 \otimes x_2 \otimes \cdots \otimes x_n \\ \dots\dots\dots\dots\dots & \dots\dots\dots\dots\dots\dots\dots \\ m_n(x_0 \otimes \cdots \otimes x_n) & = x_1 \otimes \cdots \otimes x_{n-2} \otimes x_{n-1}x_n \end{array}$$

(i.e., m_k is the product in $\mathcal{A}$ of the consecutive factors x_{k-1} and x_k). It turns out that the differential of $\Omega_u(\mathcal{A})$ has an extension, again denoted by d_u, as a differential of $T_{\mathcal{A}}(\mathcal{A}\otimes\mathcal{A})$ which is defined by

$$d_u(x_0 \otimes \cdots \otimes x_n) = \sum_{k=0}^{n+1}(-1)^k \quad x_0 \otimes \cdots \otimes x_{k-1} \otimes \mathbf{1} \otimes x_k \otimes \cdots \otimes x_n$$

for $x_i \in \mathcal{A}$ where, by convention, the first term of the summation is $\mathbf{1} \otimes x_0 \otimes \cdots \otimes x_n$ and the last term is $(-1)^{n+1}x_0 \otimes \cdots \otimes x_n \otimes \mathbf{1}$. So equipped, $T_{\mathcal{A}}(\mathcal{A}\otimes\mathcal{A})$ is a graded differential algebra, in fact a differential calculus over $\mathcal{A}$, and $\Omega_u(\mathcal{A})$ is a graded-differential subalgebra.

LEMMA 5 *The cohomologies of $T_{\mathcal{A}}(\mathcal{A}\otimes\mathcal{A})$ and of $\Omega_u(\mathcal{A})$ are trivial in the sense that one has : $H^0(T_{\mathcal{A}}(\mathcal{A}\otimes\mathcal{A})) = H^0(\Omega_u(\mathcal{A})) = \mathbb{C}$ and $H^n(T_{\mathcal{A}}(\mathcal{A}\otimes\mathcal{A})) = H^n(\Omega_u(\mathcal{A})) = 0$ for $n \geq 1$.*

Proof. Define $\delta : \mathbb{C} \to \mathcal{A}$ by $\delta(\lambda) = \lambda\mathbf{1}$, one has $d_u \circ \delta = 0$ so

$$0 \to \mathbb{C} \stackrel{\delta}{\to} \mathcal{A} \stackrel{d_u}{\to} \mathcal{A}\otimes\mathcal{A} \stackrel{d_u}{\to} \cdots \stackrel{d_u}{\to} \otimes^{n+1}\mathcal{A} \stackrel{d_u}{\to} \otimes^{n+2}\mathcal{A} \stackrel{d_u}{\to} \cdots$$

is a cochain complex with coboundary d being d_u or δ. Let ω be a linear form on $\mathcal{A}$ such that $\omega(\mathbf{1}) = 1$ and define k by $k(\mathbb{C}) = 0$ and by $k(x_0 \otimes \cdots \otimes x_n) = \omega(x_0)x_1 \otimes \cdots \otimes x_n$ for $n \geq 0$. One has $kd + dk = I$ which implies $H^n(T_{\mathcal{A}}(\mathcal{A}\otimes\mathcal{A})) = 0$ for $n \geq 1$ and $H^0(T_{\mathcal{A}}(\mathcal{A}\otimes\mathcal{A})) = H^0(\Omega_u(\mathcal{A})) = \mathbb{C}$. Then $H^n(\Omega_u(\mathcal{A})) = 0$ for $n \geq 1$ follows from the property $k(\Omega^n_u(\mathcal{A})) \subset \Omega^{n-1}_u(\mathcal{A})$ for $n \geq 1$. □

Remark 9. The graded differential algebra $C(\mathcal{A},\mathcal{A})$ of $\mathcal{A}$-valued Hochschild cochains of $\mathcal{A}$ (see in Section 4) is a differential calculus over $\mathcal{A}$. Therefore by Proposition 5 there is a unique homomorphism $\Phi : \Omega_u(\mathcal{A}) \to C(\mathcal{A},\mathcal{A})$ of graded differential algebras which induces the identity mapping of $\mathcal{A}$ onto itself. This homomorphism extends to $T_{\mathcal{A}}(\mathcal{A}\otimes\mathcal{A})$, i.e., as an homomorphism $\Phi : T_{\mathcal{A}}(\mathcal{A}\otimes\mathcal{A}) \to C(\mathcal{A},\mathcal{A})$ of graded differential algebras which is given by $\Phi(x_0 \otimes \cdots \otimes x_n)(y_1,\dots,y_n) = x_0y_1x_1\dots y_nx_n$ [55]. Notice that $\Phi(\Omega_u(\mathcal{A}))$ is contained in the graded differential subalgebra $C_0(\mathcal{A},\mathcal{A})$ of the normalized cochains of $C(\mathcal{A},\mathcal{A})$.

In Section 6 we have defined the central bimodule $\Omega^1_Z(\mathcal{A})$ to be the quotient of $\Omega^1_u(\mathcal{A})$ by the bimodule $[Z(\mathcal{A}),\Omega^1_u(\mathcal{A})]$, and the derivation $d_Z:\mathcal{A}\to\Omega^1_Z(\mathcal{A})$ to be the image of $d_u:\mathcal{A}\to\Omega^1_u(\mathcal{A})$. Let I_Z be the closed two-sided ideal of $\Omega_u(\mathcal{A})$ generated by $[Z(\mathcal{A}),\Omega^1_u(\mathcal{A})]$, i.e., the two-sided ideal generated by $[Z(\mathcal{A}),\Omega^1_u(\mathcal{A})]$ and $d_u([Z(\mathcal{A}),\Omega^1_u(\mathcal{A})])$. The space I_Z is a graded ideal which is closed and such that $I_Z\cap\Omega^1_u(\mathcal{A})=[Z(\mathcal{A}),\Omega^1_u(\mathcal{A})]$, which implies that the quotient $\Omega_Z(\mathcal{A})$ is a graded differential algebra which coincides in degree 1 with the above $\Omega^1_Z(\mathcal{A})$ and that its differential (the image of d_u) extends $d_Z:\mathcal{A}\to\Omega^1_Z(\mathcal{A})$; this differential will be also denoted by d_Z. By construction, $\Omega_Z(\mathcal{A})$ is, as a graded algebra, a quotient of the tensor algebra over $\mathcal{A}$ of the central bimodule $\Omega^1_Z(\mathcal{A})$; on the other hand, it is easily seen that tensor products over $\mathcal{A}$ of central bimodules and quotients of central bimodules are again central bimodules [35], so the $(\mathcal{A},\mathcal{A})$-bimodules $\Omega^n_Z(\mathcal{A})$ are central bimodules ($\Omega_Z(\mathcal{A})=\oplus_n\Omega^n_Z(\mathcal{A})$) and therefore the graded differential algebra $\Omega_Z(\mathcal{A})$ is a differential calculus over $\mathcal{A}$ for $\mathbf{Alg}_Z$. Proposition 5 has the following counterpart for $\Omega_Z(\mathcal{A})$.

PROPOSITION 7 *Any homomorphism φ of unital algebras of $\mathcal{A}$ into the subalgebra Ω^0 of elements of degree 0 of a graded differential algebra Ω which is such that $\varphi(z)d(\varphi(x))=d(\varphi(x))\varphi(z)$ for any $z\in Z(\mathcal{A})$ and $x\in\mathcal{A}$ (d being the differential of Ω) has a unique extension $\tilde{\varphi}_Z:\Omega_Z(\mathcal{A})\to\Omega$ as a homomorphism of graded differential algebras.*

Proof. By Proposition 5 φ has a unique extension $\tilde{\varphi}:\Omega_u(\mathcal{A})\to\Omega$ as homomorphism of graded differential algebras. On the other hand, $\varphi(z)d(\varphi(x))=d(\varphi(x))\varphi(z)$ for $z\in Z(\mathcal{A})$ and $x\in\mathcal{A}$ implies that $\tilde{\varphi}$ vanishes on $[Z(\mathcal{A}),\Omega^1_u(\mathcal{A})]$, and therefore also on I_Z since it is a homomorphism of graded differential algebras. Thus $\tilde{\varphi}$ factorizes through a homomorphism $\tilde{\varphi}_Z:\Omega_Z(\mathcal{A})\to\Omega$ of graded differential algebras which extends φ. Uniqueness is also straightforward here. □

Proposition 7 has the following corollaries.

COROLLARY 2 *For any differential calculus Ω over $\mathcal{A}$ for* $\mathbf{Alg}_Z$*, there is a unique homomorphism $j_\Omega:\Omega_Z(\mathcal{A})\to\Omega$ of differential algebras which induces the identity mapping of $\mathcal{A}$ onto itself.*

In other words $\Omega_Z(\mathcal{A})$ is universal among the differential calculi over $\mathcal{A}$ for $\mathbf{Alg}_Z$, and this universal property characterizes it (up to an isomorphism). This is why we shall refer to $\Omega_Z(\mathcal{A})$ as *the universal differential calculus over $\mathcal{A}$ for* $\mathbf{Alg}_Z$.

COROLLARY 3 *Any homomorphism $\varphi : \mathcal{A} \to \mathcal{B}$ of unital algebras mapping the center $Z(\mathcal{A})$ of $\mathcal{A}$ into the center $Z(\mathcal{B})$ of $\mathcal{B}$ has a unique extension $\Omega_Z(\varphi)$: $\Omega_Z(\mathcal{A}) \to \Omega_Z(\mathcal{B})$ as a homomorphism of graded differential algebras.*

In fact, Ω_Z is a covariant functor from the category $\mathbf{Alg}_Z$ to the category **Dif** of graded differential algebras.

In Section 2 it was pointed out that the graded center of a graded algebra is stable under the graded derivations. This implies, in particular, that the graded center $Z_{\mathrm{gr}}(\Omega)$ of a graded differential algebra Ω is a graded differential subalgebra of Ω which is graded commutative. We have defined a differential calculus over $\mathcal{A}$ for $\mathbf{Alg}_Z$ to be a graded differential algebra Ω such that $\Omega^0 = \mathcal{A}$ and such that the center $Z(\mathcal{A})$ of $\mathcal{A}(= \Omega^0)$ is contained in the center of Ω i.e., in its graded center $Z_{\mathrm{gr}}(\Omega)$ since its elements are of degree zero in Ω. It follows that if Ω is a differential calculus over $\mathcal{A}$ for $\mathbf{Alg}_Z$ then the center $Z(\mathcal{A})$ of $\mathcal{A}$ generates a graded differential subalgebra of Ω which is graded commutative and is, in fact, a graded differential subalgebra of the graded center $Z_{\mathrm{gr}}(\Omega)$ of Ω. This applies, in particular, to Ω_Z. If $\mathcal{A}$ is commutative then $\Omega_Z(\mathcal{A})$ is graded commutative since it is generated by $\mathcal{A}$, which then coincides with its center. In this case Proposition 7 has the following corollary.

COROLLARY 4 *If $\mathcal{A}$ is commutative $\Omega_Z(\mathcal{A})$ identifies canonically with the graded differential algebra $\Omega_{\mathcal{A}|\mathbb{C}}$ of Cartan–de Rham–Kähler exterior differential forms.*

Proof. Let us recall that $\Omega_{\mathcal{A}|\mathbb{C}}$ is the exterior algebra over $\mathcal{A}$ of the module $\Omega^1_{\mathcal{A}|\mathbb{C}}$ of Kähler differential, $\Lambda_{\mathcal{A}}\Omega^1_{\mathcal{A}|\mathbb{C}}$, equipped with the unique differential extending the universal derivation of $\mathcal{A}$ into $\Omega^1_{\mathcal{A}|\mathbb{C}}$. From this definition and the universality of the derivation of $\mathcal{A}$ into $\Omega^1_{\mathcal{A}|\mathbb{C}}$ (which identifies, in view of Corollary 1, with $d_Z : \mathcal{A} \to \Omega^1_Z(\mathcal{A})$) it follows that $\Omega_{\mathcal{A}|\mathbb{C}}$ is characterized by the following universal property: *Any homomorphism ψ of $\mathcal{A}$ into the subalgebra Ω^0 of the elements of degree 0 of a graded commutative differential algebra Ω has a unique extension $\tilde{\psi} : \Omega_{\mathcal{A}|\mathbb{C}} \to \Omega$ as a homomorphism of graded (commutative) differential algebras.*
Let us return to the proof of Corollary 4. Since $\Omega_Z(\mathcal{A})$ is graded commutative with $\Omega^0_Z(\mathcal{A}) = \mathcal{A}$, the above universal property implies that there is a unique homomorphism of graded differential algebras of $\Omega_{\mathcal{A}|\mathbb{C}}$ into $\Omega_Z(\mathcal{A})$ which induces the identity mapping of $\mathcal{A}$ onto itself. On the other hand, Proposition 7 (or Corollary 2) implies that there is a unique homomorphism of graded differential algebras of $\Omega_Z(\mathcal{A})$ into $\Omega_{\mathcal{A}|\mathbb{C}}$ which induces the identity of $\mathcal{A}$ onto itself. Using again these two universal properties, it follows that the above

homomorphisms are inverse isomorphisms. □

If $\mathcal{A}$ is commutative the cohomology of $\Omega_Z(\mathcal{A}) = \Omega_{\mathcal{A}|\mathbb{C}}$ is often called the de Rham cohomology ([52], [43]), in spite of it being explained in Remark 8, for $\mathcal{A} = C^\infty(M)$, $\Omega_{\mathcal{A}|\mathbb{C}}$ can be slightly different from the algebra of smooth differential forms, and that therefore there is an ambiguity. Nevertheless, $\Omega_Z(\mathcal{A})$ can be considered as a generalization of the graded differential algebra of differential forms which has the great advantage that the correspondence $\mathcal{A} \mapsto \Omega_Z(\mathcal{A})$ is functorial (Corollary 3). In contrast to the cohomology of $\Omega_u(\mathcal{A})$ (see Lemma 5), the cohomology $H_Z(\mathcal{A})$ of $\Omega_Z(\mathcal{A})$ is generally non-trivial. Since $H_Z(\mathcal{A})$ is a noncommutative generalization of the de Rham cohomology, and since, by construction, $\mathcal{A} \mapsto H_Z(\mathcal{A})$ is a covariant functor from the category $\mathbf{Alg}_Z$ to the category of graded algebras, it is natural to study the properties of this cohomology.

Let $\mathrm{Der}(\mathcal{A})$ denote the vector space of all derivations of $\mathcal{A}$ into itself. This vector space is a Lie algebra for the bracket $[\cdot,\cdot]$ defined by $[X,Y](a) = X(Y(a)) - Y(X(a))$ for $X, Y \in \mathrm{Der}(\mathcal{A})$ and $a \in \mathcal{A}$. In view of Proposition 1 (universal property of $(\Omega^1_u(\mathcal{A}), d_u)$), for each $X \in \mathrm{Der}(\mathcal{A})$ there is a unique bimodule homomorphism $i_X : \Omega^1_u(\mathcal{A}) \to \mathcal{A}$ for which $X = i_X \circ d_u$. This homomorphism of $\Omega^1_u(\mathcal{A})$ into $\mathcal{A} = \Omega^0_u(\mathcal{A})$ has a unique extension as an antiderivation of $\Omega_u(\mathcal{A}) = T_{\mathcal{A}}(\Omega^1_u(\mathcal{A}))$. This antiderivation which will be again denoted by i_X is of degree -1 (i.e., it is a graded derivation of degree -1). It is not hard to verify that $X \mapsto i_X$ is an operation of the Lie algebra $\mathrm{Der}(\mathcal{A})$ in the graded differential algebra $\Omega_u(\mathcal{A})$ (see Section 2 for the definition). The corresponding Lie derivative $L_X = i_X d_u + d_u i_X$ is for $X \in \mathrm{Der}(\mathcal{A})$ a derivation of degree 0 of $\Omega_u(\mathcal{A})$ which extends X. This operation will be referred to as *the canonical operation of* $\mathrm{Der}(\mathcal{A})$ *in* $\Omega_u(\mathcal{A})$.

Let $X \in \mathrm{Der}(\mathcal{A})$ be a derivation of $\mathcal{A}$ and let $z \in Z(\mathcal{A})$ and $\omega \in \Omega^1_u(\mathcal{A})$ one has

$$i_X([z,\omega]) = [z, i_X(\omega)] = 0$$

and

$$i_X(d([z,\omega])) = L_X([z,\omega]) = [X(z),\omega] + [z, L_X(\omega)] = [z, L_X(\omega)]$$

since $Z(\mathcal{A})$ is stable by the derivations of $\mathcal{A}$. This implies that $i_X(I_Z) \subset I_Z$ and therefore that the antiderivation i_X passes to the quotient and defines an antiderivation of degree -1 of $\Omega_Z(\mathcal{A})$ which will again be denoted by i_X. Notice that this (abuse of) notation is consistent with the one used in Proposition 3, ($\mathcal{A}$ is obviously a central bimodule). The corresponding mapping $X \mapsto i_X$ of $\mathrm{Der}(\mathcal{A})$ into the antiderivations of degree -1 of $\Omega_Z(\mathcal{A})$ is again an operation (the quotient of the one in $\Omega_u(\mathcal{A})$) which will be referred to as *the canonical operation of* $\mathrm{Der}(\mathcal{A})$ *in* $\Omega_Z(\mathcal{A})$.

Finally, if $\mathcal{A}$ is a $*$-algebra, $T_{\mathcal{A}}(\mathcal{A} \otimes \mathcal{A})$ becomes a graded differential $*$-algebra if one equips it with the involution defined by $(x_0 \otimes \cdots \otimes x_n)^* = (-1)^{\frac{n(n+1)}{2}} x_n^* \otimes \cdots \otimes x_0^*$. Since $\Omega_u(\mathcal{A})$ is stable under this involution, it is also a graded differential $*$-algebra [61]. Furthermore, $[Z(\mathcal{A}), \Omega_u^1(\mathcal{A})]$ is $*$-invariant, which implies that the involution of $\Omega_u(\mathcal{A})$ passes to the quotient and induces an involution on $\Omega_Z(\mathcal{A})$ for which $\Omega_Z(\mathcal{A})$ also becomes a graded differential $*$-algebra. More generally in this case, a differential calculus Ω over $\mathcal{A}$ will always be assumed to be equipped with an involution extending the involution of $\mathcal{A}$ and such that it is a graded differential $*$-algebra (notice that if Ω is generated by $\mathcal{A}$ such an involution is unique).

8. Diagonal and derivation-based calculi

Let $\mathcal{A}$ be a unital associative complex algebra and let $\mathcal{M}$ be an arbitrary $(\mathcal{A},\mathcal{A})$-bimodule. Then the set $\mathrm{Hom}_{\mathcal{A}}^{\mathcal{A}}(\mathcal{M},\mathcal{A})$ of all bimodule homomorphisms of $\mathcal{M}$ into $\mathcal{A}$ is a module over the center $Z(\mathcal{A})$ of $\mathcal{A}$ which will be referred to as the $\mathcal{A}$-*dual of* $\mathcal{M}$ and denoted by $\mathcal{M}^{*_{\mathcal{A}}}$, [34], [27]. Conversely, if $\mathcal{N}$ is a $Z(\mathcal{A})$-module the set $\mathrm{Hom}_{Z(\mathcal{A})}(\mathcal{N},\mathcal{A})$ of all $Z(\mathcal{A})$-module homomorphisms of $\mathcal{N}$ into $\mathcal{A}$ is canonically a $(\mathcal{A},\mathcal{A})$-bimodule which will be also referred to as the $\mathcal{A}$-*dual of* $\mathcal{N}$ and denoted by $\mathcal{N}^{*_{\mathcal{A}}}$. The $\mathcal{A}$-dual of a $Z(\mathcal{A})$-module is clearly a central bimodule over $\mathcal{A}$, so the above duality between $(\mathcal{A},\mathcal{A})$-bimodules and $Z(\mathcal{A})$-modules can be restricted to a duality between the central bimodules over $\mathcal{A}$ and the $Z(\mathcal{A})$-modules. This latter duality generalizes the duality between modules over a commutative algebra, [34], [27]. Indeed, if $\mathcal{A}$ is commutative both central bimodules over $\mathcal{A}$ and $Z(\mathcal{A})$-modules coincide with $\mathcal{A}$-modules and the above duality is then the usual duality between $\mathcal{A}$-modules. Let us come back to the general situation and let $\mathcal{M}$ be a $(\mathcal{A},\mathcal{A})$-bimodule; then one obtains by evaluation a *canonical homomorphism* of $(\mathcal{A},\mathcal{A})$-bimodule $c : \mathcal{M} \to \mathcal{M}^{*_{\mathcal{A}}*_{\mathcal{A}}}$ of $\mathcal{M}$ into *its* $\mathcal{A}$-*bidual* $\mathcal{M}^{*_{\mathcal{A}}*_{\mathcal{A}}} = (\mathcal{M}^{*_{\mathcal{A}}})^{*_{\mathcal{A}}}$.

LEMMA 6 *The following properties (a) and (b) are equivalent for a* $(\mathcal{A},\mathcal{A})$-*bimodule* $\mathcal{M}$:

(a) The canonical homomorphism $c : \mathcal{M} \to \mathcal{M}^{*_{\mathcal{A}}*_{\mathcal{A}}}$ *is injective.*

(b) $\mathcal{M}$ *is isomorphic to a sub-bimodule of* $\mathcal{A}^I$ *for some set* I.

Proof. $(a) \Rightarrow (b)$. By definition $\mathcal{M}^{*_{\mathcal{A}}*_{\mathcal{A}}}$ is a sub-bimodule of $\mathcal{A}^I$ with $I = \mathrm{Hom}_{Z(\mathcal{A})}(\mathcal{M}^{*_{\mathcal{A}}},\mathcal{A})$ so if c is injective $\mathcal{M}$ is isomorphic to a sub-bimodule of $\mathcal{M}^{*_{\mathcal{A}}*_{\mathcal{A}}}$ and therefore also to a sub-bimodule of $\mathcal{A}^I$.

$(b) \Rightarrow (a)$. Let φ be a bimodule homomorphism of $\mathcal{A}$ into itself. One has $\varphi(a) = a\varphi(1) = \varphi(1)a$ which implies $\varphi(1) \in Z(\mathcal{A})$. Conversely, any $z \in Z(\mathcal{A})$ defines a bimodule homomorphism φ of $\mathcal{A}$ into itself by setting $\varphi(a) = az$ (i.e., $\varphi(1) = z$). It follows that $\mathcal{A}^{*_\mathcal{A}} = Z(\mathcal{A})$. Let Φ be a $Z(\mathcal{A})$-module homomorphism of $Z(\mathcal{A})$ into $\mathcal{A}$. Then $\Phi(z) = z\Phi(1)$ with $\Phi(1) \in \mathcal{A}$. Conversely, any $a \in \mathcal{A}$ defines such a $Z(\mathcal{A})$-module homomorphism Φ by setting $\Phi(z) = za$ (i.e., $\Phi(1) = a$). It follows that $Z(\mathcal{A})^{*_\mathcal{A}} = \mathcal{A}$, and therefore $\mathcal{A}^{*_\mathcal{A}*_\mathcal{A}} = \mathcal{A}$. This immediately implies that if $\mathcal{M} \subset \mathcal{A}^I$ as a sub-bimodule then $c : \mathcal{M} \to \mathcal{M}^{*_\mathcal{A}*_\mathcal{A}}$ is injective. □

An $(\mathcal{A}, \mathcal{A})$-bimodule $\mathcal{M}$ satisfying the equivalent conditions of Lemma 6 will be said to be a *diagonal bimodule* over $\mathcal{A}$, [34], [35] (see also in [27]). A diagonal bimodule is central but the converse is not generally true. The $\mathcal{A}$-dual of an arbitrary $Z(\mathcal{A})$-module is a diagonal bimodule. Every sub-bimodule of a diagonal bimodule is diagonal, every product of diagonal bimodules is diagonal, and the tensor product over $\mathcal{A}$ of two diagonal bimodules is diagonal.

If $\mathcal{A}$ is commutative, a diagonal bimodule over $\mathcal{A}$ is simply an $\mathcal{A}$-module such that the canonical homomorphism in its bi-dual is injective. In particular in this case a projective module is diagonal (as a bimodule for the underlying structure).

Remark 10. It is a fortunate circumstance which is easy to verify that, for a $Z(\mathcal{A})$-module $\mathcal{N}$, the biduality does not depend on $\mathcal{A}$ but only on $Z(\mathcal{A})$. That is one has $\mathcal{N}^{*_\mathcal{A}*_\mathcal{A}} = \mathcal{N}^{**}$ and the canonical homomorphism $c : \mathcal{N} \to \mathcal{N}^{**}$ obtained by evaluation for the $\mathcal{A}$-duality reduces to the usual one for a module over the commutative algebra $Z(\mathcal{A})$.

Let $\mathcal{M}$ be a $(\mathcal{A}, \mathcal{A})$-bimodule, then the canonical image $c(\mathcal{M})$ of $\mathcal{M}$ in its $\mathcal{A}$-bidual is a diagonal bimodule. The diagonal bimodule $c(\mathcal{M})$ is the universal 'diagonalization' of $\mathcal{M}$ in the sense that it is characterized (among the diagonal bimodules over $\mathcal{A}$) by the following universal property [34], [35].

PROPOSITION 8 *For any homomorphism of $(\mathcal{A}, \mathcal{A})$-bimodules $\varphi : \mathcal{M} \to \mathcal{N}$ of $\mathcal{M}$ into a diagonal bimodule $\mathcal{N}$ over $\mathcal{A}$ there is a unique homomorphism of $(\mathcal{A}, \mathcal{A})$-bimodules $\varphi_c : c(\mathcal{M}) \to \mathcal{N}$ such that $\varphi = \varphi_c \circ c$.*

Proof. In view of the definition and Lemma 6(b) it is sufficient to prove the statement for $\mathcal{N} = \mathcal{A}^I$ for some set I, which is then equivalent to the statement for $\mathcal{N} = \mathcal{A}$. On the other hand, for $\mathcal{N} = \mathcal{A}$, $\varphi \in \mathrm{Hom}^\mathcal{A}_\mathcal{A}(\mathcal{M}, \mathcal{A}) = \mathcal{M}^{*_\mathcal{A}}$ and one has $\varphi(m) =< c(m), \varphi >= \varphi_c(c(m))$ for $m \in \mathcal{M}$ (by the definitions of $\mathcal{M}^{*_\mathcal{A}*_\mathcal{A}}$ and of the evaluation c) which defines φ_c uniquely. □

One has $c(\Omega^1_u(\mathcal{A})) = c(\Omega^1_Z(\mathcal{A}))$, and we shall denote by $\Omega^1_{\text{Diag}}(\mathcal{A})$ this diagonal bimodule and by d_{Diag} the derivation $c \circ d_u$ (or equivalently $c \circ d_Z$) of $\mathcal{A}$ into $\Omega^1_{\text{Diag}}(\mathcal{A})$.

PROPOSITION 9 *For any first order differential calculus (Ω^1, d) over $\mathcal{A}$ such that Ω^1 is diagonal, there is a unique bimodule homomorphism i_d of $\Omega^1_{\text{Diag}}(\mathcal{A})$ into Ω^1 such that $d = i_d \circ d_{\text{Diag}}$.*

Proof. In view of the above universal property of $c(\Omega^1_u(\mathcal{A}))$ the corresponding canonical homomorphism of $\Omega^1_u(\mathcal{A})$ into Ω^1 (as in Proposition 1) factorizes through a unique homomorphism $i_d : \Omega^1_{\text{Diag}}(\mathcal{A}) \to \Omega^1$. □

In other words, the derivation $d_{\text{Diag}} : \mathcal{A} \to \Omega^1_{\text{Diag}}(\mathcal{A})$ of $\mathcal{A}$ into the diagonal bimodule $\Omega^1_{\text{Diag}}(\mathcal{A})$ is universal for the derivations of $\mathcal{A}$ into diagonal bimodules over $\mathcal{A}$.

Let us recall (see Section 3) that the vector space $\text{Der}(\mathcal{A})$ of all derivations of $\mathcal{A}$ into itself is a Lie algebra and also a $Z(\mathcal{A})$-module, and that $\Omega_{\text{Der}}(\mathcal{A})$ was defined to be the graded differential subalgebra of $C_\wedge(\text{Der}(\mathcal{A}), \mathcal{A})$ generated by $\mathcal{A}$ whereas $\underline{\Omega}_{\text{Der}}(\mathcal{A})$ was defined to be the graded differential subalgebra of $C_\wedge(\text{Der}(\mathcal{A}), \mathcal{A})$ which consists of cochains of $\text{Der}(\mathcal{A})$ which are $Z(\mathcal{A})$-multilinear. Clearly $C^n_\wedge(\text{Der}(\mathcal{A}), \mathcal{A})$ is diagonal, so the first order differential calculus $(C^1_\wedge(\text{Der}(\mathcal{A}), \mathcal{A}), d)$ satisfies the conditions of Proposition 9, which implies that there is a unique bimodule homomorphism $i_d : \Omega^1_{\text{Diag}}(\mathcal{A}) \to C^1_\wedge(\text{Der}(\mathcal{A}), \mathcal{A})$ for which $d = i_d \circ d_{\text{Diag}}$.

PROPOSITION 10 *The homomorphism*

$$i_d : \Omega^1_{\text{Diag}}(\mathcal{A}) \to C^1_\wedge(Der(\mathcal{A}), \mathcal{A})$$

is injective, so by identifying $\Omega^1_{\text{Diag}}(\mathcal{A})$ with its image (by i_d) one has canonically:

$$\Omega^1_{\text{Diag}}(\mathcal{A}) = \Omega^1_{\text{Der}}(\mathcal{A}), \quad (\Omega^1_{\text{Diag}}(\mathcal{A}))^{*_\mathcal{A}} = \text{Der}(\mathcal{A})$$
$$\textit{and} \quad (\Omega^1_{\text{Diag}}(\mathcal{A}))^{*_\mathcal{A} *_\mathcal{A}} = \underline{\Omega}^1_{\text{Der}}(\mathcal{A}).$$

Proof. Applying Proposition 9 for $\Omega^1 = \mathcal{A}$ leads to the identification

$$\text{Hom}^\mathcal{A}_\mathcal{A}(\Omega^1_{\text{Diag}}(\mathcal{A}), \mathcal{A}) = \text{Der}(\mathcal{A}),$$

that is, $(\Omega^1_{\text{Diag}}(\mathcal{A}))^{*_\mathcal{A}} = \text{Der}(\mathcal{A})$ (notice that one has also $(\Omega^1_u(\mathcal{A}))^{*_\mathcal{A}} = \text{Der}(\mathcal{A})$). By definition one has

$$\underline{\Omega}^1_{\text{Der}}(\mathcal{A}) = \text{Hom}_{Z(\mathcal{A})}(\text{Der}(\mathcal{A}), \mathcal{A}),$$

that is, $\underline{\Omega}^1_{\text{Der}}(\mathcal{A}) = (\text{Der}(\mathcal{A}))^{*_\mathcal{A}}$, and therefore $(\Omega^1_{\text{Diag}}(\mathcal{A}))^{*_\mathcal{A}*_\mathcal{A}} = \underline{\Omega}^1_{\text{Der}}(\mathcal{A})$. On the other hand one has $i_d(\Omega^1_{\text{Diag}}(\mathcal{A}))=\Omega^1_{\text{Der}}(\mathcal{A})$ since $\Omega^1_{\text{Der}}(\mathcal{A})$ is generated by $\mathcal{A}$ (as bimodule). The injectivity of i_d follows from $\Omega^1_{\text{Diag}}(\mathcal{A})$ being diagonal, i.e., that the canonical homomorphism in its $\mathcal{A}$-bidual is injective. □

Notice that by definition one also has $(\bigwedge^n_{Z(\mathcal{A})} \text{Der}(\mathcal{A}))^{*_\mathcal{A}} = \underline{\Omega}^n_{\text{Der}}(\mathcal{A})$.

Let I_{Diag} be the closed two-sided ideal of $\Omega_u(\mathcal{A})$ generated by the kernel of the canonical homomorphism c of $\Omega^1_u(\mathcal{A})$ into its $\mathcal{A}$-bidual. The ideal I_{Diag} is graded such that $I_{\text{Diag}} \cap \Omega^0_u(\mathcal{A}) = 0$ and $I_{\text{Diag}} \cap \Omega^1_u(\mathcal{A}) = \text{Ker}(c)$ which implies that the quotient $\Omega^1_u(\mathcal{A})/I_{\text{Diag}}$ is a graded differential algebra which is a differential calculus over $\mathcal{A}$ and coincides in degree 1 with $c(\Omega^1_u(\mathcal{A})) = \Omega^1_{\text{Diag}}(\mathcal{A})$. This differential calculus will be referred to as the *diagonal calculus* and denoted by $\Omega_{\text{Diag}}(\mathcal{A})$. The differential of $\Omega_{\text{Diag}}(\mathcal{A})$ is the image of d_u and extends the derivation $d_{\text{Diag}} : \mathcal{A} \to \Omega^1_{\text{Diag}}(\mathcal{A})$; this differential will be also denoted by d_{Diag}. Proposition 5 and Proposition 7 have the following counterpart for $\Omega_{\text{Diag}}(\mathcal{A})$.

PROPOSITION 11 *Any homomorphism φ of unital algebras of $\mathcal{A}$ into the subalgebra Ω^0 of elements of degree 0 of a graded differential algebra Ω which is such that $d(\mathcal{A})$ spans a diagonal bimodule over $\mathcal{A}$ (for the $(\mathcal{A},\mathcal{A})$-bimodule structure on Ω^1 induced by φ) has a unique extension $\tilde{\varphi}_{\text{Diag}} : \Omega_{\text{Diag}}(\mathcal{A}) \to \Omega$ as a homomorphism of graded differential algebras.*

Proof. By Proposition 5 φ has a unique extension $\tilde{\varphi} : \Omega_u(\mathcal{A}) \to \Omega$ as homomorphism of graded differential algebras. On the other hand, the assumption means that $d : \mathcal{A} \to \tilde{\varphi}(\Omega^1_u(\mathcal{A}))$ is a derivation and that $\tilde{\varphi}(\Omega^1_u(\mathcal{A}))$ is a diagonal bimodule over $\mathcal{A}$, so in view of Proposition 9 the homomorphism $\tilde{\varphi} : \Omega^1_u(\mathcal{A}) \to \Omega^1$ factorizes through a homomorphism $\tilde{\varphi}^1_{\text{Diag}} : \Omega^1_{\text{Diag}}(\mathcal{A}) \to \Omega^1$. Thus $\tilde{\varphi}$ vanishes on $\text{Ker}(c)$ and therefore on I_Z since it is a homomorphism of graded differential algebras, so it factorizes through a homomorphism $\tilde{\varphi}_{\text{Diag}} : \Omega_{\text{Diag}}(\mathcal{A}) \to \Omega$ of graded differential algebras. Uniqueness is again straightforward. □

Thus $\Omega_{\text{Diag}}(\mathcal{A})$ is also characterized by a universal property like $\Omega_u(\mathcal{A})$ and $\Omega_Z(\mathcal{A})$ but in contrast to the cases of $\Omega_u(\mathcal{A})$ and $\Omega_Z(\mathcal{A})$ the correspondence $\mathcal{A} \mapsto \Omega_{\text{Diag}}(\mathcal{A})$ has no obvious functorial property. The reason for this is the property that the diagonal bimodules are not the bimodules for a category of algebras in the sense explained in Section 5.

Proposition 11 implies in particular that one has a unique homomorphism of graded differential algebra of $\Omega_{\text{Diag}}(\mathcal{A})$ into $\Omega_{\text{Der}}(\mathcal{A})$ which extends the identity mapping of $\mathcal{A}$ onto itself. This homomorphism $\Omega_{\text{Diag}}(\mathcal{A}) \to \Omega_{\text{Der}}(\mathcal{A})$ is surjective since $\Omega_{\text{Der}}(\mathcal{A})$ is generated by $\mathcal{A}$ as differential algebra. Furthermore, in degree 1 it is, in view of Proposition 10, a bimodule isomorphism of $\Omega^1_{\text{Diag}}(\mathcal{A})$ onto $\Omega^1_{\text{Der}}(\mathcal{A})$. However, for $m \geq 2$ the corresponding bimodule homomorphism of $\Omega^m_{\text{Diag}}(\mathcal{A})$ onto $\Omega^m_{\text{Der}}(\mathcal{A})$ is not generally injective (i.e., it has a non-trivial kernel).

For instance when $\mathcal{A}$ coincides with the algebra $M_n(\mathbb{C})$ of complex $n \times n$ matrices one has

$$\begin{aligned}\Omega_u(M_n(\mathbb{C})) &= \Omega_Z(M_n(\mathbb{C})) = \Omega_{\text{Diag}}(M_n(\mathbb{C})) \\ &\simeq C_0(M_n(\mathbb{C}), M_n(\mathbb{C})) = M_n(\mathbb{C}) \otimes T\mathfrak{sl}(n, \mathbb{C})^*,\end{aligned}$$

whereas

$$\Omega_{\text{Der}}(M_n(\mathbb{C})) = C_\wedge(\mathfrak{sl}(n, \mathbb{C}), M_n(\mathbb{C})) = M_n(\mathbb{C}) \otimes \bigwedge \mathfrak{sl}(n, \mathbb{C})^*.$$

In fact, in this case the homomorphism Φ of Remark 9 is an isomorphism which induces the isomorphism of $\Omega_u(M_n(\mathbb{C}))$ onto the differential algebra $C_0(M_n(\mathbb{C}), M_n(\mathbb{C}))$ of normalized Hochschild cochains; the latter being identical as graded algebra to the tensor product $M_n(\mathbb{C}) \otimes T\mathfrak{sl}(n, \mathbb{C})^*$ of $M_n(\mathbb{C})$ with the tensor algebra over $\mathbb{C}$ of the dual of $\mathfrak{sl}(n, \mathbb{C})$, (concerning $\Omega^1_{\text{Der}}(M_n(\mathbb{C})) = \Omega^1_u(M_n(\mathbb{C}))$ and $\Omega_{\text{Der}}(M_n(\mathbb{C})) = C_\wedge(\mathfrak{sl}(n, \mathbb{C}), M_n(\mathbb{C}))$, see in [25]).

In the case where $\mathcal{A}$ is the algebra $C^\infty(M)$ of smooth functions on a good smooth manifold (finite-dimensional paracompact, etc.) one then has $\Omega_{\text{Diag}}(C^\infty(M)) = \Omega_{\text{Der}}(C^\infty(M))$ $(= \underline{\Omega}_{\text{Der}}(C^\infty(M)))$.

It is not hard to show that the operations of the Lie algebra $\text{Der}(\mathcal{A})$ in $\Omega_u(\mathcal{A})$ and in $\Omega_Z(\mathcal{A})$ pass to the quotient to define an operation of $\text{Der}(\mathcal{A})$ in the graded differential algebra $\Omega_{\text{Diag}}(\mathcal{A})$, which will be again referred to as *the canonical operation of* $\text{Der}(\mathcal{A})$ *in* $\Omega_{\text{Diag}}(\mathcal{A})$. Furthermore, all these operations of $\text{Der}(\mathcal{A})$ pass to the quotient to define an operation of $\text{Der}(\mathcal{A})$ in $\Omega_{\text{Der}}(\mathcal{A})$ which coincides with the canonical operation of $\text{Der}(\mathcal{A})$ in $\Omega_{\text{Der}}(\mathcal{A})$ defined in Section 3.

One has the following commutative diagram of surjective homomorphisms of graded differential algebras which is also a diagram of homomorphisms of the operations of $\text{Der}(\mathcal{A})$

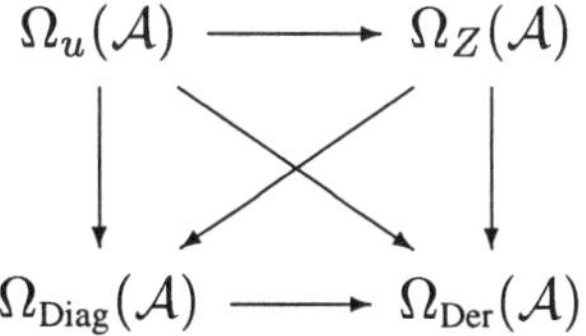

Furthermore, if $\mathcal{A}$ is a $*$-algebra there is a canonical involution on $\Omega_{\text{Diag}}(\mathcal{A})$ such that this diagram is also a diagram of graded differential $*$-algebras (the involutions of $\Omega_u(\mathcal{A}), \Omega_Z(\mathcal{A})$ and $\Omega_{\text{Der}}(\mathcal{A})$ have been defined previously in Section 7 and Section 3).

9. Noncommutative symplectic geometry and quantum mechanics

Let $\mathcal{A}$ be, as before, a unital associative complex algebra. A *Poisson bracket* on $\mathcal{A}$ is a Lie bracket which is a bi-derivation on $\mathcal{A}$ (for its associative product). That is, $(a, b) \mapsto \{a, b\}$ is a Poisson bracket if it is a bilinear antisymmetric mapping of $\mathcal{A} \times \mathcal{A}$ into $\mathcal{A}$ (i.e., a linear mapping of $\bigwedge^2 \mathcal{A}$ into $\mathcal{A}$) which satisfies

$$\{\{a, b\}, c\} + \{\{b, c\}, a\} + \{\{c, a\}, b\} = 0 \quad \text{(Jacobi identity)}$$

$$\{a, bc\} = \{a, b\}c + b\{a, c\} \quad \text{(derivation property)}$$

for any elements a, b, c of $\mathcal{A}$. Equipped with such a Poisson bracket, $\mathcal{A}$ is referred to as a *Poisson algebra* [38].

There is a lot of classical commutative Poisson algebras, for instance the symmetric algebra $S(\mathfrak{g})$ (over $\mathbb{C}$) of a (complex) Lie algebra $\mathfrak{g}$, the algebra $C^\infty(M)$ of smooth functions on a symplectic manifold, etc.. For a noncommutative algebra $\mathcal{A}$ a generic type of Poisson bracket $\{\cdot, \cdot\}$ is obtained by setting for $a, b \in \mathcal{A}$

$$\{a, b\} = \frac{i}{\hbar}[a, b],$$

where $[a, b]$ denotes the commutator in $\mathcal{A}$, i.e., $[a, b] = ab - ba$, and where $\hbar \in \mathbb{C}$ is any non-zero complex number. We have put an $i \in \mathbb{C}$ in front of the right hand side of the above formula in order that in the case where $\mathcal{A}$ is a $*$-algebra *the Poisson bracket is real*, i.e., satisfies $\{a, b\}^* = \{a^*, b^*\}$, whenever $\hbar$ is real. The reason why the Poisson brackets proportional to the commutator are generic (in the noncommutative case) is connected to the following lemma [38].

LEMMA 7 *Let $\mathcal{A}$ be a Poisson algebra, then one has $[a, b]\{c, d\} = \{a, b\}[c, d]$ and more generally $[a, b]x\{c, d\} = \{a, b\}x[c, d]$ for any elements a, b, c, d and x of $\mathcal{A}$.*

Proof. The first identity is obtained by developing $\{ac, bd\}$ in two different orders by using the bi-derivation property. The second (more general, since $\mathbb{1} \in \mathcal{A}$) identity is obtained by replacing c by xc in the first identity, by developing and by using again the first identity. □

For more details concerning the 'generic side' of Poisson brackets proportional to the commutator we refer to [38]. We simply observe here that this is the type of Poisson bracket which occurs in quantum mechanics.

Our aim is now to develop a (noncommutative) generalization of symplectic structures related to the above Poisson brackets. One should start from a notion of differential form, i.e., from a differential calculus Ω over $\mathcal{A}$. Since for a Poisson bracket $x \mapsto \{a, x\}$ is an element of Der$(\mathcal{A})$ for any $a \in \mathcal{A}$, it is natural to assume that one has an operation $X \mapsto i_X$ of the Lie algebra Der$(\mathcal{A})$ in the graded differential algebra Ω. Furthermore, we wish to take into account the structure of $Z(\mathcal{A})$-module of Der$(\mathcal{A})$ so we require that Ω is a central bimodule over $\mathcal{A}$ and that $X \mapsto i_X$ is a $Z(\mathcal{A})$-linear mapping of Der$(\mathcal{A})$ into $\mathrm{Der}^{-1}_{\mathrm{gr}}(\Omega)$. Notice that this $Z(\mathcal{A})$-linearity is well defined since Ω central is equivalent to $Z(\mathcal{A}) \subset Z^0_{\mathrm{gr}}(\Omega)$ (see in Section 2 for the notations). Having such a differential calculus one defines a homomorphism λ of Ω into $\underline{\Omega}_{\mathrm{Der}}(\mathcal{A})$ by setting $\lambda(\omega)(X_1, \dots, X_n) = i_{X_n} \dots i_{X_1}\omega$ for $\omega \in \Omega^n$. That this defines a homomorphism of graded differential algebra of Ω into $C_\wedge(\mathrm{Der}(\mathcal{A}), \mathcal{A})$ follows from the general properties of operations, whereas the property that the image of λ is contained in $\underline{\Omega}_{\mathrm{Der}}(\mathcal{A})$ follows from the $Z(\mathcal{A})$-linearity. It turns out that even if one uses a general differential calculus Ω for the symplectic structures, the only relevant parts for the corresponding Poisson structures are the images by λ in $\underline{\Omega}_{\mathrm{Der}}(\mathcal{A})$ (see, e.g., in [38]). One is then led to the definitions of [26], or more precisely to the following slight generalizations [27].

An element ω of $\underline{\Omega}^2_{\mathrm{Der}}(\mathcal{A})$ will be said to be *non-degenerate* if, for any $x \in \mathcal{A}$, there is a derivation $\mathrm{Ham}(x) \in \mathrm{Der}(\mathcal{A})$ such that one has $\omega(X, \mathrm{Ham}(x)) = X(x)$ for any $X \in \mathrm{Der}(\mathcal{A})$. Notice that if ω is non-degenerate then $X \mapsto i_X\omega$ is an injective linear mapping of Der$(\mathcal{A})$ into $\underline{\Omega}^1_{\mathrm{Der}}(\mathcal{A})$, but that the converse is not true; the condition for ω to be non-degenerate is stronger than the injectivity of $X \mapsto i_X\omega$. If M is a manifold an element $\omega \in \underline{\Omega}^2_{\mathrm{Der}}(C^\infty(M))$ is an ordinary 2-form on M, and it is non-degenerate in the above sense if and only if the 2-form ω is non-degenerate in the classical sense (i.e., everywhere non-degenerate).

Let $\omega \in \underline{\Omega}^2_{\mathrm{Der}}(\mathcal{A})$ be non-degenerate, then for a given $x \in \mathcal{A}$ the derivation $\mathrm{Ham}(x)$ is unique and $x \mapsto \mathrm{Ham}(x)$ is a linear mapping of $\mathcal{A}$ into Der$(\mathcal{A})$.

A closed non-degenerate element ω of $\underline{\Omega}^2_{\mathrm{Der}}(\mathcal{A})$ will be called *a symplectic structure for $\mathcal{A}$.*

LEMMA 8 *Let ω be a symplectic structure for $\mathcal{A}$ and define an antisymmetric bilinear bracket on $\mathcal{A}$ by $\{x,y\} = \omega(\mathrm{Ham}(x), \mathrm{Ham}(y))$. Then $(x,y) \mapsto \{x,y\}$ is a Poisson bracket on $\mathcal{A}$.*

Proof. One has $\{x, yz\} = \{x,y\}z + y\{x,y\}$ for $x,y,z \in \mathcal{A}$. Furthermore, one has the identity

$$d\omega(\mathrm{Ham}(x), \mathrm{Ham}(y), \mathrm{Ham}(z)) = -\{x,\{y,z\}\} - \{y,\{z,x\}\} - \{z,\{x,y\}\},$$

which implies the Jacobi identity since $d\omega = 0$. □

Let ω be a symplectic structure for $\mathcal{A}$, then one has

$$[\mathrm{Ham}(x), \mathrm{Ham}(y)] = \mathrm{Ham}(\{x,y\}),$$

i.e., Ham is a Lie algebra homomorphism of $(\mathcal{A}, \{,\})$ into Der($\mathcal{A}$). We shall refer to the above bracket as *the Poisson bracket associated to the symplectic structure* ω. If $\mathcal{A}$ is a $*$-algebra and if, furthermore, ω is real, i.e., $\omega = \omega^*$, then this Poisson bracket is real and $\mathrm{Ham}(x^*) = (\mathrm{Ham}(x))^*$ for any $x \in \mathcal{A}$.

An algebra $\mathcal{A}$ equipped with a symplectic structure will be referred to as a *symplectic algebra*. Thus symplectic algebras are particular Poisson algebras.

Remark 11. Let $\mathcal{A}$ be an arbitrary Poisson algebra with Poisson bracket $(x,y) \mapsto \{x,y\}$; one defines a linear mapping Ham : $\mathcal{A} \to \mathrm{Der}(\mathcal{A})$ by $\mathrm{Ham}(x)(y) = \{x,y\}$, (i.e., $\mathrm{Ham}(x) = \{x,\cdot\}$), for $x,y \in \mathcal{A}$. In this general setting one also has the identity $[\mathrm{Ham}(x), \mathrm{Ham}(y)] = \mathrm{Ham}(\{x,y\})$ since it is equivalent to the Jacobi identity for the Poisson bracket.

If M is a manifold a symplectic structure for $C^\infty(M)$ is just a symplectic form on M. Since there are manifolds which do not admit symplectic form, one cannot expect that an arbitrary $\mathcal{A}$ admits a symplectic structure.

Assume that $\mathcal{A}$ has a trivial center $Z(\mathcal{A}) = \mathbb{C}\mathbb{1}$ and that all its derivations are inner (i.e., of the form $\mathrm{ad}(x), x \in \mathcal{A}$). Then one defines an element ω of $\underline{\Omega}^2_{\mathrm{Der}}(\mathcal{A})$ by setting $\omega(\mathrm{ad}(ix), \mathrm{ad}(iy)) = i[x,y]$. It is easily seen that ω is a symplectic structure for which one has $\mathrm{Ham}(x) = \mathrm{ad}(ix)$ and $\{x,y\} = i[x,y]$. If furthermore $\mathcal{A}$ is a $*$-algebra then this symplectic structure is real ($\omega = \omega^*$). Although a little tautological, this construction is relevant for quantum mechanics.

Let $\mathcal{A}$ be, as above, a complex unital $*$-algebra with a trivial center and only inner derivations, and assume that there exists a linear form τ on $\mathcal{A}$ which is central, i.e., $\tau(xy) = \tau(yx)$, and normalized by $\tau(\mathbb{1}) = 1$. Then one defines an element $\theta \in \underline{\Omega}^1_{\text{Der}}(\mathcal{A})$ by $\theta(\text{ad}(ix)) = x - \tau(x)\mathbf{1}$. One has $(d\theta)(\text{ad}(ix), \text{ad}(iy)) = i[x, y]$, i.e., $\omega = d\theta$, so in this case the symplectic form ω is exact. As examples of such algebras one can take $\mathcal{A} = M_n(\mathbb{C})$, (a factor of type I_n), with $\tau = (1/n)$ trace, or $\mathcal{A} = \mathcal{R}$, a von Neumann algebra which is a factor of type II_1 with τ equal to the normalized trace. The algebra $M_n(\mathbb{C})$ is the algebra of observables of a quantum spin $s = (n-1)/2$ while $\mathcal{R}$ is the algebra used to describe the observables of an infinite assembly of quantum spins; two typical types of quantum systems with no classical counterpart.

Let us now consider the C.C.R. algebra (canonical commutation relations) $\mathcal{A}_{\text{CCR}}$ [26]. This is the complex unital $*$-algebra generated by two Hermitian elements q and p satisfying the relation $[q, p] = i\hbar\mathbf{1}$. This algebra is the algebra of observables of the quantum counterpart of a classical system with one degree of freedom. We keep here the positive constant $\hbar$ (the Planck constant) in the formula for comparison with classical mechanics, although the algebra for $\hbar \neq 0$ is isomorphic to the one with $\hbar = 1$. We restrict attention to one degree of freedom to simplify the notations, but the discussion extends easily to a finite number of degrees of freedom. This algebra has again only inner derivations and a trivial center so $\omega(\text{ad}(\frac{i}{\hbar}x), \text{ad}(\frac{i}{\hbar}y)) = \frac{i}{\hbar}[x, y]$ defines a symplectic structure for which $\text{Ham}(x) = \text{ad}(\frac{i}{\hbar}x)$ and $\{x, y\} = \frac{i}{\hbar}[x, y]$ which is the standard quantum Poisson bracket. In this case one can express ω in terms of the generators q and p and their differentials [26], [27]:

$$\omega = \sum_{n \geq 0} \left(\frac{1}{i\hbar}\right)^n \frac{1}{(n+1)!} [\dots[dp, \underbrace{p], \dots, p}_{n}][\dots[dq, \underbrace{q], \dots, q}_{n}]$$

Notice that this formula is meaningful, because if one inserts two derivations $\text{ad}(ix), \text{ad}(iy)$ in it only a finite number of terms contribute to the sum. In contrast to the preceding case, here the symplectic form is not exact, i.e., it corresponds to a non-vanishing element of $H^2(\underline{\Omega}_{\text{Der}}(\mathcal{A}_{\text{CCR}}))$ which is therefore non-trivial. This was guessed in [26] on the basis of the non-existence of a finite trace (i.e., central linear form) on $\mathcal{A}_{\text{CCR}}$ and finally proved in [38]. For $\hbar = 0$, q and p commute and the algebra reduces to the algebra of complex polynomial functions on the phase space $\mathbb{R}^2$. Furthermore, the limit of $\{x, y\} = \frac{i}{\hbar}[x, y]$ at $\hbar = 0$ reduces to the usual classical Poisson bracket as well known and, by using the above formula, one sees that the formal limit of ω at $\hbar = 0$ is $dpdq$. This limit is, however, very singular since the limit algebra is the algebra of complex polynomials in two indeterminates, the limit symplectic form is exact and not every derivation is Hamiltonian in contrast to what happens for $\mathcal{A}_{\text{CCR}}$ (i.e., for $\hbar \neq 0$).

10. Theory of connections

Throughout this section $\mathcal{A}$ is a unital associative complex algebra and Ω is a differential calculus over $\mathcal{A}$ that is a graded differential algebra such that $\Omega^0 = \mathcal{A}$ with differential denoted by d.

Let $\mathcal{M}$ be a left $\mathcal{A}$-module; a Ω-*connection* on $\mathcal{M}$ (or simply a *connection* on $\mathcal{M}$ if no confusion arises) is a linear mapping $\nabla : \mathcal{M} \to \Omega^1 \otimes_{\mathcal{A}} \mathcal{M}$ such that one has

$$\nabla(am) = a\nabla(m) + d(a) \otimes_{\mathcal{A}} m$$

for any $a \in \mathcal{A}$ and $m \in \mathcal{M}$, ($\Omega^1 \otimes_{\mathcal{A}} \mathcal{M}$ being equipped with its canonical structure of left $\mathcal{A}$-module). One extends ∇ to $\Omega \otimes_{\mathcal{A}} \mathcal{M}$ by setting $\nabla(\omega \otimes_{\mathcal{A}} m) = (-1)^n \omega\nabla(m) + d(\omega) \otimes_{\mathcal{A}} m$ for $\omega \in \Omega^n$ and $m \in \mathcal{M}$ ($\Omega \otimes_{\mathcal{A}} \mathcal{M}$ is canonically a left Ω-module). It then follows from the definitions that ∇^2 is a left Ω-module endomorphism of $\Omega \otimes_{\mathcal{A}} \mathcal{M}$ which implies that its restriction $\nabla^2 : \mathcal{M} \to \Omega^2 \otimes_{\mathcal{A}} \mathcal{M}$ to $\mathcal{M}$ is a homomorphism of left $\mathcal{A}$-modules; this homomorphism is called the *curvature* of the connection ∇.

Not every left $\mathcal{A}$-module admits a connection. If $\mathcal{M}$ is the free $\mathcal{A}$-module $\mathcal{A} \otimes E$, where E is some complex vector space, then $\nabla = d \otimes I_E$ is a connection on $\mathcal{A} \otimes E$ which has a vanishing curvature (such a connection with zero curvature is said to be flat). If $\mathcal{M} \subset \mathcal{A} \otimes E$ is a direct summand of a free $\mathcal{A}$-module $\mathcal{A} \otimes E$ and if $P : \mathcal{A} \otimes E \to \mathcal{M}$ is the corresponding projection, then $\nabla = P \circ (d \otimes I_E)$ is a connection on $\mathcal{M}$. Thus a projective module admits (at least one) a connection. In the case where Ω is the universal differential calculus $\Omega_u(\mathcal{A})$ the converse is also true: It was shown in [22] that a (left) $\mathcal{A}$-module admits a $\Omega_u(\mathcal{A})$-connection if and only if it is projective.

One defines in a similar manner Ω-connections on right modules. Namely if $\mathcal{N}$ is a right $\mathcal{A}$-module, a Ω-connection on $\mathcal{N}$ is a linear mapping $\nabla : \mathcal{N} \to \mathcal{N} \otimes_{\mathcal{A}} \Omega^1$ such that $\nabla(na) = \nabla(n)a + n \otimes_{\mathcal{A}} d(a)$ for any $n \in \mathcal{N}$ and $a \in \mathcal{A}$.

Let $\mathcal{M}$ be a left $\mathcal{A}$-module, then its dual $\mathcal{M}^*$ (i.e., the set of left $\mathcal{A}$-module homomorphisms of $\mathcal{M}$ into $\mathcal{A}$) is a right $\mathcal{A}$-module. We denote by $\langle m, n\rangle \in \mathcal{A}$ the evaluation of $n \in \mathcal{M}^*$ on $m \in \mathcal{M}$. Let ∇ be a Ω-connection on $\mathcal{M}$, then one defines a unique linear mapping ∇^* of $\mathcal{M}^*$ into $\mathcal{M}^* \otimes_{\mathcal{A}} \Omega^1$ by setting (with obvious notations)

$$\langle m, \nabla^*(n)\rangle = d(\langle m, n\rangle) - \langle \nabla(m), n\rangle$$

for any $m \in \mathcal{M}$ and $n \in \mathcal{M}^*$. It is easy to verify that ∇^* is a Ω-connection on the right module $\mathcal{M}^*$ which will be referred to as *the dual connection* of

∇. One defines in the same way the dual connection of a connection on a right module.

Our aim is now to recall the definitions of Hermitian modules over a $*$-algebra $\mathcal{A}$ and of Hermitian connections. We assume that $\mathcal{A}$ is a $*$-algebra such that the convex cone $\mathcal{A}^+$ generated by the a^*a ($a \in \mathcal{A}$) is a strict cone, i.e., such that $\mathcal{A}^+ \cap (-\mathcal{A}^+) = 0$. This property is satisfied for instance by $*$-algebras of operators in Hilbert spaces. A *Hermitian structure* on a right $\mathcal{A}$-module $\mathcal{M}$ [14] is a sesquilinear mapping $h : \mathcal{M} \times \mathcal{M} \to \mathcal{A}$ such that one has:

$$(i) \quad h(ma, nb) = a^* h(m,n) b, \ \ \forall m, n \in \mathcal{M} \text{ and } \forall a, b \in \mathcal{A};$$

$$(ii) \quad h(m,m) \in \mathcal{A}^+, \ \ \forall m \in \mathcal{M} \text{ and } h(m,m) = 0 \Rightarrow m = 0.$$

A right $\mathcal{A}$-module $\mathcal{M}$ equipped with a Hermitian structure will be referred to as a *Hermitian module* over $\mathcal{A}$. If $\mathcal{M}$ is a Hermitian module over $\mathcal{A}$, a *Hermitian connection* on $\mathcal{M}$ is a connection ∇ on the right $\mathcal{A}$-module $\mathcal{M}$ such that one has

$$d(h(m,n)) = h(\nabla m, n) + h(m, \nabla n)$$

for any $m, n \in \mathcal{M}$ with obvious notations. We have chosen to define Hermitian structures on right modules for notational reasons (we prefer the convention of physicists for sesquilinearity, i.e., linearity in the second argument); one can define similarly Hermitian structures and connections for left modules.

Let $\mathcal{M}$ be a right $\mathcal{A}$-module. The group $\text{Aut}(\mathcal{M})$ of all module automorphisms of $\mathcal{M}$ acts on the affine space of all connections on $\mathcal{M}$ via $\nabla \mapsto \nabla^U = U \circ \nabla \circ U^{-1}, U \in \text{Aut}(\mathcal{M})$, (canonically one has $\text{Aut}(\mathcal{M}) \subset \text{Aut}(\mathcal{M} \otimes_{\mathcal{A}} \Omega^1)$). If, furthermore, $\mathcal{A}$ is a $*$-algebra as above and if h is a Hermitian structure on $\mathcal{M}$ then the subgroup of $\text{Aut}(\mathcal{M})$ of all automorphisms U which preserve h, i.e., such that $h(Um, Un) = h(m,n)$ for $m, n \in \mathcal{M}$ will be denoted by $\text{Aut}(\mathcal{M}, h)$ and called the *gauge group*, whereas its elements will be called *gauge transformations*; it acts on the real affine space of Hermitian connections on $\mathcal{M}$.

As pointed out before, one-sided modules are not sufficient and one needs bimodules for many reasons. Firstly, in the case where $\mathcal{A}$ is a $*$-algebra one needs $*$-bimodules to formulate and discuss reality conditions [34], [18], [27] (see also in the introduction). Secondly, a natural noncommutative generalization of linear connections should be connections on Ω^1, since Ω is taken as an analog of differential forms, but this is a $(\mathcal{A}, \mathcal{A})$-bimodule in an essential way. Thirdly, in order to have an analog of local couplings one needs to have a tensor product over $\mathcal{A}$, since the latter is the noncommutative version of the local tensor product of tensor fields. In short, one needs a theory of connections for bimodules, and any of the above quoted problems shows that one-sided connections on bimodules (i.e., on bimodules considered as left or right modules) are of no help. The difficulty of defining a Ω-connection on a $(\mathcal{A},\mathcal{A})$-bimodule $\mathcal{M}$ lies in that a left $\mathcal{A}$-module connection on $\mathcal{M}$ sends $\mathcal{M}$ into $\Omega^1 \otimes_{\mathcal{A}} \mathcal{M}$ whereas a right $\mathcal{A}$-module connection on $\mathcal{M}$ sends $\mathcal{M}$ into $\mathcal{M} \otimes_{\mathcal{A}} \Omega^1$. A solution of this problem adapted to the case where $\mathcal{M} = \Omega^1$ has been given in [56], and generalized in [32] for arbitrary $(\mathcal{A},\mathcal{A})$-bimodules on the basis of an analysis of first order differential operators in bimodules. This led to the following definition [32].

Let $\mathcal{M}$ be a $(\mathcal{A},\mathcal{A})$-bimodule; a *left bimodule Ω-connection* on $\mathcal{M}$ is a left $\mathcal{A}$-module Ω-connection ∇ on $\mathcal{M}$ for which there is a bimodule homomorphism $\sigma : \mathcal{M} \otimes_{\mathcal{A}} \Omega^1 \to \Omega^1 \otimes_{\mathcal{A}} \mathcal{M}$ such that

$$\nabla(ma) = \nabla(m)a + \sigma(m \otimes_{\mathcal{A}} d(a))$$

for any $a \in \mathcal{A}$ and $m \in \mathcal{M}$. Clearly σ is then unique under these conditions. One defines similarly a *right bimodule Ω-connection* on $\mathcal{M}$ to be a right $\mathcal{A}$-module Ω-connection ∇ on $\mathcal{M}$ for which there is a bimodule homomorphism $\sigma : \Omega^1 \otimes_{\mathcal{A}} \mathcal{M} \to \mathcal{M} \otimes_{\mathcal{A}} \Omega^1$ such that

$$\nabla(am) = a\nabla(m) + \sigma(d(a) \otimes_{\mathcal{A}} m)$$

for any $a \in \mathcal{A}$ and $m \in \mathcal{M}$. When no confusion arises about Ω and about 'left/right' we simply refer to this notion as *bimodule connection*.

In the case where $\mathcal{M}$ is the bimodule Ω^1 itself, a left bimodule Ω-connection is just the first part of the proposal of [56] for the definition of linear connections in noncommutative geometry; the second part of the proposal of [56] relates σ and the product $\Omega^1 \otimes_{\mathcal{A}} \Omega^1 \to \Omega^2$, so it makes sense only for $\mathcal{M} = \Omega^1$, which is necessary to define the generalization of torsion.

It has been shown in [7] (Appendix A of [7]) that on general grounds the above definition is just what is needed to define tensor products over $\mathcal{A}$ of bimodule connections and of left (right) bimodule connections with left (right) module connections. In fact, let ∇' be a left bimodule connection on the bimodule $\mathcal{M}'$ and let ∇'' be a connection on a left module $\mathcal{M}''$. Then one defines a connection ∇ on the left module $\mathcal{M}' \otimes_{\mathcal{A}} \mathcal{M}''$ by setting

$$\nabla = \nabla' \otimes_{\mathcal{A}} I_{\mathcal{M}''} + (\sigma' \otimes_{\mathcal{A}} I_{\mathcal{M}''}) \circ (I_{\mathcal{M}'} \otimes_{\mathcal{A}} \nabla''),$$

where $\sigma' : \mathcal{M}' \otimes_{\mathcal{A}} \Omega^1 \to \Omega^1 \otimes_{\mathcal{A}} \mathcal{M}'$ is the bimodule homomorphism corresponding to ∇'. If, furthermore, $\mathcal{M}''$ is a $(\mathcal{A},\mathcal{A})$-bimodule and if ∇'' is a left bimodule connection with corresponding bimodule homomorphism

$\sigma'' : \mathcal{M}'' \otimes_{\mathcal{A}} \Omega^1 \to \Omega^1 \otimes_{\mathcal{A}} \mathcal{M}''$, then ∇ is also a left bimodule connection with corresponding bimodule homomorphism σ given by

$$\sigma = (\sigma' \otimes_{\mathcal{A}} I_{\mathcal{M}''}) \circ (I_{\mathcal{M}'} \otimes_{\mathcal{A}} \sigma'')$$

of $\mathcal{M}' \otimes_{\mathcal{A}} \mathcal{M}'' \otimes_{\mathcal{A}} \Omega^1$ into $\Omega^1 \otimes_{\mathcal{A}} \mathcal{M}' \otimes_{\mathcal{A}} \mathcal{M}''$.

Let $\mathcal{M}$ be a $(\mathcal{A}, \mathcal{A})$-bimodule and let $\mathcal{M}^*$ denote the dual of $\mathcal{M}$ considered as a left $\mathcal{A}$-module. Then $\mathcal{M}^*$ is a right $\mathcal{A}$-module as dual of a left $\mathcal{A}$-module, but it is in fact a bimodule if one defines the left action $m' \mapsto am'$ of $\mathcal{A}$ on $\mathcal{M}^*$ by $\langle m, am' \rangle = \langle ma, m' \rangle$ for any $m \in \mathcal{M}, a \in \mathcal{A}, m' \in \mathcal{M}^*$. If ∇ is a left bimodule Ω-connection on $\mathcal{M}$ then one verifies that ∇^* is a right bimodule Ω-connection on $\mathcal{M}^*$ [7] (Appendix B of [7]). Notice that this kind of duality between bimodules is different of the $\mathcal{A}$-duality between bimodules over $\mathcal{A}$ and modules over $Z(\mathcal{A})$ discussed in Section 8.

When $\mathcal{A}$ is a $*$-algebra there is also a generalization of Hermitian forms on $(\mathcal{A}, \mathcal{A})$-bimodules which has been introduced on [57] and called *right Hermitian forms* in [34] which is adapted for tensor products over $\mathcal{A}$. If $\mathcal{M}$ is a $(\mathcal{A}, \mathcal{A})$-bimodule, then a right Hermitian form on $\mathcal{M}$ is a Hermitian form h on $\mathcal{M}$ considered as a right $\mathcal{A}$-module which is such that for the left multiplication by $a \in \mathcal{A}$ one has $h(m, an) = h(a^*m, n)$. One can then define the notion of right Hermitian bimodule connection, (which is, in particular, a right bimodule connection).

We now explain the relation between the above notion of bimodule connection and the theory of first order operators in bimodules. Let $\mathcal{A}$ and $\mathcal{B}$ be unital associative complex algebras and let $\mathcal{M}$ and $\mathcal{N}$ be two $(\mathcal{A}, \mathcal{B})$-bimodules. We denote by l_a the left multiplication by $a \in \mathcal{A}$ in $\mathcal{M}$ and in $\mathcal{N}$ and we denote by r_b the right multiplication by $b \in \mathcal{B}$ in $\mathcal{M}$ and in $\mathcal{N}$. A linear mapping D of $\mathcal{M}$ into $\mathcal{N}$ which is such that one has $[[D, l_a], r_b] = 0$ for any $a \in \mathcal{A}$ and $b \in \mathcal{B}$ is called *a first-order operator* or *an operator of order 1* of $\mathcal{M}$ into $\mathcal{N}$ [16]. Notice that homomorphisms of left $\mathcal{A}$-modules of $\mathcal{M}$ into $\mathcal{N}$ as well as homomorphisms of right $\mathcal{B}$-modules of $\mathcal{M}$ into $\mathcal{N}$ are first-order operators of $\mathcal{M}$ into $\mathcal{N}$. The structure of first-order operators is given by the following theorem [32].

THEOREM 4 *Let $\mathcal{M}$ and $\mathcal{N}$ be two $(\mathcal{A}, \mathcal{B})$-bimodules and let D be a first order operator of $\mathcal{M}$ into $\mathcal{N}$. Then there is a unique $(\mathcal{A}, \mathcal{B})$-bimodule homomorphism $\sigma_L(D)$ of $\Omega^1_u(\mathcal{A}) \otimes_{\mathcal{A}} \mathcal{M}$ into $\mathcal{N}$ and there is a unique $(\mathcal{A}, \mathcal{B})$-bimodule homomorphism $\sigma_R(D)$ of $\mathcal{M} \otimes_{\mathcal{B}} \Omega^1_u(\mathcal{B})$ into $\mathcal{N}$ such that one has:*

$$D(amb) = aD(m)b + \sigma_L(D)(d_u a \otimes m)b + a\sigma_R(D)(m \otimes d_u b)$$

for any $m \in \mathcal{M}$, $a \in \mathcal{A}$ *and* $b \in \mathcal{B}$.

For the proof and further informations see in [32]. It is clear that $\sigma_L(D)$ and $\sigma_R(D)$ are the appropriate generalization of the notion of symbol in this setting. We shall refer to them as the *left* and the *right universal symbol of D* respectively.

Remark 12. The converse of Theorem 4 is also true. More precisely, let (Ω^1_L, d_L) be a first order differential calculus over $\mathcal{A}$, let (Ω^1_R, d_R) be a first order differential calculus over $\mathcal{B}$ and let $D : \mathcal{M} \to \mathcal{N}$ be a linear mapping then any of the following condition (1) or (2) implies that D is a first-order operator.

(1) There is a $(\mathcal{A}, \mathcal{B})$-bimodule homomorphism $\sigma_L : \Omega^1_L \otimes_{\mathcal{A}} \mathcal{M} \to \mathcal{N}$ such that

$$D(am) = aD(m) + \sigma_L(d_L(a) \otimes m), \ \ \forall m \in \mathcal{M} \text{ and } \forall a \in \mathcal{A}$$

(2) There is a $(\mathcal{A}, \mathcal{B})$-bimodule homomorphism $\sigma_R : \mathcal{M} \otimes_{\mathcal{B}} \Omega^1_R \to \mathcal{N}$ such that

$$D(mb) = D(m)b + \sigma_R(m \otimes d_R(b)), \ \ \forall m \in \mathcal{M} \text{ and } \forall b \in \mathcal{B}.$$

Let $\mathcal{M}$ be a $(\mathcal{A}, \mathcal{A})$-bimodule and let ∇ be a left $\mathcal{A}$-module Ω-connection on $\mathcal{M}$. It is obvious that ∇ is a first-order operator of the $(\mathcal{A}, \mathcal{A})$-bimodule $\mathcal{M}$ into the $(\mathcal{A}, \mathcal{A})$-bimodule $\Omega^1 \otimes_{\mathcal{A}} \mathcal{M}$. It follows therefore from the above theorem that there is a unique $(\mathcal{A}, \mathcal{A})$-bimodule homomorphism $\sigma_R(\nabla)$: $\mathcal{M} \otimes_{\mathcal{A}} \Omega^1_u(\mathcal{A}) \to \Omega^1 \otimes_{\mathcal{A}} \mathcal{M}$ such that one has

$$\nabla(ma) = \nabla(m)a + \sigma_R(\nabla)(m \otimes_{\mathcal{A}} d_u(a))$$

for any $m \in \mathcal{M}$ and $a \in \mathcal{A}$. Therefore, ∇ is a left bimodule Ω-connection on $\mathcal{M}$ if and only if $\sigma_R(\nabla)$ factorizes through a $(\mathcal{A}, \mathcal{A})$-bimodule homomorphism $\sigma : \mathcal{M} \otimes_{\mathcal{A}} \Omega^1 \to \Omega^1 \otimes_{\mathcal{A}} \mathcal{M}$ as $\sigma_R(\nabla) = \sigma \circ (I_{\mathcal{M}} \otimes i_d)$, where $I_{\mathcal{M}}$ is the identity mapping of $\mathcal{M}$ onto itself and i_d is the unique $(\mathcal{A}, \mathcal{A})$-bimodule homomorphism of $\Omega^1_u(\mathcal{A})$ into Ω^1 such that $d = i_d \circ d_u$ (see Proposition 1). This implies in particular that any left $\mathcal{A}$-module $\Omega_u(\mathcal{A})$-connection is a left bimodule $\Omega_u(\mathcal{A})$-connection.

In the case of the derivation-based differential calculus there is an easy natural way of defining connections on left and right modules and on central bimodules over $\mathcal{A}$, [34]. We describe it in the case of central bimodules (for left and for right modules, just forget multiplications on the right and on the left, respectively). Let $\mathcal{M}$ be a central bimodule over $\mathcal{A}$, i.e., a $\mathcal{A}$-bimodule for

$\mathbf{Alg}_Z$, *a (derivation-based) connection on* $\mathcal{M}$ is a linear mapping ∇, $X \mapsto \nabla_X$, of Der($\mathcal{A}$) into the linear endomorphisms of $\mathcal{M}$ such that

$$\nabla_{zX}(m) = z\nabla_X(m), \nabla_X(amb) = a\nabla_X(m)b + X(a)mb + amX(b)$$

for any $m \in \mathcal{M}$, $X \in \text{Der}(\mathcal{A})$, $z \in Z(\mathcal{A})$ and $a, b \in \mathcal{A}$. One verifies that such a connection on the central bimodule $\mathcal{M}$ is a bimodule $\underline{\Omega}_{\text{Der}}(\mathcal{A})$-connection on $\mathcal{M}$ in the previous sense with a well defined σ (modulo some technical problems of completion of the tensor products $\underline{\Omega}^1_{\text{Der}}(\mathcal{A}) \otimes_{\mathcal{A}} \mathcal{M}$ and $\mathcal{M} \otimes_{\mathcal{A}} \underline{\Omega}^1_{\text{Der}}(\mathcal{A})$). The interest of this formulation is that curvature is straightforwardly defined and is a bimodule homomorphism [34]. We refer to [34] (and also to [27]) for more details and, in particular, for the relation with $\mathcal{A}$-duality. Furthermore, in this frame the notion of reality on connections is obvious. Assume that $\mathcal{A}$ is a $*$-algebra and that $\mathcal{M}$ is a central bimodule which is a $*$-bimodule over $\mathcal{A}$ then a (derivation-based) connection ∇ on $\mathcal{M}$ will be said to be *real* if one has $\nabla_X(m^*) = (\nabla_X(m))^*$ for any $m \in \mathcal{M}$ and any $X \in \text{Der}_{\mathbb{R}}(\mathcal{A})$, i.e., $X \in \text{Der}(\mathcal{A})$ with $X = X^*$.

The notion of reality in the general frame of bimodule Ω-connections is slightly more involved and will not be discussed here.

11. Classical Yang–Mills–Higgs models

An aspect, with no counterpart in ordinary differential geometry, of the theory of Ω-connections on $\mathcal{A}$-modules for a differential calculus Ω which is not graded commutative is the generic occurrence of inequivalent Ω-connections with vanishing curvature (on a fixed $\mathcal{A}$-module). By taking as the algebra $\mathcal{A}$ the algebra of functions on space-time with values in some algebra $\mathcal{A}_0$, i.e., $\mathcal{A} = C^\infty(\mathbb{R}^{s+1}) \otimes \mathcal{A}_0$, this led to classical Yang–Mills–Higgs models based on noncommutative geometry in which the Higgs field is the part of the connection which is in the 'noncommutative directions'.

In the following we display the case of Ω-connections on right modules over the algebra $\mathcal{A} = C^\infty(\mathbb{R}^{s+1}) \otimes M_n(\mathbb{C})$ of smooth $M_n(\mathbb{C})$-valued functions on $\mathbb{R}^{s+1}$ for $\Omega = \Omega_{\text{Der}}(\mathcal{A})$.

Let us first describe the situation for $\mathcal{A} = M_n(\mathbb{C})$. The derivations of $M_n(\mathbb{C})$ are all inner, so the complex Lie algebra $\text{Der}(M_n(\mathbb{C}))$ reduces to $\mathfrak{sl}(n)$ and the real Lie algebra $\text{Der}_{\mathbb{R}}(M_n(\mathbb{C}))$ reduces to $\mathfrak{su}(n)$. As already mentioned in Section 8, one has

$$\Omega_{\text{Der}}(M_n(\mathbb{C})) = C_\wedge(\text{Der}M_n(\mathbb{C}), M_n(\mathbb{C})) = C_\wedge(\mathfrak{sl}(n), M_n(\mathbb{C})),$$

as can be shown directly [25] and as also follows from the formulas below. Let $E_k, k \in \{1, 2, \ldots, n^2 - 1\}$ be a base of self–adjoint traceless $n \times n$ matrices. The $\partial_k = \mathrm{ad}(iE_k)$ form a basis of real derivations, i.e., a basis of $\mathrm{Der}_{\mathbb{R}}(M_n(\mathbb{C})) = \mathfrak{su}(n)$. One has $[\partial_k, \partial_\ell] = C^m_{k\ell}\partial_m$, the $C^m_{k\ell}$ are the corresponding structure constants of $\mathfrak{su}(n)$, (or $\mathfrak{sl}(n)$). Define $\theta^k \in \Omega^1_{\text{Der}}(M_n(\mathbb{C}))$ by $\theta^k(\partial_\ell) = \delta^k_\ell \mathbf{1}$.The following formulas give a presentation of the graded differential algebra $\Omega_{\text{Der}}(M_n(\mathbb{C}))$ [28], [26]:

$$\begin{aligned}
E_k E_\ell &= g_{k\ell}\mathbf{1} + (S^m_{k\ell} - \tfrac{i}{2}C^m_{k\ell})E_m, \\
E_k \theta^\ell &= \theta^\ell E_k, \\
\theta^k \theta^\ell &= -\theta^\ell \theta^k, \\
dE_k &= -C^m_{k\ell}E_m\theta^\ell, \\
d\theta^k &= -\tfrac{1}{2}C^k_{\ell m}\theta^\ell\theta^m,
\end{aligned}$$

where $g_{k\ell} = g_{\ell k}$, $S^m_{k\ell} = S^m_{\ell k}$ are real, $g_{k\ell}$ are the components of the Killing form of $\mathfrak{su}(n)$, and $C^m_{k\ell} = -C^m_{\ell k}$ are, as above, the (real) structure constants of $\mathfrak{su}(n)$.Formula giving the dE_k can be inverted and one has $\theta^k = -(i/n^2)\ g^{\ell m}g^{kr}E_\ell E_r dE_m$, where $g^{k\ell}$ are the components of the inverse matrix of $(g_{k\ell})$. The element $\theta = E_k\theta^k$ of $\Omega^1_{\text{Der}}(M_n(\mathbb{C}))$ is real, $\theta = \theta^*$, and independent of the choice of the E_k, in fact we have already met θ in Section 9: $\theta(\mathrm{ad}(iA)) = A - (1/n)\mathrm{tr}(A)\mathbf{1}$ and $\omega = d\theta$ is the natural symplectic structure for $M_n(\mathbb{C})$. Furthermore, θ is invariant, $L_X\theta = 0$, and any invariant element of $\Omega^1_{\text{Der}}(M_n(\mathbb{C}))$ is a scalar multiple of θ. We call θ the *canonical invariant element* of $\Omega^1_{\text{Der}}(M_n(\mathbb{C}))$. One has

$$dM = i[\theta, M], \quad \forall M \in M_n(\mathbb{C})$$
$$d(-i\theta) + (-i\theta)^2 = 0.$$

The $*$–algebra $M_n(\mathbb{C})$ is simple with only one irreducible representation in $\mathbb{C}^n$. A general finite right–module (which is projective) is the space $M_{Kn}(\mathbb{C})$ of $K \times n$–matrices with right action of $M_n(\mathbb{C})$. One has $\mathrm{Aut}(M_{Kn}(\mathbb{C})) = GL(K)$ with left matrix multiplication. The module $M_{Kn}(\mathbb{C})$ is naturally Hermitian with $h(\Phi, \Psi) = \Phi^*\Psi$ where Φ^* is the $n \times K$ matrix Hermitian conjugate to Φ. The gauge group is then the unitary group $U(K)(\subset GL(K))$. Here, there is a natural origin $\overset{0}{\nabla}$ in the space of connections given by $\overset{0}{\nabla}\Phi = -i\Phi\theta$ where $\Phi \in M_{Kn}(\mathbb{C})$ and where θ is the canonical invariant element of $\Omega^1_{\text{Der}}(M_n(\mathbb{C}))$. The fact that this defines a connection follows from

$$\overset{0}{\nabla}(\Phi M) = (\overset{0}{\nabla}\Phi)M + \Phi i[\theta, M]$$

and from the above expression of dM for $M \in M_n(\mathbb{C})$. This connection is Hermitian and its follows from the above expression for $d\theta$ that its curvature vanishes, i.e., $(\overset{0}{\nabla})^2 = 0$. Any connection ∇ is of the form $\nabla\Phi = \overset{0}{\nabla}\Phi + A\Phi$ where $A = A_k\theta^k$ with $A_k \in M_K(\mathbb{C})$ and $A\Phi$ means $A_k\Phi\otimes\theta^k$. The connection ∇ is Hermitian if and only if the A_k are anti-Hermitian i.e., $A_k^* = -A_k$. The curvature of ∇ is given by $\nabla^2\Phi = F\Phi$ $(= F_{k\ell}\Phi \otimes \theta^k\theta^\ell)$ with

$$F = \frac{1}{2}([A_k, A_\ell] - C^m_{k\ell}A_m)\theta^k\theta^\ell.$$

Thus $\nabla^2 = 0$ if and only if the A_k form a representation of the Lie algebra $\mathfrak{sl}(n)$ in $\mathbb{C}^K$ and two such connections are in the same $\mathrm{Aut}(M_{Kn}(\mathbb{C}))$–orbit if and only if the corresponding representations of $\mathfrak{sl}(n)$ are equivalent. This implies that the *gauge orbits of flat* $(\nabla^2 = 0)$ *Hermitian connections are in one–to–one correspondence with unitary classes of representations of* $\mathfrak{su}(n)$ *in* $\mathbb{C}^K$, [28]. For instance if $n = 2$, these orbits are labelled by the number of partitions of the integer K.

We now come to the case $\mathcal{A} = C^\infty(\mathbb{R}^{s+1}) \otimes M_n(\mathbb{C})$. Let x^μ, $\mu \in \{0, 1, \dots, s\}$, be the canonical coordinates of $\mathbb{R}^{s+1}$. One has $\Omega_{\mathrm{Der}}(C^\infty(\mathbb{R}^{s+1}) \otimes M_n(\mathbb{C})) = \Omega_{\mathrm{Der}}(C^\infty(\mathbb{R}^{s+1})) \otimes \Omega_{\mathrm{Der}}(M_n(\mathbb{C}))$ so one can split the differential as $d = d' + d''$ where d' is the differential along $\mathbb{R}^{s+1}$ and d'' is the differential of $\Omega_{\mathrm{Der}}(M_n(\mathbb{C}))$. A typical finite projective right module is $C^\infty(\mathbb{R}^{s+1}) \otimes M_{Kn}(\mathbb{C})$. This is an Hermitian module with Hermitian structure given by $h(\Phi, \Psi)(x) = \Phi(x)^*\Psi(x)$, $(x \in \mathbb{R}^{s+1})$. As a $C^\infty(\mathbb{R}^{s+1})$–module, this module is free (of rank $K.n$), so $d'\Phi$ is well defined for $\Phi \in C^\infty(\mathbb{R}^{s+1}) \otimes M_{Kn}(\mathbb{C})$. In fact, $d'\Phi(x) = \frac{\partial\Phi}{\partial x^\mu}(x)dx^\mu$. A connection on the $C^\infty(\mathbb{R}^{s+1}) \otimes M_n(\mathbb{C})$–module $C^\infty(\mathbb{R}^{s+1}) \otimes M_{Kn}(\mathbb{C})$ is of the form $\nabla\Phi = d'\Phi - i\Phi\theta + A\Phi$ with $A = A_\mu dx^\mu + A_k\theta^k$, where the A_μ and the A_k are $K \times K$ matrix valued functions on $\mathbb{R}^{s+1}$ (i.e., elements of $C^\infty(\mathbb{R}^{s+1}) \otimes M_K(\mathbb{C})$) and where $A\Phi(x) = A_\mu(x)\Phi(x)dx^\mu + A_k(x)\Phi(x)\theta^k$. Such a connection is Hermitian if and only if the $A_\mu(x)$ and the $A_k(x)$ are anti-Hermitian, $\forall x \in \mathbb{R}^{s+1}$. The curvature of ∇ is given by $\nabla^2\Phi = F\Phi$ where

$$\begin{aligned} F &= \tfrac{1}{2}(\partial_\mu A_\nu - \partial_\nu A_\mu + [A_\mu, A_\nu])dx^\mu dx^\nu \\ &\quad +(\partial_\mu A_k + [A_\mu, A_k])dx^\mu\theta^k \\ &\quad +\tfrac{1}{2}([A_k, A_\ell] - C^m_{k\ell}A_m)\theta^k\theta^\ell. \end{aligned}$$

The connection ∇ is flat (i.e., $\nabla^2 = 0$) if and only if each term of the above formula vanishes, which implies that ∇ is gauge equivalent to a connection for which one has $A_\mu = 0$, $\partial_\mu A_k = 0$ and $[A_k, A_\ell] = C^m_{k\ell}A$. Furthermore, two such connections are equivalent if and only if the corresponding representations of $\mathfrak{su}(n)$ in $\mathbb{C}^K$ (given by the constant $K \times K$–matrices A_ℓ) are equivalent. So, again, *the gauge orbits of flat Hermitian connections are in one-to-one correspondence with the unitary classes of (anti-Hermitian) representations of* $\mathfrak{su}(n)$ *in* $\mathbb{C}^K$. Again, in the case $n = 2$, the number of such orbits is the number of partitions of the integer K, i.e.,

$$\text{card}\left\{(n_r) | \sum_r n_r.r = K\right\}.$$

If we consider $\mathbb{R}^{s+1}$ as the $(s+1)$–dimensional space-time and if we replace the algebra of smooth functions on $\mathbb{R}^{s+1}$ by $C^\infty(\mathbb{R}^{s+1}) \otimes M_n(\mathbb{C})$ which we interpret as the algebra of 'smooth functions on a noncommutative generalized space-time'. It is clear from the above expression for the curvature that the generalization of the (Euclidean) Yang–Mills action for a Hermitian connection ∇ on $C^\infty(\mathbb{R}^{s+1}) \otimes M_{Kn}(\mathbb{C})$ is

$$\begin{aligned} \|F\|^2 &= \int d^{s+1}x \,\text{tr}\Big\{\tfrac{1}{4}\sum(\partial_\mu A_\nu - \partial_\nu A_\mu + [A_\mu, A_\nu])^2 \\ &\qquad + \tfrac{1}{2}\sum(\partial_\mu A_k + [A_\mu, A_k])^2 \\ &\qquad + \tfrac{1}{4}\sum([A_k, A_\ell] - C^m_{k\ell}A_m)^2\Big\}, \end{aligned}$$

where the metrics of space-time is $g_{\mu\nu} = \delta_{\mu\nu}$ and where the basis E_k of Hermitian traceless $n \times n$ matrices is chosen in such a way that $g_{k\ell} = \delta_{k\ell}$, i.e., $\text{tr}(E_k E_\ell) = n\delta_{k\ell}$. This can be more deeply justified by introducing the analog of the Hodge involution on $\Omega_{\text{Der}}(M_n(\mathbb{C}))$, the analog of the integration of elements of $\Omega^{n^2-1}_{\text{Der}}(M_n(\mathbb{C}))$ (essentially the trace), and by combining these operations with the corresponding one on $\mathbb{R}^{s+1}$ to obtain a scalar product on $\Omega_{\text{Der}}(C^\infty(\mathbb{R}^{s+1}) \otimes M_n(\mathbb{C}))$, etc.. See in [28], [29] for more details.

The above action is the Yang–Mills action on the noncommutative space corresponding to $C^\infty(\mathbb{R}^{s+1}) \otimes M_n(\mathbb{C})$. However, it can be interpreted as the action of a field theory on the $(s+1)$-dimensional space–time $\mathbb{R}^{s+1}$. At first sight this field theory consists of a $U(K)$-Yang–Mills potential $A_\mu(x)$ minimally coupled with scalar fields $A_k(x)$ with values in the adjoint representation which interact among themselves through a quartic potential. The action is positive and vanishes for $A_\mu = 0$ and $A_k = 0$, but it also vanishes on other gauge orbits. Indeed, $\|F\|^2 = 0$ is equivalent to $F = 0$, so the gauge orbits on which the action vanishes are labelled by unitary classes of representations of $\mathfrak{su}(n)$ in $\mathbb{C}^K$. By the standard semi-heuristic argument, these gauge orbits are interpreted as different vacua for the corresponding quantum theory. To specify a quantum theory one has to choose one and to translate the fields in order that the zero of these translated fields corresponds to the chosen vacuum (i.e., is the corresponding zero of the action). The variables A_μ, A_k are thus adapted to the specific vacuum φ_0 corresponding to the trivial representation $A_k = 0$ of $\mathfrak{su}(n)$. If one chooses the vacuum φ_α corresponding to a representation $\overset{\alpha}{R}_k$ of $\mathfrak{su}(n)$ (i.e., one has $[\overset{\alpha}{R}_k, \overset{\alpha}{R}_\ell] = C^m_{k\ell}\overset{\alpha}{R}_n)$ one must instead use the variables A_μ and $\overset{\alpha}{B}_k = A_k - \overset{\alpha}{R}_k$. Making this change of variable one observes that components of A_μ become massive and that the $\overset{\alpha}{B}_k$ have different masses; the whole mass spectrum depends on α. This is very analogous to the Higgs mechanism. Here, however, the gauge invariance is not broken, the non-invariance of the mass terms of the A_μ is compensated by the gauge transformation of the $\overset{\alpha}{B}_k$ becoming inhomogeneous (they are components of a connection). Nevertheless, from the point of view of the space-time interpretation this is the Higgs mechanism and the A_k are Higgs fields.

The above models were the first ones of classical Yang–Mills–Higgs models based on noncommutative geometry. They certainly admit a natural supersymmetric extension since there is a natural extension of the derivation-based calculus to graded matrix algebras [42]. There is also another extension of the above calculus where $C^\infty(\mathbb{R}^{s+1}) \otimes M_n(\mathbb{C})$ is replaced by the algebra $\Gamma\mathrm{End}(E)$ of smooth sections of the endomorphisms bundle of a (non-trivial) smooth vector bundle E (of rank n) admitting a volume over a smooth ($(s+1)$-dimensional) manifold [33].

The use of the derivation-based calculus makes the above models quite rigid. By relaxing this, i.e., by using other differential calculi Ω, other models based on noncommutative geometry which are closer to the classical version of the standard model have been constructed [15], [19], [21]. Furthermore, there is an elegant way of combining the introduction of the (spinors) matter fields with the differential calculus and the metric [16] as well as with the reality conditions [18] in noncommutative geometry (and also with the action principles [11]). Within this general set-up one can probably absorb any classical model of gauge theory.

A problem arises for the quantization of these classical models based on noncommutative geometry. Namely, is it possible to keep something of the noncommutative geometrical interpretation of these classical models at the quantum level? The best would be to find some B.R.S. symmetry [3] ensuring that (perturbative) quantization does not spoil the correspondence with noncommutative geometry. Unfortunately no such symmetry has been discovered up to now. As long as no progress is obtained on this problem the noncommutative geometrical interpretation of the gauge theory with Higgs field must be taken with some circumspection, in spite of its appealing features.

12. Conclusion : Further remarks

Concerning the noncommutative generalization of differential geometry the point of view more or less explicit here is that the data are encoded in an algebra $\mathcal{A}$ which plays the role of the algebra of smooth functions. This is why although we have described various notions in terms of an arbitrary differential calculus Ω, we have studied in some detail specific differential calculi 'naturally' associated with $\mathcal{A}$ (i.e., which do not depend on other data than $\mathcal{A}$ itself) such as the universal differential calculus $\Omega_u(\mathcal{A})$, the generalization $\Omega_Z(\mathcal{A})$ of the Kähler exterior forms, the diagonal calculus $\Omega_{\text{Diag}}(\mathcal{A})$ and the derivation-based calculus. There are other possibilities, for instance some authors consider that the data are encoded in a graded differential algebra which plays the role of the algebra of smooth differential forms, e.g., [54]. This latter point of view can be taken into account here by using an arbitrary differential calculus Ω.

In all the above points of view the generalization of differential forms is provided by a graded differential algebra. This is not always so natural. For instance, it was shown in [46] (see also [47]) that the subspace $[\Omega_u(\mathcal{A}), \Omega_u(\mathcal{A})]_{\text{gr}}$ of graded commutators in $\Omega_u(\mathcal{A})$ is stable by d_u and that the cohomology of the cochain complex $\Omega_u(\mathcal{A})/[\Omega_u(\mathcal{A}), \Omega_u(\mathcal{A})]$ is closely related to the cyclic homology (it is contained in the reduced cyclic homology), and is also in several respects a noncommutative version of de Rham cohomology. This complex $\Omega_u(\mathcal{A})/[\Omega_u(\mathcal{A}), \Omega_u(\mathcal{A})]$ (which is generally not a graded algebra) is sometimes called *the noncommutative de Rham complex* [58]. It is worth noticing that for $\mathcal{A}$ noncommutative there is no tensor product over $\mathcal{A}$ between $\mathcal{A}$-modules (i.e., no analog of the tensor product of vector bundles), and that therefore the Grothendieck group $K_0(\mathcal{A})$ (of classes of projective $\mathcal{A}$-modules) has no product. Thus for $\mathcal{A}$ noncommutative $K_0(\mathcal{A}) \otimes \mathbb{C}$ is not an algebra, and therefore there is no reason for a cochain complex such that its cohomology is a receptacle for the image of the Chern character of $K_0(\mathcal{A}) \otimes \mathbb{C}$ to be a graded algebra.

Also we have not described here the approach to the differential calculus and to the metric aspects in noncommutative geometry based on generalized Dirac operators (spectral triples) [16], [17], [18] as well as the related supersymmetric approach of [39], see in O. Grandjean's lectures. In these notes we did not introduce specifically generalizations of linear connections, and *a fortiori* not generalizations of Riemannian structures.

Finally, we have not discussed differential calculus for quantum groups, i.e., bicovariant differential calculus [61]. In the spirit of Section 2 let us define a *graded differential Hopf algebra* to be a graded differential algebra

$\mathfrak{A}$ which is also a graded Hopf algebra with coproduct Δ such that $\Delta : \mathfrak{A} \to \mathfrak{A} \otimes \mathfrak{A}$ is a homomorphism of graded differential algebras (i.e., in particular, the differential d of $\mathfrak{A}$ satisfies the graded co-Leibniz rule), with counit ε such that $\varepsilon \circ d = 0$ and with antipode S homogeneous of degree 0 such that $S \circ d = d \circ S$. If $\mathfrak{A}$ is a graded differential Hopf algebra then the subalgebra $\mathfrak{A}^0$ of elements of degree 0 of $\mathfrak{A}$ is an ordinary Hopf algebra, i.e., a quantum group, and $\mathfrak{A}$ is a bicovariant differential calculus over $\mathfrak{A}^0$. Notice that if G is a Lie group then the graded differential algebra $\Omega(G)$ of differential forms on G is, in fact, a graded differential Hopf algebra which is graded commutative (in order to be correct, one has to complete the tensor product in the definition of the coproduct or to use, instead of $\Omega(G)$, the graded differential subalgebra of forms generated by the representative functions on G).

References

[1] A. Alekseev, E. Meinrenken, The non-commutative Weil algebra, math-DG/9903052 to appear in *Inv. Math.*

[2] Y. André, Différentielles non-commutatives et théorie de Galois différentielle ou aux différences, prépublication 221, Institut de Mathématiques de Jussieu, 1999.

[3] C. Becchi, A. Rouet, R. Stora, Renormalization models with broken symmetries, in *Renormalization Theory* (Erice 1975), G. Velo and A.S. Wightman (eds), D. Reidel, Dordrecht, 1976.

[4] R. Bott, L.W. Tu, *Differential forms in algebraic topology*, Springer-Verlag 1982.

[5] S. Boukraa, The BRS algebra of a free minimal differential algebra, *Nucl. Phys.* **B303** (1988), 237–259.

[6] N. Bourbaki, *Algèbre I*, Chapitre III. Paris, Hermann 1970.

[7] K. Bresser, F. Müller-Hoissen, A. Dimakis, A. Sitarz, Noncommutative geometry of finite groups, *J. of Physics A (Math. and General)* **29** (1996), 2705–2735.

[8] A. Cap, A. Kriegl, P.W. Michor, J. Vanzura, The Frölicher-Nijenhuis bracket in non commutative differential geometry, *Acta Math. Univ. Comenianae* **LXII** (1993), 17–49.

[9] H. Cartan, *Notion d'algèbre différentielle; application aux groupes de Lie et aux variétés où opère un groupe de Lie and*
La trangression dans un groupe de Lie et dans un espace fibré principal, Colloque de topologie (Bruxelles 1950), Paris, Masson 1951.

[10] H. Cartan, S. Eilenberg, *Homological algebra*, Princeton University Press 1973.

[11] A. Chamseddine, A. Connes, The spectral action principle, *Commun. Math. Phys.* **186** (1997), 731–750.

[12] A. Connes, Une classification des facteurs de type III, *Ann. Scient. E.N.S. 4ème Série* **t.6** (1973), 133–252.

[13] A. Connes, Non-commutative differential geometry, *Publ. IHES* **62** (1986), 257–360.

[14] A. Connes, C^* algèbres et géométrie différentielle, *C.R. Acad. Sci. Paris*, **290**, Série A (1980), 599–604.

[15] A. Connes, Essay on physics and noncommutative geometry, in *The interface of mathematics and particles physics*, pp. 9–48, Oxford Univ. Press 1990.

[16] A. Connes, *Non-commutative geometry*, Academic Press, 1994.

[17] A. Connes, Geometry from the spectral point of view, *Lett. Math. Phys.* **34** (1995), 203–238.

[18] A. Connes, Noncommutative geometry and reality, *J. Math. Phys.* **36** (1995), 6194–6231.

[19] A. Connes, J. Lott, Particle models and noncommutative geometry, *Nucl. Phys.* **B18** Suppl. (1990), 29–47.

[20] R. Coquereaux, Noncommutative geometry and theoretical physics, *J. Geom. Phys.* **6** (1989), 425–490.

[21] R. Coquereaux, *Higgs fields and superconnections*, in Differential Geometric Methods in Theoretical Physics, Rapallo 1990 (C. Bartocci, U. Bruzzo, R. Cianci, eds), Lecture Notes in Physics **375**, Springer Verlag 1991.

R. Coquereaux, R. Häußling, F. Scheck, Algebraic connections on parallel universes, *Int. J. Mod. Phys.* **A10** (1995), 89–98.

[22] J. Cuntz, D. Quillen, Algebra extensions and nonsingularity, *J. Amer. Math. Soc.* **8** (1995), 251–289.

[23] P.A.M. Dirac, On quantum algebras, *Proc. Camb. Phil. Soc.* **23** (1926), 412.

[24] M. Dubois-Violette, The Weil-BRS algebra of a Lie algebra and the anomalous terms in gauge theory, *J. Geom. Phys.* **3** (1987), 525–565.

[25] M. Dubois-Violette, Dérivations et calcul différentiel non commutatif, *C.R. Acad. Sci. Paris,* **307**, Série I (1988), 403–408.

[26] M. Dubois-Violette, Non-commutative differential geometry, quantum mechanics and gauge Theory, in *Differential Geometric Methods in Theoretical Physics*, Rapallo 1990 (C. Bartocci, U. Bruzzo, R. Cianci, eds), Lecture Notes in Physics **375**, Springer-Verlag 1991.

[27] M. Dubois-Violette, Some aspects of noncommutative differential geometry, in *Contemporary Mathematics* **203** (1997), 145–157.

[28] M. Dubois-Violette, R. Kerner, J. Madore, Non-commutative differential geometry of matrix algebras, *J. Math. Phys.* **31** (1990), 316–322.

[29] M. Dubois-Violette, R. Kerner, J. Madore, Gauge bosons in a noncommutative geometry, *Phys. Lett.* **B217** (1989), 485–488.

M. Dubois-Violette, R. Kerner, J. Madore, Classical bosons in a noncommutative geometry, *Class. Quantum Grav.* **6** (1989), 1709–1724.

M. Dubois-Violette, R. Kerner, J. Madore, Non-commutative differential geometry and new models of gauge theory, *J. Math. Phys.* **31** (1990), 323–330.

[30] M. Dubois-Violette, A. Kriegl, Y. Maeda, P.W. Michor, In preparation.

[31] M. Dubois-Violette, T. Masson, Basic cohomology of associative algebras, *Journal of Pure and Applied Algebra,* **114** (1996), 39–50.

[32] M. Dubois-Violette, T. Masson, On the first order operators in bimodules, *Lett. Math. Phys.* **37** (1996), 467–474.

[33] M. Dubois-Violette, T. Masson, $SU(n)$-gauge theories in noncommutative differential geometry, *J. Geom. Phys.* **25** (1998), 104–118.

[34] M. Dubois-Violette, P.W. Michor, Connections on central bimodules in noncommutative geometry, *J. Geom. Phys.* **20** (1996), 218–232.

[35] M. Dubois-Violette, P.W. Michor, Dérivations et calcul différentiel noncommutatif. II, *C.R. Acad. Sci. Paris*, **319**, Série I (1994), 927–931.

[36] M. Dubois-Violette, P.W. Michor, More on the Frölicher-Nijenhuis bracket in noncommutative differential geometry, *Journal of Pure and Applied Algebra,* **121** (1997), 107–135.

[37] M. Dubois-Violette, M. Talon, C.M. Viallet, B.R.S. algebras. Analysis of consistency equations in gauge theory, *Commun. Math. Phys.* **102** (1985), 105–122.

M. Dubois-Violette, M. Henneaux, M. Talon, C.M. Viallet, General solution of the consistency equation, *Phys. Lett.* **B289** (1992), 361–367.

[38] D.R. Farkas, G.. Letzter, Ring theory from symplectic geometry, *Journal of Pure and Applied Algebra,* **125** (1998), 155–190.

[39] J. Fröhlich, O. Grandjean, A. Recknagel, Supersymmetric quantum theory, non-commutative geometry, and gravitation, in *Quantum Symmetries*, Les Houches 1995 (A. Connes, K. Gawedzki, J. Zinn-Justin, eds), Elsevier 1998.

[40] M. Gerstenhaber, The cohomology structure of an associative ring, *Ann. Math.* **78** (1963), 267–287.

[41] W. Greub, S. Halperin, R. Vanstone, *Connections, curvature, and cohomology*, Vol. III, Academic Press 1976.

[42] H. Grosse, G. Reiter, Graded differential geometry of graded matrix algebras, math-ph/9905018 to appear in *J. Math. Phys.*.

[43] D. Husemoller, *Lectures on cyclic homology*, Springer-Verlag 1991.

[44] N. Jacobson, *Structure and representations of Jordan algebras*, American Mathematical Society 1968.

[45] N. Jacobson, *Basic algebra II*, second edition, Freeman and Co., New York 1989.

[46] M. Karoubi, Homologie cyclique des groupes et algèbres, *C.R. Acad. Sci. Paris* **297**, Série I (1983), 381–384.

[47] M. Karoubi, Homologie cyclique et K-théorie, *Astérisque* **149** (SMF), 1987.

[48] D. Kastler, Lectures on Alain Connes' noncommutative geometry and applications to fundamental interactions, in *Infinite dimensional geometry, noncommutative geometry, operator algebras, fundamental interactions*; Saint-François, Guadeloupe 1993 (R. Coquereaux, M. Dubois-Violette, P. Flad, eds), World Scientific 1995.

[49] J.L. Koszul, Homologie et cohomologie des algèbres de Lie, *Bull. Soc. Math. Fr.* **78** (1950), 65–127.

[50] J.L. Koszul, *Fibre bundles and differential geometry*, Tata Institute of Fundamental Research, Bombay, 1960.

[51] G. Landi, *An introduction to noncommutative spaces and their geometries*, Springer-Verlag, 1997.

[52] J.-L. Loday, *Cyclic homology*, Springer-Verlag, New York 1992.

[53] J. Madore, *Noncommutative differential geometry and its physical applications*, Cambridge University Press 1995.

[54] G. Maltsiniotis, Le langage des espaces et des groupes quantiques, *Commun. Math. Phys.* **151** (1993), 275–302.

[55] T. Masson, *Géométrie non commutative et applications à la théorie des champs*, Thesis, Orsay 1995.

[56] J. Mourad, Linear connections in noncommutative geometry, *Class. Quantum Grav.* **12** (1995), 965–974.

[57] M. Rieffel, Projective modules over higher dimensional noncommutative tori, *Can. J. Math.* **XL2** (1988), 257–338.

[58] P. Seibt, *Cyclic homology of algebras*, World Scientific Publishing Co., 1987.

[59] D. Sullivan, Infinitesimal computations in topology, *Publ. IHES* **47** (1977), 269–331.

[60] C.A. Weibel, *An introduction to homological algebra*, Cambridge University Press 1994.

[61] S.L.Woronowicz, Differential calculus on compact matrix pseudogroups (quantum groups), *Commun. Math. Phys.* **122** (1989), 125–170.

Mathematical Physics Studies

1. F.A.E. Pirani, D.C. Robinson and W.F. Shadwick: *Local Jet Bundle Formulation of Bäcklund Transformations.* 1979 ISBN 90-277-1036-8
2. W.O. Amrein: *Non-Relativistic Quantum Dynamics.* 1981 ISBN 90-277-1324-3
3. M. Cahen, M. de Wilde, L. Lemaire and L. Vanhecke (eds.): *Differential Geometry and Mathematical Physics.* 1983 Pb ISBN 90-277-1508-4
4. A.O. Barut (ed.): *Quantum Theory, Groups, Fields and Particles.* 1983 ISBN 90-277-1552-1
5. G. Lindblad: *Non-Equilibrium Entropy and Irreversibility.* 1983 ISBN 90-277-1640-4
6. S. Sternberg (ed.): *Differential Geometric Methods in Mathematical Physics.* 1984 ISBN 90-277-1781-8
7. J.P. Jurzak: *Unbounded Non-Commutative Integration.* 1985 ISBN 90-277-1815-6
8. C. Fronsdal (ed.): *Essays on Supersymmetry.* 1986 ISBN 90-277-2207-2
9. V.N. Popov and V.S. Yarunin: *Collective Effects in Quantum Statistics of Radiation and Matter.* 1988 ISBN 90-277-2735-X
10. M. Cahen and M. Flato (eds.): *Quantum Theories and Geometry.* 1988 ISBN 90-277-2803-8
11. Bernard Prum and Jean Claude Fort: *Processes on a Lattice and Gibbs Measures.* 1991 ISBN 0-7923-1069-1
12. A. Boutet de Monvel, Petre Dita, Gheorghe Nenciu and Radu Purice (eds.): *Recent Developments in Quantum Mechanics.* 1991 ISBN 0-7923-1148-5
13. R. Gielerak, J. Lukierski and Z. Popowicz (eds.): *Quantum Groups and Related Topics.* Proceedings of the First Max Born Symposium. 1992 ISBN 0-7923-1924-9
14. A. Lichnerowicz, *Magnetohydrodynamics: Waves and Shock Waves in Curved Space-Time.* 1994 ISBN 0-7923-2805-1
15. M. Flato, R. Kerner and A. Lichnerowicz (eds.): *Physics on Manifolds.* 1993 ISBN 0-7923-2500-1
16. H. Araki, K.R. Ito, A. Kishimoto and I. Ojima (eds.): *Quantum and Non-Commutative Analysis.* Past, Present and Future Perspectives. 1993 ISBN 0-7923-2532-X
17. D.Ya. Petrina: *Mathematical Foundations of Quantum Statistical Mechanics.* Continuous Systems. 1995 ISBN 0-7923-3258-X
18. J. Bertrand, M. Flato, J.-P. Gazeau, M. Irac-Astaud and D. Sternheimer (eds.): *Modern Group Theoretical Methods in Physics.* Proceedings of the Conference in honour of Guy Rideau. 1995 ISBN 0-7923-3645-3
19. A. Boutet de Monvel and V. Marchenko (eds.): *Algebraic and Geometric Methods in Mathematical Physics.* Proceedings of the Kaciveli Summer School, Crimea, Ukraine, 1993. 1996 ISBN 0-7923-3909-6

Mathematical Physics Studies

20. D. Sternheimer, J. Rawnsley and S. Gutt (eds.): *Deformation Theory and Symplectic Geometry.* Proceedings of the Ascona Meeting, June 1996. 1997
ISBN 0-7923-4525-8
21. G. Dito and D. Sternheimer (eds.): *Conférence Moshé Flato 1999.* Quantization, Deformations, and Symmetries, Volume I. 2000
ISBN 0-7923-6540-2 / Set: 0-7923-6542-9
22. G. Dito and D. Sternheimer (eds.): *Conférence Moshé Flato 1999.* Quantization, Deformations, and Symmetries, Volume II. 2000
ISBN 0-7923-6541-0 / Set: 0-7923-6542-9
23. Y. Maeda, H. Moriyoshi, H. Omori, D. Sternheimer, T. Tate and S. Watamura (eds.): *Noncommutative Differential Geometry and Its Applications to Physics.* Proceedings of the Workshop at Shonan, Japan, June 1999. 2001
ISBN 0-7923-6930-0

Kluwer Academic Publishers – Dordrecht / Boston / London